ANNOUNCING

PLENUM/ROSETTA EDITIONS

Plenum, one of the world's leading scientific and technical publishers, is pleased to announce its new paperback line—*plenum/rosetta editions.* Those advanced-level, scientific works in chemistry, biology, physics, etc., written or edited by leading authorities in each field, which are most suitable for school use will be published under the new imprint.

IR
Theory and Practice of Infrared Spectroscopy

Nelson L. Alpert
Instrumentation Consultant
Stamford, Connecticut

William E. Keiser
Optical Technology Division
Perkin–Elmer Corporation
Danbury, Connecticut

and

Herman A. Szymanski
Dean of the College
Alliance College
Cambridge Springs, Pennsylvania

A PLENUM / ROSETTA EDITION

Library of Congress Cataloging in Publication Data

Alpert, Nelson L. 1925–
IR: theory and practice of infrared spectroscopy.
"Plenum/Rosetta edition."
1. Infra-red spectrometry. I. Keiser, William E., joint author. II. Szymanski, Herman A., joint author. III. Title.
[QC457.A68 1973] 535'.842 73-12968
ISBN 0-306-20001-5 (pbk.)

A Plenum/Rosetta Edition
Published by Plenum Publishing Corporation
227 West 17th Street, New York, N.Y. 10011

First paperback printing 1973
Based on the second edition, published in 1970

United Kingdom edition published by Plenum Press, London
A Division of Plenum Publishing Company
Davis House (4th Floor), 8 Scrubs Lane, Harlesden, London, NW10 6SE, England

Printed in the United States of America

PREFACE

The first edition of this text was written primarily by one of the present authors (HAS), with a chapter on instrumentation contributed by a second (NLA). The volume was well received, and to keep the text up-to-date a second edition was planned. For this second edition, a third author (WEK) was invited, whose background complemented that of the other two. Each of the authors was assigned several chapters as his primary task while the complete manuscript remained the secondary responsibility of all three. It is hoped that this approach has resulted in a work that is even more thorough than the first edition in covering the basic concepts of infrared spectroscopy.

NELSON L. ALPERT
WILLIAM E. KEISER
HERMAN A. SZYMANSKI

PREFACE TO THE FIRST EDITION

My experience with the many infrared spectroscopy institutes held at Canisius College and many discussions with both beginners and experienced practitioners in infrared spectroscopy have convinced me that there is a need for an introductory text devoted entirely to infrared spectroscopy, a text which can be utilized even by those who approach this study with only a limited background. This volume sprang from that conviction. It is intended for all who wish to use infrared spectroscopy in research – especially chemists doing structural work – in routine control work, in industrial development, or in medical applications or those military applications where it is employed as an analytical tool.

Except for the chapter on theory, the text material can be easily assimilated even by students with only the equivalent of a two-year technical degree. While it is primarily intended as a textbook for courses at either the upper undergraduate or the graduate level, this volume should also prove valuable as a reference book in the infrared laboratory.

Because I feel that the basic principles of instrument design should be understood by everyone working in infrared spectroscopy, I asked Dr. Nelson Alpert of the Perkin-Elmer Corporation to prepare the chapter on instrumentation. Dr. Alpert's wide experience in instrument design qualifies him exceptionally well for the task of writing a description of the design concepts that underlie *all* instruments in such a way that the description will not be outdated by the development of new instrument designs. I feel that he has acquitted himself admirably.

Perhaps the most difficult chapter to present was that concerned with the theory of infrared spectroscopy. To achieve a balance between a completely theoretical presentation which could be understood only by those at the graduate level and a presentation which would be so simple in approach as to have no value, I attempted to select the topics which the reader will most frequently encounter in the current literature and to explain the principles upon which each concept is based. Early drafts of this chapter were as long as the entire volume is now, and I finally decided to omit several important but not absolutely necessary topics, such as the calculation of thermodynamic parameters from observed spectra. The theory of band intensity is discussed only briefly, and group theory is introduced only to illustrate some of its general applications.

I have tried to connect the qualitative aspects of group frequencies with their theoretical foundations and therefore suggest that the reader attempting to understand the interpretation of spectra combine the study of Chapter 5, which deals with qualitative analysis, with that of the theory chapter, Chapter 4.

With deep gratitude and pleasure I acknowledge the assistance of many people who have helped to make this book possible. They include Dr. Forrest F. Cleveland of the Illinois Institute of Technology and Dr. Richard Stanton of Canisius College, who made suggestions concerning the chapter on theory; Dr. Raymond Annino of Canisius College and Mr. Abram Davis of the Hooker Chemical Company, who made suggestions concerning the chapter on quantitative analysis; Dr. Frank Bajer of the Hooker Chemical Company and Dr. Ronald Erickson of Canisius College, who made suggestions concerning the chapter on qualitative analysis; and Fr. Paul McCarthy and Fr. James Ruddick of Canisius College, who made valuable suggestions for material in several chapters. Finally, many of the spectra were run by Dr. William Keiser of the Perkin-Elmer Corporation, who also offered helpful suggestions for the chapter on laboratory techniques and sample preparation.

HERMAN A. SZYMANSKI

November 1963
Buffalo, New York

In writing the chapter on instrumentation, my prime objective was to focus on information which (a) relates to operation of an instrument in obtaining useful results and (b) contributes to an understanding of the instrument so as to enhance its utility.

I wish to take this opportunity to acknowledge the constructive suggestions of Dr. Van Zandt Williams and Dr. Robert C. Gore of the Perkin-Elmer Corporation. Finally, I am grateful to Dr. Herman A. Szymanski for inviting me to participate in this project.

NELSON L. ALPERT

November 1963
Norwalk, Connecticut

CONTENTS

Chapter 8

CHAPTER 1

Introduction to Infrared Spectroscopy

Infrared spectroscopy can be described as the use of instrumentation in measuring a physical property of matter, and the relating of the data to chemical composition. The instruments used are called infrared spectrophotometers, and the physical property measured is the ability of matter to absorb, transmit, or reflect infrared radiation.

During the past several decades this technique has become increasingly important and useful in the qualitative and quantitative analysis of materials, and its utilization is spreading to such other areas as layer-thickness measurements, reflectivities, and refractive index. It is a technique that finds a wide variety of uses both in industrial analytical laboratories and in research laboratories of all types, as it furnishes information that is useful in qualitative and quantitative analysis, in the calculation of various physical constants, in the determination of the structure of compounds, and in many other areas.

Infrared analysis can be used for almost any type of sample as long as the material is composed of or contains compounds (rather than pure elements). It is a nondestructive type of analysis (the sample can normally be recovered for other use), and is useful for microsamples (down to the sub-microgram range). In its earlier years, infrared spectroscopy was used primarily for organic materials, but, especially since the advent of long-wavelength instrumentation, it has been found to be equally useful for the analysis of inorganic compounds.

This book was written primarily for the new man in infrared spectroscopy, the student and the inexperienced worker, but should also be useful to the more experienced, especially to those whose experience has been limited to one small segment of the field. To cover infrared technology in a comprehensive or detailed manner would require a book of unmanageable size; so, except for the chapters dealing with interpretation of spectra, the material covered was limited to broad, general descriptions or discussions of instrumentation, accessories, general sample-handling techniques, and quantitative analysis, and to a library of reference spectrograms. The chapters on qualitative analysis cover the theory of interpretation of data, as well as a discussion on how the information provided by the spectrogram is interpreted and used in practice.

1.1. UNITS

An infrared spectrogram is a two-dimensional presentation on a sheet of paper of the absorption characteristics of a molecule. These absorption characteristics, appearing on the spectrogram as bands or peaks, can be described in terms of three variables: position, intensity, and shape. The first two of these can be expressed in numbers, while the third (band shape) is usually expressed in words.

Position. Position is the point on the x-axis or abscissa where the band appears; it always expressed numerically. Four different parameters are used to describe position: energy (in ergs), frequency (in sec^{-1}), wavelength (in microns), and wavenumber (in cm^{-1}). The units used on the spectrogram will depend on instrument design, and if one of the other units is required for proper work-up, the data must be appropriately converted. This conversion can easily be accomplished using the equations given below. The first two of the above parameters, energy and frequency, give numbers that are unwieldly and not easily comprehended; they are therefore not used for normal presentation. Current instrumentation presents the data either in wavelength or wavenumber units, the latter being preferred. A survey of the literature will show that while both types of unit have been used in the past, wavelength units were preferred for many years.

The units in question can be converted as follows.

Energy and frequency are related by

$$E = h\nu \tag{1-1}$$

where

E = energy in ergs

h = Planck's constant (6.67×10^{-27} erg-sec)

ν = frequency in sec^{-1}

Frequency and wavelength are related by

$$\lambda\nu = c \tag{1-2}$$

where

λ = wavelength in cm

ν = frequency in sec^{-1}

c = velocity of light (3×10^{10} cm/sec)

Wavelength and wavenumber are related by

$$\bar{\nu} = 1/\lambda \tag{1-3}$$

where

$$\lambda = \text{wavelength in cm}$$

$$\bar{\nu} = \text{wavenumber in cm}^{-1}$$

Equation (1-3) is usually written as

$$\mu = 10{,}000/\bar{\nu} \tag{1-4}$$

Where λ is now the wavelength in microns, the unit normally used in infrared measurements.

Wavelength can be expressed in several units, all of which are related to the centimeter (cm). The unit that is used in the presentation depends on the region of the electromagnetic spectrum in which the work is being done. In the X-ray region the unit is the angstrom (Å); in the ultraviolet region it is either the angstrom or the millimicron (mμ); in the visible and near-infrared regions it is the millimicron; and, as mentioned above, in the infrared region it is the micron (μ). The relationship of these units to each other and to the cm is as follows:

$$10\ \text{Å} = 1\ \text{m}\mu$$

$$1000\ \text{m}\mu = 1\ \mu$$

$$1000\ \mu = 1\ \text{mm}$$

$$1 \times 10^4\ \mu = 1\ \text{cm}$$

The following example showing the calculations required and the size of the numbers involved should help to clarify the relationship of these units to each other.

Example

Calculate the energy associated with a 10 μ band as well as its frequency and wavenumber.

$$10\ \mu = 1 \times 10^{-3}\ \text{cm}$$

The frequency is calculated from (1-2):

$$\lambda\nu = c$$

or

$$\nu = c/\lambda$$

$$= \frac{3 \times 10^{10}\ \text{cm/sec}}{10^{-3}\ \text{cm}} = 3 \times 10^{13}\ \text{sec}^{-1}$$

From this number we can calculate the energy using 1-1

$$
\begin{aligned}
E &= h\nu \\
&= 6.67 \times 10^{-27}\ \text{erg-sec} \times 3 \times 10^{13}\ \text{sec}^{-1} \\
&= 2 \times 10^{-13}\ \text{erg}
\end{aligned}
$$

Using (1-3) or (1-4) we can calculate the wavenumber (remembering to use the proper units for λ in these equations).

Using (1-3) and $\lambda = 1 \times 10^{-3}$ cm, we get

$$\lambda = 1/\bar{\nu}$$

or

$$
\begin{aligned}
\bar{\nu} &= 1/\lambda \\
&= \frac{1}{1 \times 10^{-3}\ \text{cm}} \\
&= 1 \times 10^{3}\ \text{cm}^{-1} = 1000\ \text{cm}^{-1}
\end{aligned}
$$

Using (1-4) and $\lambda = 10\ \mu$

$$
\begin{aligned}
\bar{\nu} &= 10{,}000/\lambda \\
&= 10{,}000/10 \\
&= 1000\ \text{cm}^{-1}
\end{aligned}
$$

If these calculations are repeated for wavelengths of 1 μ and 100 μ the results will show that as the wavelength increases, the energy, frequency, and wavenumber decrease, while a decrease in wavelength causes the energy, frequency, and wavenumber to increase.

In practice the wavenumber is usually called "frequency," and the symbol ν is used instead of $\bar{\nu}$. This is not as confusing as it may appear, as the size of the number will indicate which type of frequency is involved; true frequency in sec^{-1} is a much larger number (by several orders of magnitude) than the wavenumber in cm^{-1}.

Intensity. Intensity is a measure of the quantity of energy absorbed by the sample and is determined from the y-axis, or ordinate data. This parameter can be expressed in numbers or in words. In the former case, the number is valid only for the instrument, the operating conditions, the sample composition, and the sample path length for which it was obtained. In the latter case, the intensity is said to be weak, medium, or strong, and depends to a lesser degree on the same factors.

If the intensity is in numbers it may be expressed either as percent transmittance, or in absorbance units. If the energy, or radiant power, impinging on the sample at some frequency is P_0, and the energy transmitted by the sample is some fraction of this, P, then the transmittance T can be expressed as

$$T = P/P_0 \tag{1-5}$$

and percent transmittance as

$$\%T = 100\, P/P_0 \tag{1-6}$$

Frequently, however, especially for quantitative analysis, it is desirable to express this data in absorbance units rather than in percent transmittance. The absorbance is given by

$$A = \log 1/T = \log P_0/P \tag{1-7}$$

Instruments are usually designed to record the transmitted energy in linear percent transmittance, but they can also record it in absorbance units on a logarithmic scale. The operator can choose the type of presentation by placing the correct paper on the recorder. For most purposes, nonlinear absorbance is as useful as, and in many cases even more useful than percent transmittance. Whichever way the data is recorded, the 100%-transmittance (zero-absorbance) line is at the top of the chart paper, and though the absorptions are called peaks they appear as valleys. In some articles, and on some instruments constructed abroad, this presentation is reversed, and the peaks appear in the opposite direction (here the 0%-transmittance or infinite-absorbance lines are at the top of the chart). Usually, chart paper is available in both linear percent-transmittance and nonlinear absorbance scales for both linear wavelength and linear wavenumber abscissas.

Shape. As stated above, the third of these descriptive variables is given in words and not in numbers, although some attempts have been made to reduce the qualitative nature of this description by assigning letters or numbers to the various types of band shapes. The band shape is usually described as broad, narrow, sharp, etc., and operator opinion plays a large role in these descriptions. The situation is further complicated by the fact that a band on a linear wavelength chart will usually not have the same shape as the same band on a linear wave number chart.

1.2. REGIONS OF ELECTROMAGNETIC RADIATION

Although this book is concerned only with the infrared region of the electromagnetic spectrum, it will be useful to look at the entire spectrum normally used in analysis. The spectrum is divided into several regions, each of which is used for a different purpose, and each of which furnishes a different type of information. These are the X-ray, ultraviolet, visible, near-infrared, infrared, far-infrared, and microwave regions. This division of the spectrum is arbitrary, and the boundaries

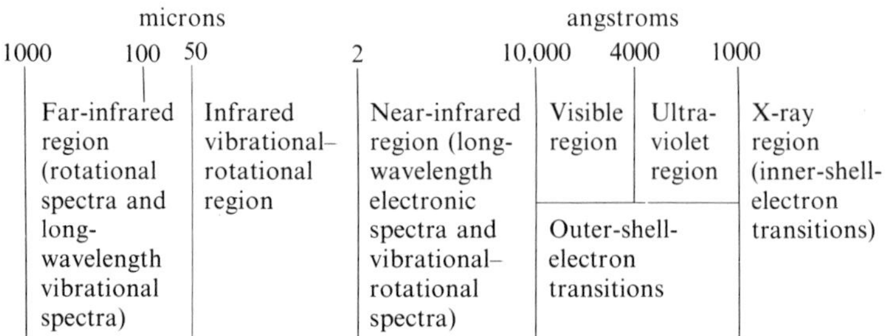

Figure 1-1. The electromagnetic spectrum.

are not well defined. Figure 1-1 is one schematic representation of the spectrum from the X-ray through part of the far-infrared regions.

The calculations of the example in Section 1.1 can be extended to both shorter (X-ray region) and longer (far-infrared region) wavelengths. These will show that the wavelengths in the X-ray region are associated with the highest energy, while the reverse is true of the wavelengths in the far-infrared region. Matter can interact with radiation only when the energy of the radiation matches an energy-level transition in an atom or molecule. In matter the higher energy levels are associated with the interior electrons of the atoms, hence X-rays are used to determine these internal-shell electron transitions; ultraviolet and visible spectroscopy deal with valence-shell electron transitions; while near-infrared measurements can deal either with similar electronic transitions (an extension of the visible) or with rotational-vibrational transitions (an extension of infrared). Infrared spectroscopy is concerned with rotational and vibrational energy transitions, with the 2–15 μ region being known as the fundamental region (Chapter 4). The far infrared is primarily a region of rotational energy transitions, although several low-energy vibrational transitions interact with radiation in this region.

In practice there is some overlap among these regions, and depending on the instrumentation available it is frequently possible to go some distance beyond the usual boundary into the next region. In the laboratory it is possible to find instruments for use in each region, but in some cases the same instrument will cover two or three regions; the usual instrument of the latter variety is the ultraviolet, visible, and near-infrared combination.

1.3. THE INFRARED SPECTRUM OF A COMPOUND

Infrared spectra are normally recorded on chart paper as a graph of intensity (absorbance or percent transmittance) *versus* position (wavelength or wavenumber), although it is possible to present them in other ways (e.g., on an oscilloscope screen). The recordings are usually called spectrograms. Figures 1-2 and 5-24 show spectrograms of polystyrene run on two commercial instruments; one of the instruments presents the data linearly in wavelength units, while the other presents it linearly in wavenumber units. (Figures 1–2 and 5–24 will be found in a pocket on the inside of the back cover). Comparison of these spectrograms will show how the appearance of the bands changes between two types of presentation.

Not all matter is capable of producing an infrared spectrum (e.g., metals do not). To interact with infrared radiation the molecule must have a permanent dipole moment and must vibrate about a bond (changing the bond length or angle), or rotate about an axis perpendicular to the bond. It is the interaction of radiation with these vibrations and rotations that gives rise to the absorption bands appearing on the spectrogram.

The spectra of most compounds show a number of bands, the wavelengths or frequencies at which they appear supplying the primary data used to identify the compound. It is frequently assumed that this band-frequency information is all that is required for compound identification, but experience shows that intensity and band shape must also be considered. Lists of absorption frequencies or wavelengths normally do not furnish sufficient information for positive identification. However, the information available from the abscissa is the most important in qualitative analysis. The proper use of this information is discussed in Chapter 5.

While the data from the ordinate have some use in qualitative analysis, it is primarily used in quantitative work. The abscissa data are absolute within the accuracy of the instrument, but ordinate data are relative, and depend on the instrument used, the operating conditions, the sample composition, the sample path length, and the concentration. The use of these data in quantitative analysis is discussed in Chapter 6.

1.4. ATMOSPHERIC ABSORPTIONS

One further item that should be brought to the attention of the spectroscopist is the effect of the atmosphere on instrument operation and the resultant data. Unless the instrument is operated in a special atmosphere or in a vacuum, atmospheric absorption may present problems. (See Figure 2-7.)

Nitrogen, oxygen, and the noble gases have no infrared absorptions, and therefore do not require further discussion. Carbon dioxide and water vapor, always present in the normal atmosphere, do absorb infrared radiation, and may present serious problems. The atmosphere in which the instrument is operated may also contain other materials (particulate or gaseous) which may cause not only operating problems, but also instrument deterioration.

In single-beam instruments all the atmospheric absorptions will appear as bands on the spectrogram, with their intensities varying with changes in concentration. When the atmospheric bands are totally absorbing, it is impossible to determine whether or not the sample has any absorption at these frequencies; when the atmospheric bands are only partially absorbing, it may be difficult to determine the presence of weak sample bands in the same regions. The variation in the intensities of the atmospheric bands makes quantitative analysis almost impossible at these frequencies. The best solution to this problem is to purge the instrument or enclose it in a housing that can be purged. It may also be possible to place trays of adsorbing material in the various sections of the instrument, to eliminate these gases.

Double-beam instruments are designed to minimize some of these problems. The lengths of the sample and reference beams are equal, so that, as the atmospheric concentrations should be uniform throughout the instrument, the detector will not see any difference in the energy of the two beams. This means that there will be no signal, and the pen will not record the bands. However, even though the bands do not appear, the atmosphere still removes energy from the system, and the remaining energy may be insufficient to operate the instrument properly. The most appropriate remedy is the same as for single-beam instruments—purge or adsorb the unwanted gases.

CHAPTER 2

Instruments

Use of an infrared spectrophotometer in good operating condition does not in itself assure accurate results. The validity of the spectra also depends on the sample-handling techniques, which will be discussed in Chapter 7, and on the selection of proper instrument operating variables, which is a primary concern of this chapter. To aid in understanding the operating variables and their interrelations we shall first examine the basis of operation of one type of widely used infrared spectrophotometer. This will be followed by a discussion of the operating variables and their qualitative and quantitative interdependence. The remainder of the chapter will describe the components and features of infrared spectrophotometers.

2.1. DESCRIPTION

A block diagram delineating the essential subsystems of a double-beam spectrophotometer is shown in Figure 2-1. "Double-beam" refers to an instrument system which compares the radiant energy transmitted by the unknown sample to that transmitted through a known reference, which may be air. Thus, in the block diagram, infrared energy from the

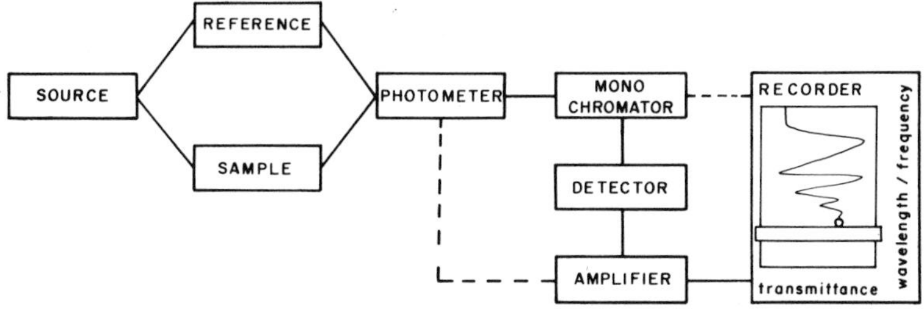

Figure 2-1. Block diagram of a double-beam spectrophotometer system.

source illuminates both the *sample* and the *reference*. The amount of energy transmitted by them is compared in the *photometer*. In order to determine the spectral variation of this comparison, the energy from the photometer is dispersed by the *monochromator* so that only a narrow spectral region is transmitted to the *detector* at any time. The *amplifier* operates on the signal from the detector so as to present the sample-to-reference comparison from the photometer on the ordinate of the *recorder*. In order for the recorder to plot this information as a function of position in the spectrum, the abscissa of the recorder is coupled to the monochromator.

The schematic optical diagram in Figure 2-2 provides an additional basis for the discussion of operating variables (Section 2.2). This diagram demonstrates the salient features of the general type of double-beam spectrophotometers found in most analytical laboratories. Details and information relative to other types of instrument systems will be presented in Section 2.3B.

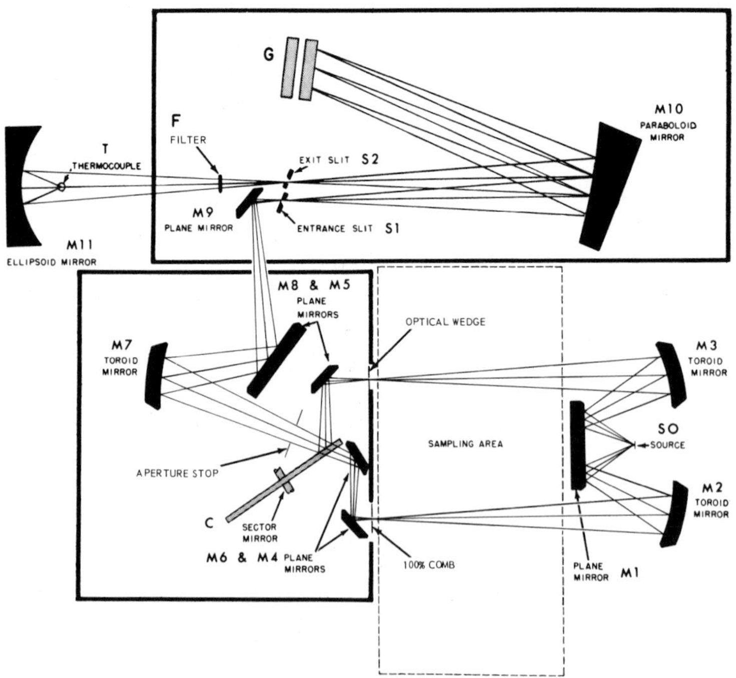

Figure 2-2. Schematic optical diagram of a double-beam infrared spectrophotometer. (Courtesy of Perkin–Elmer Corporation.)

The energy radiated by the source *SO* is split into sample and reference beams by the plane mirror *M*1. Mirror *M*2 focuses the sample beam on a comb-shaped device used to adjust the 100% transmittance level. Since the photometer compares sample- and reference-beam transmittances, the setting of this 100% comb is relative and a matter of convenience. Mirror *M*3 focuses the reference beam on the optical wedge, whose function is explained below.

The sample and reference beams are recombined at the rotating sector mirror *C* into a single beam consisting of alternate pulses of reference- and sample-beam radiation. This section of the instrument is called the photometer; here the pulsating radiation may be used to indicate the difference in energy between the sample and reference beams. In addition, the photometer contains the means of equalizing the energy in the two beams in any wavelength interval isolated by the monochromator, which follows in the optical system and is described below. The means of obtaining optical null is an optical-wedge type of attenuator actuated by the servosystem of the instrument so as to adjust its position in the reference beam to equalize the energy in the two beams. The position of the recorder pen in the instrument at all times indicates the position of the wedge. It therefore indicates the transmission of the sample relative to the reference at the wavelength for which the instrument is set.

Mirror *M*7 refocuses the alternate pulses of reference- and sample-beam energy on the entrance slit *S*1 of the monochromator. The monochromator performs three functions basic to the operation of an infrared spectrophotometer:

1. it disperses the radiation into its component wavelengths;
2. it selects the particular wavelength of radiation to be transmitted to the detector;
3. it maintains approximately constant energy at the detector at all wavelengths when no sample is in the instrument.

The energy which passes through the entrance slit is collimated (all rays made parallel to each other) by mirror *M*10. The collimated beam proçeeds to the dispersing element, which in this case is the grating *G*. The grating returns the beam to *M*10, which refocuses the beam on the exit slit *S*2. This monochromator system is called a Littrow configuration. The angle of the grating relative to the collimated beam determines the wavelength of the radiation which passes through *S*2. In scanning, the grating is rotated by a cam so as to move the spectral band across the exit slit at a predetermined rate. Thus the energy reaching the detector changes in sequence from one wavelength to the next.

The width of the slit openings determines both the width of the spectral interval and the amount of energy which passes through the exit slit. The slit width is varied by a second cam coupled to the wavelength cam to maintain constant energy throughput as the wavelength is scanned with no sample in either beam. This energy throughput, therefore, determines the resolution of the instrument at each wavelength. The spectral interval passed by a given exit slit width is called the *spectral slit width.*

The energy which passes through the exit slit is then concentrated onto the detector T by the mirror $M11$.

2.2. OPERATING VARIABLES

The range of operating variables at the analyst's command depends on the particular instrument in use. However, serious work should not be undertaken on any type of infrared spectrophotometer without an understanding of the basic variables and their interrelations. The three basic variables are: (1) spectral resolution; (2) photometric (quantitative) accuracy; and (3) scanning speed. Since all three variables are interrelated, any change in the condition of one will influence the others. For example, when higher resolution or photometric accuracy are desired, it is necessary to spend a longer time scanning the spectrum. If a fast scan is desired, this necessitates a loss in either resolution or photometric accuracy or both.

2.2A. Resolution

One of the important properties of an infrared spectrophotometer is its resolution—i.e., its ability to distinguish between neighboring spectral regions. The ability to resolve neighboring bands is closely related to the spectral slit width expressed in units of wavenumber ($\Delta\nu$) or of wavelength ($\Delta\lambda$). As stated above, the spectral slit width is the small spectral region isolated by the exit slit of the monochromator. As the term is generally used, high resolution is associated with small spectral slit width.

In addition to the direct effect the spectral slit width and resolution have on the ability to separate neighboring bands in a spectrum, there is also a commonly experienced but often unrecognized effect on the recorded band depth. Unless the resolution of the instrument is very much less than the natural width of a particular band, its recorded absorbance will be less than its true absorbance. A demonstration of this effect on two bands of cyclohexane is shown in Figure 2-3. Because the 903 cm^{-1} band has a narrower natural width than the 861 cm^{-1} band, it is more strongly affected by variations in the spectral slit width, to the point where the relative band depths actually reverse.

At any given point in the spectrum the spectral slit width increases linearly with the physical slit width. However, the power available at

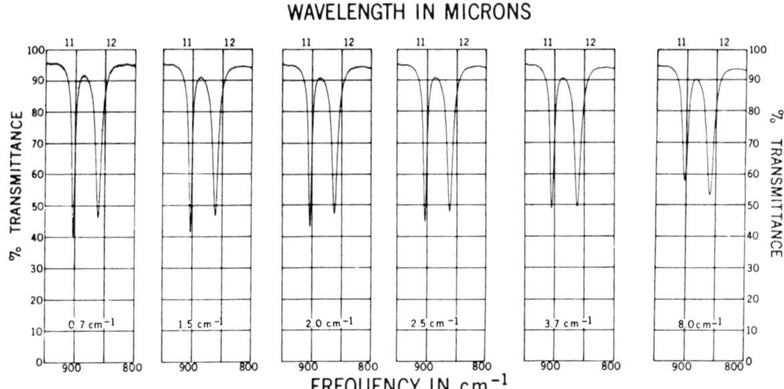

Figure 2-3. The 861 and 903 cm^{-1} bands of cyclohexane at various spectral slit widths. (Courtesy of Beckman Instruments.)

the detector is proportional to the square of the physical slit width. The reason for this is that the radiation source emits not just a single spectral line but a spectral continuum. Thus, not only does widening the entrance slit allow a proportionally greater amount of energy to enter the monochromator but, in addition, widening the exit slit allows a proportionally greater amount of energy to reach the detector. Since in most spectrophotometers the entrance and exit slits operate synchronously at very nearly equal widths, the energy reaching the detector is proportional to the square of the width of either slit.

To improve resolution the slits must be narrowed, but when the slit widths are reduced the energy on the detector is also reduced, in proportion to the square of the slit opening. Unfortunately, the sources of noise which limit the precise readability of the spectral data remain constant as the slits are narrowed. To compensate for the loss in signal power the gain in the system must be increased; this, in turn, raises the recorded noise level proportionally. To maintain a reasonable level of noise on the recorded spectrum the speed of response of the pen system must be made slower, thereby smoothing the noise peaks. To compensate for the slower pen response, a longer scanning time is required for the pen to follow the spectrum accurately.

The ability of a dispersing element (discussed in Section 2.3C) to provide high resolution, or high energy in a given spectral band pass, is dependent on its angular dispersion—i.e., on its ability to spread a spectral region over a large angle. In general, gratings have a much higher angular dispersion than prisms, and consequently make possible operation with a higher resolution or with an enhanced signal energy. The latter improvement permits a lower system gain, which results in lower recorded noise and enables use of a faster pen speed and faster response time.

In general, infrared spectrophotometers are limited in ultimate performance by energy. More specifically, the ultimate limitation is the relationship of the detector signal, generated by the available energy, to the natural noise generated in the instrument. The noise in a well designed instrument depends primarily on the inherent thermal noise generated within the detector element, with only a small additional contribution from the amplifier system. Although this noise may be smoothed out by use of a slower system response time, there are obvious practical limits to this procedure. However, some alleviation of this performance limitation may be gained by use of recently developed computer techniques in smoothing and averaging the noise (Section 2.6).

It is important to understand what instrumental factors affect this energy limitation. The energy which reaches the detector is proportional to the transmittance efficiency (T) of the optical elements in the system; to the square of the slit width (W^2), as mentioned above; to the angular slit height (h/F), the height divided by the focal length of the monochromator; to the normal cross-sectional area (A) of the beam incident on the dispersing element; to the energy B radiated by the source in the spectral interval $\Delta\lambda$ ($B_{\Delta\lambda}$); and to the angular dispersion of the dispersing element ($d\theta/d\lambda$).

This may be restated as a formula:

$$E \propto T\, W^2(h/F)AB_{\Delta\lambda}(d\theta/d\lambda) \qquad (2\text{-}1)$$

The above relationship enables the analyst to compare the relative energies for equal spectral slit widths in different instruments.

The narrower one can make the slit width, the better the resolution, except that in the extreme we encounter optical limitations, which include image aberrations and, when the slit width becomes comparable to the wavelength, image smearing. When the latter effect occurs we speak of having reached the "diffraction limit." Obviously, in order to obtain high resolution at a reasonable energy or signal level, it is necessary to optimize the remaining terms in the equation.

The angular slit-height term is not generally at the disposal of the analyst. The designer of the instrument has predetermined this to avoid too small a value. The designer must also avoid too large an angular slit height, since this would tend to introduce aberrations and other disadvantages, cancelling out any possible gain in energy.

In some instruments the radiation from the source may be controlled by the analyst over a limited range. However, he must be aware that any increase in the output of the source over its nominal rating shortens its useful life.

The angular dispersion of the monochromator may not be arbitrarily varied by the analyst, except by selection of the dispersing element which best suits the needs of the particular problem.

2.2B. Photometric Accuracy

Photometric accuracy is the accuracy with which the pen indicates the true transmittance of the sample. It enables the analyst to convert band-depth measurements to accurate quantitative results.

Five major factors determine photometric accuracy:

1. the inherent accuracy of the photometric system employed;
2. noise;
3. dynamic response of the system;
4. sampling limitations;
5. false radiation.

Although the *inherent accuracy of a spectrophotometric system* depends on its design, a knowledge of the factors which contribute to this inherent accuracy may be useful in maintaining an instrument in its best operating condition. In any system the optical alignment interacts with the photometric accuracy in some way. One common cause of this is the nonuniformity of response of detectors over their sensitive areas. As a result, photometric accuracy can be achieved only with a specific alignment.

In the commonly used double-beam optical null system employing a variable-aperture optical wedge or comb, the source is usually imaged on the comb and reimaged on the slits, which in turn are focused on the detector (see Figure 2-2). Thus the uniformity of the source image on the wedge is important. In addition, the double-beam recombination optics in the photometer section must carefully superimpose the image from the sample beam on the image from the reference beam. Within reasonable limits, the alignment of the monochromator does not directly affect photometric accuracy. However, the alignment of the detector optics does have a fundamental effect on photometric accuracy.

In the single-beam instrument, photometric accuracy to a great extent is dependent on the precise linearity of the electronic amplifiers, as well as on the stability of these amplifiers between the time that the 100% level is established, the time that the zero line is checked, and the time during which the spectrum is obtained. Another factor is the presence of atmospheric absorption bands or solvent bands in the vicinity of the band undergoing a quantitative measurement. This is one of the inherent limitations in the accuracy of single-beam systems which is subject to some control by the analyst.

In electronic ratio-recording systems as well as single-beam systems, a slide-wire is used in the output recording system. Therefore, the slide-wire itself may become a limiting factor in determining photometric accuracy.

The second main factor determining photometric accuracy, *noise*, may generally be controlled by the analyst. The deepest point of the band (maximum absorption), which is of greatest interest in quantitative measurements, can vary by as much as the peak-to-peak noise level; hence, photometric precision is limited by this factor. In the more versatile spectrophotometers it is possible to slow down the response of the system, thereby smoothing out and averaging the noise fluctuations and decreasing their peak-to-peak variations. However, if this is done more time must be spent in recording the spectrum to compensate for the slower response of the system.

The third factor, *dynamic response*, refers to the ability of the system to approximate, under dynamic scanning conditions, the shape of the spectrum which would be obtained by tedious point-to-point measurements under "static" conditions. Dynamic response can be quite troublesome to an unwary operator because it does not exhibit itself in such an obvious manner as noise.

The prime factor influencing dynamic response is the gain of the spectrophotometric system. The gain determines the amplitude of the signal which actuates the servomotor when the sample and reference beams are unequal in energy, and it must be carefully set at a proper value. If it is much too high, the recorder pen will break into oscillation, a condition analogous to the booming resonance which sometimes affects loudspeaker systems. If the gain is only moderately excessive, the spectrum will appear superficially normal. However, the system may tend to overshoot, particularly on sharp bands, thus indicating fallacious band depths. If the gain is set too low, the system cannot respond accurately to the spectral information, an effect similar to the loss of appreciation of the finer details of a musical masterpiece when the gain of a hi-fi set is too low.

In addition, dynamic response depends on the establishment of a suitable compatibility between the response time of the system and the time spent scanning the spectrum. If the time spent at each bit of available spectral information is not many times the response time of the system, the recorder pen will not respond accurately and some information will necessarily be lost through distortion of the true band shape. In slowing down the rate of scan, however, we encounter a point of diminishing returns, which depends on the balance between the resolution desired, the photometric accuracy required, and the patience of the operator. Except for practical limitations in the available instrument and operator time, it is impossible to err in the direction of slower scans.

The effect of scanning time and concomitant dynamic response on the appearance of a spectrum is demonstrated in Figure 2-4. As the

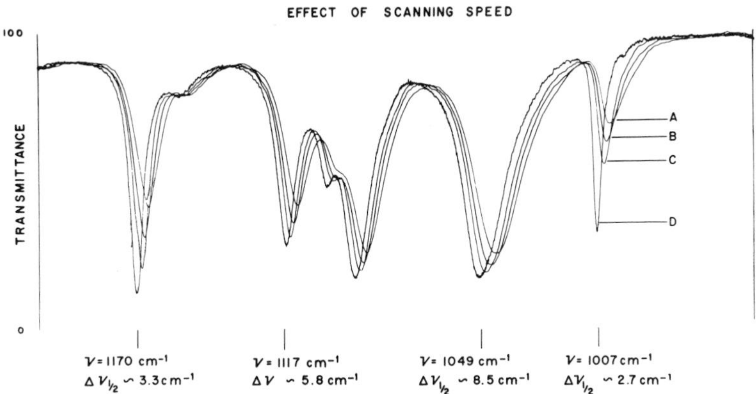

Figure 2-4. Effect of scanning speed on the appearance of a recorded spectrum. Effective times for full scan (4000–550 cm^{-1}): A, 8 min; B, 11 min; C, 16 min; D, 3 hr.

scan time is made too fast for maintaining proper dynamic response, the absorption bands become distorted with regard to peak position, depth of the peak, and band shape. It will be noted that the effect is greater for the sharper peaks.

Sampling limitations, the fourth factor listed, cover a host of problems. Among factors requiring close attention are: (1) control of sample and cell-window purity; (2) avoidance of fingerprints or other dirt on sample-cell windows; (3) choice of proper sample thickness to optimize accuracy; (4) in the case of vapors, filling at a pressure consistent with pressure-broadening requirements; (5) complete filling of liquid cells; (6) accurate positioning of the sample cell in the beam; (7) solvent compensation; (8) solvent interaction effects; (9) reradiation of energy from hot samples; (10) polarization effects; and (11) scattering of energy from the sample. This list is by no means complete, but it will serve to indicate some of the pitfalls that must be avoided in the quest for quantitative accuracy.

The final factor, *false radiation*, refers to the spectral impurity of the energy reaching the detector. This is quite commonly referred to as scattered light or stray light. Energy from a spectral region other than the interval nominally isolated by the monochromator will cause incorrect indications of band depth. In the extreme case of a totally absorbing band, the bottom of the band will not approach zero transmittance, but will be offset from zero by an amount equal to the false radiation. (In fact, this indicates one method for measuring false radiation.) Interpolating from this extreme case it may be concluded that, when false radiation is a factor in measurements, the indicated band depths will always be less than the true band depth.

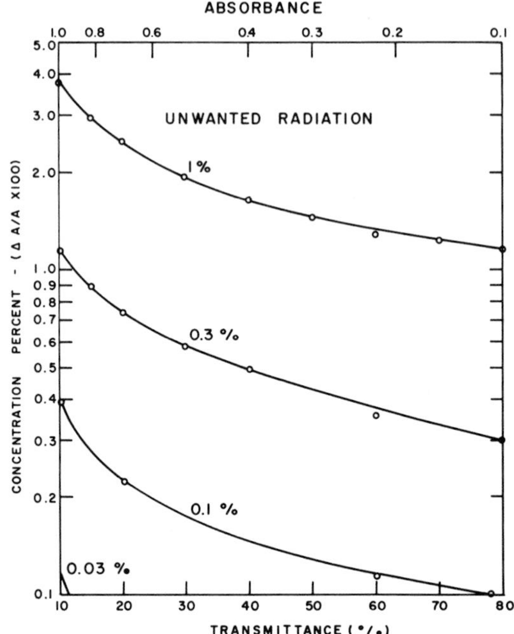

Figure 2-5. Error in quantitative results due to unwanted radiation.

The effect of false radiation on the quantitative accuracy obtained with an instrument is shown in Figure 2-5. The range of values of false radiation selected for these calculations includes values commonly found in laboratory instruments. In contrast to the other factors which have been discussed as having an influence on photometric accuracy, this factor of false radiation may be corrected for arithmetically, at least to a significant degree, provided carefully obtained data on false radiation are available.

2.2C. Quantitative Relations Between Operating Variables

From the discussion above we may deduce useful quantitative relationships between variables which may be at the disposal of the operator. The emphasis here will be on the interdependence of pairs of variables. Any attempt to combine these in a single, all-encompassing expression leads at best to a quasimathematical relation, which would tend to becloud the decision making process.

Resolution may be expressed in terms of the smallest frequency difference $\Delta\nu$ between neighboring bands which can be distinguished from each other. Except near the inherent limits of resolution $\Delta\nu$ is proportional to the physical slit width W. A change in slit width for greater or lesser resolution implies a change in scan time. To adopt a concept from information theory: if one wishes to maintain the same dynamic response, the scan time is proportional to the number of *bits of information* in the spectral region being scanned. The narrower the resolution, the greater the number of information bits. Therefore, the scan time T is inversely proportional to the resolution or slit width. These relations may be stated as

$$W = C_1 \Delta \nu \tag{2-2}$$

$$TW = C_2 \tag{2-3}$$

where C_1 and C_2 are constants.

However, there are other effects of changing the slit width, as indicated in equation (2-1). Specifically, the signal reaching the detector is proportional to the square of the slit width. To maintain the same dynamic response the gain of the amplifier must be changed inversely proportional to the signal—i.e., inversely proportional to W^2. As previously described, the recorded noise N is directly proportional to this gain. From this we conclude:

$$NW^2 = C_3 \tag{2-4}$$

where C_3 is another proportionality constant.

In the above discussion, N is a dependent variable. Within limits the noise is subject to operator control through the response time τ of the pen or amplifier. The longer the response time the lower the noise level, because of the smoothing effect on noise peaks of the slower response. Quantitatively this relationship, which involves a square root dependence, is expressed as follows:

$$N\tau^{1/2} = C_4 \tag{2-5}$$

When the response time is changed this interacts with the dynamic response of the system and must be compensated by a proportional change in scan time, thus:

$$T = C_5 \tau \tag{2-6}$$

Example.

Starting with a resolution $\Delta\nu_1 = 2\ \text{cm}^{-1}$, corresponding to a slit width $W_1 = 0.2$ mm, which required a scan time across the spectrum of $T_1 = 12$ min with a recorded noise level of $N_1 = 1\%$ and a pen response time $\tau_1 = 2$ sec, if the resolution is improved by a factor of 2, what scan time is required to maintain the same dynamic response, and what does the noise level become?

$$(\Delta\nu)_2 = \tfrac{1}{2}(\Delta\nu)_1 = 1\ \text{cm}^{-1}$$

From (2-2) $\quad W_2 = \tfrac{1}{2}W_1 = 0.1$ mm

From (2-3) $\quad T_2 = 2T_1 = 24$ min

From (2-4) $\quad N_2 = 4N_1 = 4\%$

Since a noise level of 4 % interferes with quantitative measurements, what changes must be made to reduce the noise level to the starting value of 1 % without degrading resolution? That is:

$$N_3 = \tfrac{1}{4}N_2 = N_1 = 1\,\%$$

From (2-5) $\qquad \tau_3 = 16\tau_2 = 16\tau_1 = 32 \text{ sec}$

From (2-6) $\qquad T_3 = 16T_2 = 32T_1 = 384 \text{ min } (6 \text{ h } 24 \text{ min})$

The above example illustrates the drastic steps which are required in order to obtain high-quality, high-resolution spectra. One compensating factor is that when high resolution is required, it is generally not necessary to scan a broad spectral region. In other words, one is seeking only to resolve a particular band system. However, this still necessitates careful and often tedious work.

The example indicates one other useful relationship between a pair of variables. In the case where we wish to maintain constant noise level we may note that:

$$T(\Delta\nu)^5 = C_6(N) \tag{2-7}$$

where C_6 is a constant for a given noise level N.

2.3. COMPONENTS OF INFRARED SPECTROPHOTOMETERS

This section will describe the fundamental components which go into infrared spectrophotometers. The material is presented so as to provide the reader with a sufficient knowledge and comprehension to make optimum use of his spectrophotometer, but it is not our aim to turn him into an instrument designer. For those who may wish to delve more deeply, references are given at the end of this chapter.

2.3A. Sources

Absorption spectrophotometers require the use of continuous sources of radiation—i.e., sources which radiate energy over the entire band of spectral interest, without sharp discontinuities created by emission lines or self-absorption bands. In the infrared portion of the spectrum, all of the commonly used sources are incandescent solids. The useful sources all approximate the theoretical performance of a blackbody radiator; it is of course impossible for any to exceed it. The radiation efficiency of any real object compared to an ideal blackbody is called *emissivity.*

Blackbody Radiators. According to the well-known Planck distribution law, the spectral distribution of energy emitted by an ideal blackbody is determined solely by the temperature of the radiating element. Figure 2-6 shows this distribution for a source at 1500°K,

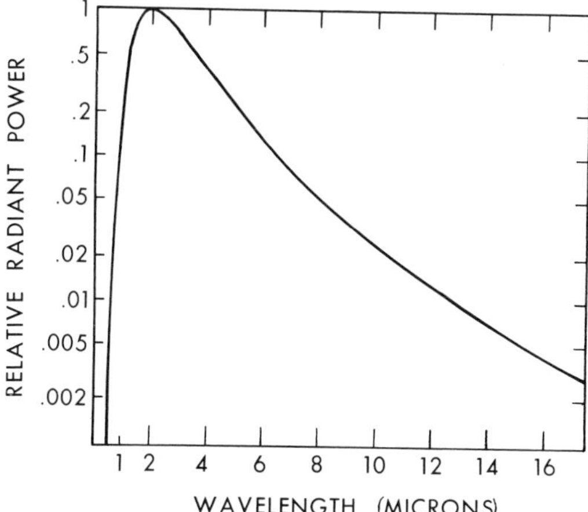

Figure 2-6. Spectral distribution of energy emitted by a 1500°K blackbody source.

which is the range in which commonly used sources are generally operated.

Certain key properties of this figure should be noted. With an energy peak at 1.85 μ the energy distribution drops off more sharply on the short-wavelength side than on the long-wavelength side. Even with this slower dropoff at longer infrared wavelengths, the radiant power at 15 μ, for example, is only roughly 1/200 as great as the peak power. The slit width of the spectrophotometer must be adjusted to compensate for this wide variation of available radiant power. In most of the spectrophotometers used by analytical chemists, the slit widths are automatically programmed to compensate for this as well as for other factors, so as to maintain a constant reference energy signal over the entire spectral range of the instrument.

In view of the fact that the distribution curve in Figure 2-6 is affected only by temperature, the effect of varying the temperature should be considered next. First, if the temperature is raised, the peak both shifts toward shorter wavelengths and grows in magnitude. Consequently, a greater percentage of the energy irradiating the sample is in the near-infrared and visible regions, outside the spectral region of interest in an infrared instrument. Of more direct significance is the variation of blackbody radiation with temperature at any given wavelength. Near the peak of the blackbody curve the radiant energy is proportional to T^5, where T is the absolute temperature. However, for wavelengths that are long compared to the peak wavelength, the radiant intensity is very nearly linearly proportional to T. Therefore, the potential gain of utilizing sources significantly hotter than 1500°K is meager relative to the accompanying disadvantages of more complex source control systems, shorter source life, greater heating effects on the sample, and similar factors.

It should be noted that all bodies radiate to all others. Thus, any warm object within an infrared spectrophotometer may become a source radiating more energy than it receives from its surroundings. The optical path in an instrument is always carefully shielded from radiation from hot electronic components. Potentially troublesome, however, are warm samples which, of necessity, are directly in the optical path. For example, a sample at 77°C (350°K, 171°F) exhibits a radiation peak at 8.4 μ whose intensity is 1.7% of that due to a 1500°K source at the same wavelength. Because of the slower decrease with wavelength of the 350°K source compared to the 1500°K source, the former radiates about 6% as much energy as the latter at 15 μ.

Sources Commonly Used. Analytical instruments most commonly use either a Nernst glower, a Globar, or a wire-coil source. In the near infrared, tungsten lamps such as the type used for projectors are frequently adapted.

The *Nernst glower* is made from rare-earth oxides. It is generally shaped as a cylinder up to a few millimeters in diameter and a few centimeters long, and is fitted with platinum leads. Thus it is conveniently shaped for focusing efficiently on a monochromator entrance slit. It generally operates in the 1400–1600°K range. Except for a deficiency in its emissivity below 5 μ, which is partly compensated by the proximity to its peak radiance, the Nernst glower is an efficient radiator. In utilizing a Nernst glower one must take into account its large negative temperature coefficient of resistance. At room temperature its resistance is so high that it is not feasible to heat it by passing a current through it. Instead, instruments employing Nernst glowers provide an indirect means of preheating the glower to a dull red temperature, after which the direct heating takes over. This extra step in starting a Nernst glower is frequently avoided by leaving the source on a low standby current when the instrument is not in use. The current is sufficiently low so as not to deteriorate the life of the source, but high enough to avoid the preheating cycle. Nernst glowers do not deteriorate on exposure to the atmosphere but are subject to mechanical distortion, which is their most severe practical limitation.

A *Globar* is made from silicon carbide or carborundum. It operates generally in a slightly lower temperature range than a Nernst glower. For wavelengths shorter than about 7 μ it has a significantly higher emissivity than a Nernst glower. Beyond this the two are comparable except for a drop in the Globar emissivity between 10 and 14 μ. Globars are not troubled by the negative temperature coefficient and the susceptibility to mechanical distortion of the Nernst glower. The most serious problem in working with Globars is that large thermal gradients are produced around the electrical contacts, frequently necessitating water cooling to avoid arcing problems. In some cases Globars have been made with a large diameter in the vicinity of the contacts and a smaller cross section in the area radiating to the slit. This minimizes the problem and in some cases reduces the requirement for water cooling to more convenient air cooling. In addition, it should be noted that Globar rods are generally of larger diameter than Nernst glowers, thus requiring a larger electrical power input for a given radiant energy through the entrance slit. However, the large diameter of a Globar makes it easier to illuminate wider slits uniformly with a Globar than with a Nernst glower.

Different types of *incandescent wire sources* have been incorporated in analytical infrared spectrophotometers. Their greatest limitation is reaching a sufficiently high temperature without excessively shortening their usable lifetime. Nichrome coils have been utilized up to a temperature of about 1100°K without unduly shortening their life. Another variation of a coil source is a rhodium or platinum coil packed in a sealed ceramic tube. This type of source operates at about 1500°K, with a usable life of approximately 1000 hours.

2.3B. Photometers and Photometric Systems

The purpose of a spectrophotometer is to record the spectral variation of energy transmitted through the sample. The photometric system is the means of establishing the variation and relating it to a reference, e.g., the energy transmitted through the sample in regions of no absorbance.

Single-Beam Systems. The earliest work in infrared spectroscopy utilized single-beam systems. Although still prevalent in some applications, primarily related to infrared physical measurements, these systems play only an infrequent role in analytical chemistry.

In contemporary single-beam systems the radiation is interrupted at a frequency compatible with the detector (e.g., 13 Hz†) to utilize ac amplifier systems. These have many advantages over the dc amplifiers which would otherwise be required, including increased stability and freedom from drift. Nevertheless, the stability requirements of the source and the amplifiers in a single-beam system exceed those of the double-beam systems to be described below.

In order to measure transmittance bands in a single-beam system it is necessary to scan the region at least twice—with and without the sample cell. To ensure that amplifiers and source are adequately stable, it is preferable to record the background scanned without the sample

† Hz, or hertz, is the presently accepted nomenclature for cycles/second.

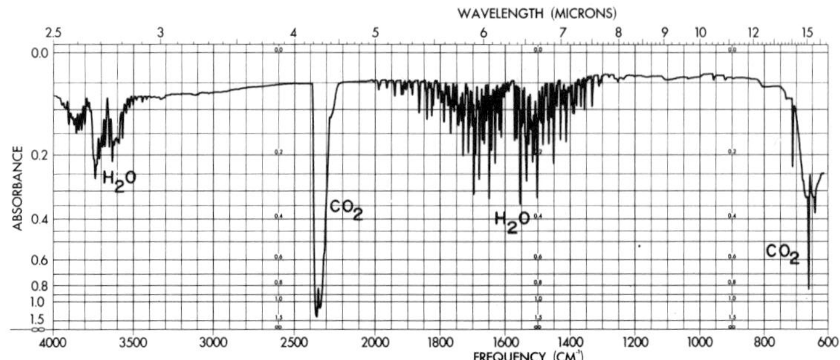

Figure 2-7. Atmospheric absorption bands.

both before and after the sample spectrum. Care must be exercised in regions where there are atmospheric absorption bands. Depending on the accuracy desired, it may be necessary to purge the optical path of water vapor and CO_2 in order to do quantitative analyses. In fact, even for qualitative analyses on moderate- or high-resolution instruments, it may be necessary to purge the optical path, unless the analyst can work in spectral regions free from atmospheric bands. Figure 2-7 shows the absorption spectrum of air of roughly one-meter path length. The regions most troublesome because of atmospheric absorption are: 3740 cm^{-1} (2.67 μ; H_2O), 2350 cm^{-1} (4.25 μ; CO_2), 1820–1330 cm^{-1} (5.5–7.5 μ; H_2O), and 670 cm^{-1} (14.98 μ; CO_2).

Double-Beam Optical Null Systems. Virtually all instruments used for analytical chemical applications utilize a double-beam optical null photometric system. In these instruments an electrooptical servosystem continually attenuates the energy in the reference beam so that there is no net signal difference between the reference beam and the sample beam. The recording pen indicates the position of the reference beam attenuator and therefore the relative transmittance of the sample.

Because the servosystem always drives to a null signal, the spectrum is basically independent of time or spatial changes in source intensity, amplifier drift, and most important of all, atmospheric absorption bands. The optical systems for this method of photometry are designed to ensure symmetry of the reference and sample beams with respect to path length, number of reflections, etc. However, the operator of such an instrument must keep in mind the fact that insertion of a sample cell may introduce a sufficient asymmetry between the reference and sample beams so that atmospheric absorption bands may appear in the spectrum. The degree to which these appear depends on the difference in absorption path length introduced, on the concentration of CO_2 and H_2O vapor in this path difference, and on the resolution of the instrument. The bands in question are very narrow and are naturally deeper or more pronounced with high resolution.

This phenomenon, called *uncompensation*, may look like noise and may be properly interpreted by noting the wavelength regions in which it appears. The same effect may arise from a totally different cause—i.e., too fast a scan speed. This may be readily understood by taking into account the fact that the double-beam optical null system is a time-shared system in which the detector receives energy from the reference beam half the time and from the sample beam the other half. When the instrument is scanning, the successive views of the sample and reference beams by the detector will actually be in adjoining regions of the spectrum rather than at the same spectral position. If one is scanning too rapidly through atmospheric absorption bands, the detector will be sensitive to the difference in absorbed energy corresponding to the small change in wavelength between successive impulses from the sample beam and the reference beam. Under these conditions the servosystem will attempt to follow the contour of the atmospheric absorption bands and the pen will trace a miniature version of this spectrum. However, if the scan rate is sufficiently slow, the change in energy from the reference half of the cycle to the sample half of the cycle will be so small that it will not be detected by the servosystem. This effect will also be more pronounced with high resolution, because of the accompanying increased band depths. This discussion of uncompensated effects shows that even though double-beam systems tend to counteract many of the effects of atmospheric absorption bands, the only way to eliminate their deleterious effects is to purge the optical system with dry N_2 or air. The higher the resolution of the instrument, the more scrupulously this rule must be followed.

Figure 2-8 shows a schematic diagram of an optical null system. Some of the key components in this system may now be described.

The *sector mirror C* alternately reflects reference-beam energy and transmits sample-beam energy through the remainder of the system. The rate at which a spectrophotometer can be scanned is essentially limited by the speed of this sector, which in turn is determined by the speed of response of the detector. If the latter is a thermocouple or a metal bolometer, the most commonly used detectors in analytical instrumentation, the chopping frequency is generally in the 10–13 Hz range.

The *synchronous rectifier S* is mechanically or electrically coupled to the sector mirror. It converts the amplified low-frequency output of the detector to direct current. The rectifier is phased with the optical chopper mirror so that the polarity of the rectified output indicates the condition of unbalance of the optical null system. That is, one polarity indicates more energy in the reference beam than in the sample beam,

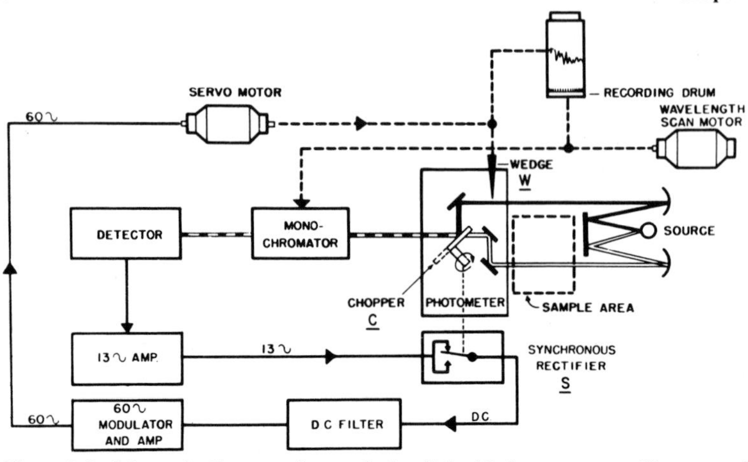

Figure 2-8. Schematic diagram of an optical null double-beam system. (Courtesy of Perkin–Elmer Corporation.)

and the opposite polarity indicates more energy in the sample beam than in the reference beam. A balanced null signal, of course, indicates equal energy in the two beams. The polarity of the synchronously rectified voltage determines the direction in which the servomotor drives the optical attenuator.

The *reference beam attenuator* or *wedge W* is driven by the servosystem so as to maintain a null energy balance between the reference and sample beams. The attenuator itself is generally a device with an open area which varies linearly with position. For example, it commonly resembles a thin comb with a small number of very long V-shaped teeth. The openings are precision-etched in a very thin metal sheet to ensure the shape of the opening and the ability to position it in a uniform portion of the beam. These two—i.e., the uniform variation of the open area with position, and the constancy of the energy distribution over the open area of the attenuator—are the most important factors influencing the photometric accuracy of the system. The attenuator usually consists of three to five open V's, so as to average over small nonuniformities in the cross section of the image on it or in the local sensitivity of the detector surface. The fact that the thickness of the attenuator may be only 0.003 in. makes it a delicate component, highly susceptible to damage by pencils, etc.

In most spectrophotometers the source is imaged on the attenuator, whence it is focused on the slits, which are reimaged on the detector, as in Figure 2-2. An alternate position for mounting the attenuator is at a pupil image (defined below), where the beam is also quite uniform. The attenuator may take forms other than a comb, such as that of a venetian blind, or a guillotine-like device.

In the last two paragraphs concerning the position of the optical attenuator in the system, we have indicated the requirement for beam uniformity. It may be of some interest to the reader to be aware of some of the properties of optical systems in spectrophotometers which relate to this. If a white card is placed in the beam of an instrument to observe the energy distribution across the beam, one will generally observe a bright central area surrounded by a dimmer area in which the energy drops gradually to zero. Exceptions to this observation will occur at fundamental image points, i.e., at the images of the pupil or of the slit. An image of the slit is characterized by the fact that there is a one-to-one correspondence between any point on the slit and a particular point along the image. The *pupil image* is a conjugate image point and may be defined by the fact that points along the slit have no corresponding image points on the image of the pupil. More specifically, each point on the slit is uniformly spread out over the entire pupil image, and each point on the pupil image is uniformly spread over the entire slit image.

How may we find these image points in an instrument? In the absence of a detailed optical layout, determination of the sharpness and uniformity of the beam on a white card is a helpful technique. However, many of the image points may be located easily on the basis of the following general principles. In general, to minimize area requirements, the source and the detector are located at the image of the slit. In addition, to minimize the size of the required sample, an image of the slit is usually located in the sample space or very close to one end of the sample space. In a monochromator the slit is generally located at the focal point of a paraboloid or sphere, so that the energy to and from the dispersing element is collimated with all rays in the beam parallel to each other. This places the dispersing element at, or very close to, the pupil of the system. Some instruments utilize an aperture stop situated at a pupil image in front of the monochromator as one means of controlling scattered radiation. The aperture stop delineates the beam so that no extreme rays strike the dispersing element on the edge only to be scattered around without having been properly dispersed and reimaged on the exit slit.

With these principles in mind we may summarize some of the implications of the location of the optical attenuator. Obviously it must be placed either at a slit image or at a pupil image, where the beam is uniform enough so that there is a simple relationship between open areas of the attenuator and the transmitted energy. If it is at a slit image, precautions must be taken to minimize or compensate for variations in radiant power over the surface of the source and for variations in sensitivity over the used area of detector. Conversely, if the attenuator is at a pupil image, one must be wary of surface imperfections on the dispersing element (scratching, fogging, etc.) and of inhomogeneities in transmitting optics near a pupil image (e.g., the prism).

Having examined the working basis and requirements of a double-beam optical null system, we shall next examine some of the properties of such systems of which the analyst should be aware. The accuracy of calibration of these systems is generally quite constant until the open area of the attenuator becomes very small. For very small openings it is virtually impossible to ensure precise calibration. The principal cause for this is the effect of slit width. If the slit is very narrow, the zero position of the attenuator is sharply defined. As the slit widens, the attenuator will have to move farther to stop the energy and produce zero signal. There is an analogous effect near 100%.

When a spectrum is recorded, quantitative results can be derived from it only if the zero and 100% levels are established. We must first define what is meant by these terms before the method of measurement becomes clear.

The *100% level* is the pen position corresponding to zero absorption in the sample at the wavelength being analyzed. Note that this is not the level indicated in a record made with no sample or sample cell in either beam. The latter is frequently referred to as the P_0 line and is characterized by smooth, small variations throughout the working region of a properly operating instrument. However, the P_0 line neglects factors such as reflection losses at cell windows and absorption at the wavelength under study in the cell window, in the solvent, or in any material in the beam other than the sample being analyzed.

One technique commonly used to determine the true 100% level at the position of a particular band is to interpolate this level from the level of the spectrum in nonabsorbent regions close to this analytical wavelength. To ensure accurate interpretation of the data it is necessary to record a preliminary curve of everything involved in the spectrum except the sample itself, preferably recording this on the same chart as the spectrum of the sample. For a solid sample this preliminary curve is the same as the P_0 line. If a KBr pressed disk is being used, more accurate results would be obtained if the preliminary curve were recorded with a pure KBr disk without sample, mounted in the same holder as that to be used with the KBr disk with sample. For a liquid sample, a run with the same cell filled with the solvent would satisfy the requirement. The same holds true for a gas or vapor, where a carrier gas, if any, should be taken into account for completeness.

A second technique for determining the 100% line is differential analysis. If this is adopted, a cell (or disk) devoid of sample but otherwise identical with that being inserted in the sample beam (including solvent or carrier gas, if any) is placed in the reference beam. Thus, the only difference between the two beams is the sample material itself. It should be noted that a liquid cell with sample cannot be accurately compared with a liquid cell containing no sample at all because of the difference in reflection losses within the two cells. This differential method has the advantage of closely approximating the symmetry condition, which is so important to double-beam optical null systems. However, if the solvent has bands of high absorbance itself, the reference cell may remove excessive amounts of energy from the system in these regions of the spectrum and cause fallacious results due to sluggish operation of the null system. This may be handled with greater confidence if a system employing automatic gain control (see Section 2.4D) is utilized.

Measurement of the *zero level* is not as simple as one might expect. The complication arises from the inherent lack of energy of a double-beam optical null system at zero transmittance, where both beams are blocked. Under these conditions it is possible for the optical attenuator to drift or coast below its zero position, since the system has no means of returning the errant attenuator to the true zero. In a well-operating instrument there are three potential causes for fallacious zero reading: a too rapid approach to zero, an improper electrical balance, and scattered radiation.

Too Rapid Approach to Zero. When the attenuator reaches zero too rapidly, the momentum of the system causes the attenuator to move below zero. There are well-defined techniques to overcome this difficulty. One is to block both beams under conditions where the attenuator is transmitting a significant amount of energy and then unblock the reference beam a very small amount in order to introduce a creeping down-scale motion of the attenuator. When the attenuator comes to rest near zero, the reference beam must be completely unblocked. Another is to scan the absorption edge of an optical material slowly. If the material is sufficiently thick, the zero level will be well defined and the approach to zero under normal scan conditions will be slow enough to avoid the drift-below-zero problem.

Improper Electrical Balance. The electrical balance control in an optical null system equalizes the reference and sample signal phases, including the effects of spurious pickup signals within the instrument. A negative unbalance will drive the attenuator below zero when both beams are blocked. A positive unbalance will tend to stop the attenuator before it reaches the true zero. The balance control is set for zero drift under zero-energy conditions, i.e., with both beams blocked. Therefore, patience must be exercised to avoid the problem of momentum drift discussed above.

Scattered or Unwanted Radiation. This is energy of wavelengths different from the small spectral interval under observation which nevertheless reaches the detector. Thus, even when there is total attenuation at the wavelength of interest, the scattered or unwanted radiation will cause a residual signal which prevents the attenuator from indicating true zero. Consequently, the zero level of the system should be determined under conditions where the percentage of scattered radiation is known to be trivial. The sources of scattered radiation are discussed in Section 2.5, and the effects of scattered radiation in quantitative work are considered in Chapter 6.

Ratio-Recording Systems. Ratio-recording systems have found only very limited application in analytical infrared systems, and the discussion here will be accordingly brief.

In the double-beam ratio-recording system the signals from the sample and reference beams are separated within the amplification system, and the output circuit records the ratio of the sample-beam signal to the reference-beam signal. There are many variations of ratio-recording systems, but most of them depend on utilizing only half of the available energy in each beam through some method of space sharing, as contrasted to the time sharing of full energy in optical null systems. For example, in the Halford–Savitzky type of ratio-recording system, the beam from the source is split so that the top half goes through the sample beam and the bottom half through the reference beam. The two beams are chopped 90° out of phase, and the phase-sensitive amplifiers operate on the two signals separately.

Because infrared systems are inherently energy-limited, any photometric system which requires inefficient use of the available energy would be less desirable. In addition to this factor, ratio-recording systems are also characterized by stringent requirements on the electronic amplifying systems. Ratio-recording systems have been commonly used in ultraviolet- and visible-range spectrophotometers, which distinguish themselves from infrared spectrophotometers by the general lack of energy limitations, availability of high-speed detectors and associated amplifiers, and less frequent occurrence of narrow absorption bands.

However, there are generally real advantages to ratio recording systems which, for some applications, more than compensate for the energy disadvantage. The zero is "live" and easily determined in a ratio-recording system. Also, the optical system generally involves light chopping before, not after, the sample. In such a system, a hot sample produces a constant background signal at the detector, which is easily separated electronically from the chopped light signal transmitted by the sample. In other words, ratio-recording systems generally require no correction for sample temperature.

2.3C. Dispersing Elements

Up to a few years ago prisms were the usual dispersing elements of infrared instruments. Now grating instruments have reached a state of development where they have become commonplace. With the large variety of instrumentation available to the analyst today, a wise selection of the best instrument for a particular problem or combination of problems depends particularly on a familiarity with the properties of the dispersing elements.

The dispersing element spreads out the spectral continuum of energy radiated by the source so that only narrow regions of the spectrum pass through the exit slit to the detector. The dispersing element determines both the resolution of the instrument through its angular dispersion characteristics and the spectral range of the instrument by virtue of its inherent physical properties. It is the object of this section to summarize the properties of the dispersing elements in most common use in infrared spectrophotometers.

Prisms. A prism placed in a light beam refracts (i.e., bends) the light which is incident at an angle different from the normal. A prism monochromator analogous to the grating monochromator included in Figure 2-2 is shown in Figure 2-9. The use of a prism as a dispersing element depends on the fact that the refractive index changes with wavelength so that different wavelengths are refracted by different amounts. Also, the variation of n with λ is a single-valued, unique function for the material of the prism.

The two prime requisites of a dispersing prism are that it have a high transmission in the applicable spectral region and that its angular dispersion be sufficient. As indicated in Section 2.2A, both the transmission efficiency and the angular dispersion directly affect the signal-to-noise ratio for a given resolution. However, these requisites are not independent of each other. Angular dispersion depends directly on $dn/d\lambda$, i.e., the variation of the index of refraction with wavelength. This derivative is called the *dispersion*. The nature of the physical mechanism of the transmission of radiation is such that the dispersion

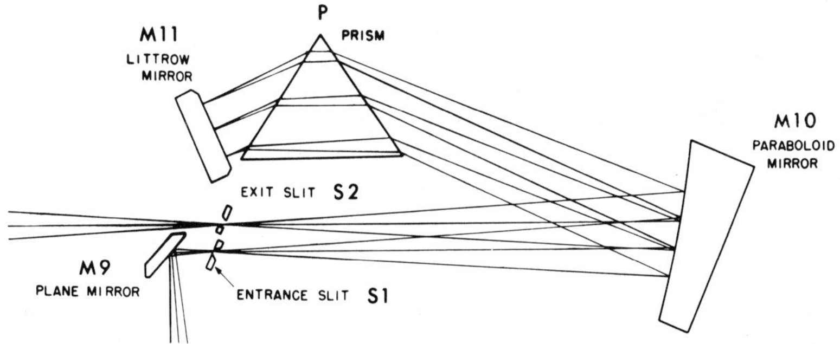

Figure 2-9. Littrow-type prism monochromator. (Courtesy of Perkin–Elmer Corporation.)

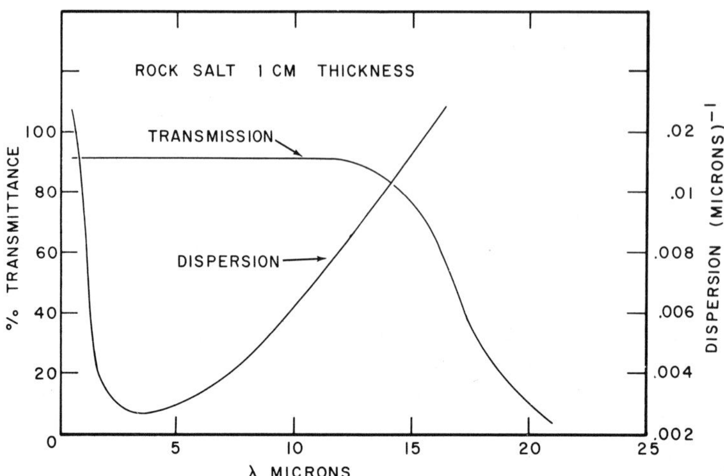

Figure 2-10. Transmission and dispersion of NaCl.

is greatest near a region of absorption. In Figure 2-10, where the transmission and dispersion of rock salt (NaCl) are shown, the interdependence of these variables may be noted in the vicinity of 15 μ. From the figure it is apparent why NaCl is efficiently usable as a prism material in the infrared only between roughly 5 and 15 μ, even though the material is quite transparent at wavelengths much shorter than 5 μ.

Rock salt has proved to be the most generally useful prism material in infrared analytical instrumentation. Its dispersion is high in a spectral region particularly rich in absorption bands—5 to 15 μ (2000 to 667 cm^{-1})—and continues to be adequate to supply additional information to even shorter wavelengths. Some instruments with rock salt prisms scan as far as 1 μ (10,000 cm^{-1}). More commonly though 2.5 μ (4000 cm^{-1}) is the short-wavelength limit of such instruments because of a change in sample-cell requirements in that region.

The most versatile prism instruments provide for relatively simple interchange of prism material, both to optimize the performance in a given range and to extend the efficient performance range of the instrument to shorter and longer wavelengths. Table 2-I lists the important properties of the prism and window materials commonly used in analytical instrumentation. Most of these materials must be artificially grown crystals to be adequately pure and homogeneous for optical use. They are also generally notable for varying degrees of undesirable properties, such as sensitivity to moisture and to scratching. CaF_2, which is widely used, forms one exception to this; it is insoluble in

TABLE 2-I. Properties of Infrared Optical Materials

Material	Long-wavelength limit,* μ	Refractive index	Cold-water solubility g/100 g H_2O
Fused silica (SiO_2)	4.5	1.45 at 1.0 μ	0
Artificial sapphire (Al_2O_3)	6.5	1.76 at 1.0 μ	9.8×10^{-5}
Magnesium fluoride (MgF_2)	7.5	1.38 at 3.3 μ	0.0076
Lithium fluoride (LiF)	8	1.38 at 2.0 μ	0.27
† Irtran 1 (MgF_2)	9	1.34 at 3.3 μ	0.007
Irtran 5 (MgO)	9.5	1.66 at 3.3 μ	0
Calcium fluoride (CaF_2)	11	1.42 at 2.0 μ	0.0017
Irtran 3 (CaF_2)	11.5	1.39 at 3.3 μ	0.0017
† T-12 (BaF_2/CaF_2)	12	1.41 at 3.3 μ	0.20
Arsenic-trisulfide glass (As_2S_3)	13	2.40 at 3.3 μ	0
Barium fluoride (BaF_2)	13	1.46 at 2.0 μ	0.17
Irtran 2 (ZnS)	14.5	2.20 at 3.3 μ	0
Silicon (Si)	15	3.4 at 10 μ	0
Irtran 4 (ZnSe)	21.8	2.40 at 3.3 μ	0
Sodium Chloride (NaCl)	22	1.50 at 9.0 μ	35.7 at 0°C
Germanium (Ge)	23	4 at 2 μ	0
Silver chloride (AgCl)	25	1.98 at 10 μ	0
Potassium chloride (KCl)	26	1.46 at 10 μ	34.7
Irtran 6 (CdTe)	28	2.69 at 3.3 μ	0
KRS-6 (TlBr–TlCl)	30	2.18 at 10 μ	0.32
Potassium bromide (KBr)	33	1.53 at 10 μ	54
Silver bromide (AgBr)	35	2.23 at 0.67 μ	0
Potassium iodide (KI)	40	1.62 at 10 μ	127
KRS-5 (TlBr–TlI)	40	2.37 at 10 μ	0.05
Cesium bromide (CsBr)	42	1.66 at 10 μ	124
Cesium iodide (CsI)	55	1.74 at 10 μ	44

* All of the listed materials transmit at all shorter infrared wavelengths to 1 μ or beyond, with the exception of Si (1.2 μ), Ge (1.8 μ), and Irtran 6 (2 μ).

† Irtran refers to microcrystalline hot pressed materials developed by Eastman Kodak Company. T-12 is a two-phase polycrystalline material developed by Harshaw Chemical Company.

water and is very hard. Selection of materials in the above table for use as prisms in a spectrophotometer depends both on physical properties and on the cost for the relatively large size required for this application. Because most of the materials suitable for use as prisms are very soluble in water and will tend to fog (lose their polish) if subjected to condensation, infrared spectrophotometers generally operate with the prism at a temperature higher than its surroundings. With even the most exacting care, moisture-sensitive prisms will always tend to fog with time. While they can be repolished, the techniques involved are such that the task should generally not be undertaken in a chemical laboratory. Another physical property requiring attention is the variation of the refractive index with temperature. Errors in wavelength calibration due to this effect may be avoided within reasonable temperature limits by use of a bimetallically controlled mechanism, which is incorporated in most prism instruments.

In addition to the absorption losses referred to above, there are also reflection losses which are subtracted from the optical beam at each surface of incidence on the prism or on any other transmitting optical part. When a beam is normally incident on a surface of refractive index n, a certain fraction of the energy, given by the formula $(n - 1)^2/(n + 1)^2$, is reflected. If we call this ratio the reflection coefficient R, the energy transmitted after reflection loss at a single surface is $T_1 = 1 - R$. A window has two surfaces, and its transmission after reflection losses at both surfaces is $T_2 \cong (1 - R)^2$. This is plotted in Figure 2–11 as a function of the index of refraction. This figure also shows the corresponding transmission $T_4 \cong (1 - R)^4$ of, for example, the two windows of a sample cell. These curves neglect the absorption within the material and the effect of the index of refraction of a sample in contact with the cell windows, as well as multiple internal reflections.

The normal-incidence values plotted in Figure 2–11 may be used as a first approximation for prisms, for which the angle of incidence necessarily differs from 0°. Taking into account the fact that in most prism monochromators the beam is reflected back through the prism

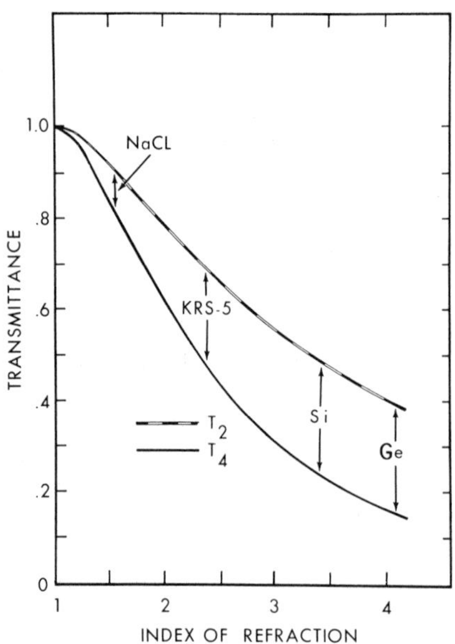

Figure 2-11. Effect of reflection losses as a function of refractive index.

a second time before returning to the collimator (Littrow system), there are four surfaces involved, with reflection losses inherent at each of the four. As may be seen in Figure 2-10, this severely detracts from the usefulness of such materials as KRS-5 as prism materials.

When the incidence of the beam is not normal to the surface, the reflection coefficient is larger than R. Another phenomenon also exhibits itself under these conditions, in that the reflection coefficient at high angles of incidence is increased differently for the two polarization components of the beam. The result is that the prism produces a small amount of polarization, which is nevertheless great enough so that it cannot be neglected in spectra of optically inhomogeneous samples. This includes many organic solids, particularly long-chain hydrocarbons. The degree of polarization caused by a 60° NaCl prism, in a standard monochromator, is such that the horizontal component is 1.51 times as great as the vertical component (electric vector vertical). The degree of polarization, which has been defined as $(E_H - E_v)/(E_H + E_v)$, is therefore 20%. This factor varies only slightly with wavelength.

Gratings. Gratings used in the infrared are of the plane-reflection type, which consists generally of a glass blank with an aluminum coating containing many closely and precisely spaced parallel grooves. The theory of light dispersion by a grating may be pursued in any good textbook on optics. Suffice it to say here that the incident light is dispersed by the phenomenon of diffraction, which is fundamentally a wave interference effect among the light rays from the long series of grooves.

If one is to take full advantage of a grating spectrophotometer, it is desirable to become familiar with the fundamental properties of radiation diffracted from a grating. First of all, in contrast to a prism: (a) the spectral distribution of radiation from a grating follows simple trigonometric laws; and (b) the energy diffracted at a given angle is not of a single wavelength but consists of a multiplicity of wavelengths. Both of these unique properties are demonstrated in the grating equation, which for a reflection grating may be stated as

$$mN\lambda = 2 \sin \theta \cos \delta \qquad \text{or} \qquad \bar{\nu}/mN = \csc \theta / 2 \cos \delta \tag{2-8}$$

Figure 2-12 sketches the geometry involved: θ is the mean of the angles of incidence and diffraction, measured relative to the grating normal; δ is the difference between the incident or the diffracted angle and the mean angle θ; m is the number of grooves per unit length; λ is the wavelength; $\bar{\nu}$ is the wavenumber; and N is the order number. The first unique property mentioned above follows from the fact that for each given order N, λ is proportional to $\sin \theta$ and $\bar{\nu}$ is proportional to $\csc \theta$, since the other quantities in the equation are all constants of

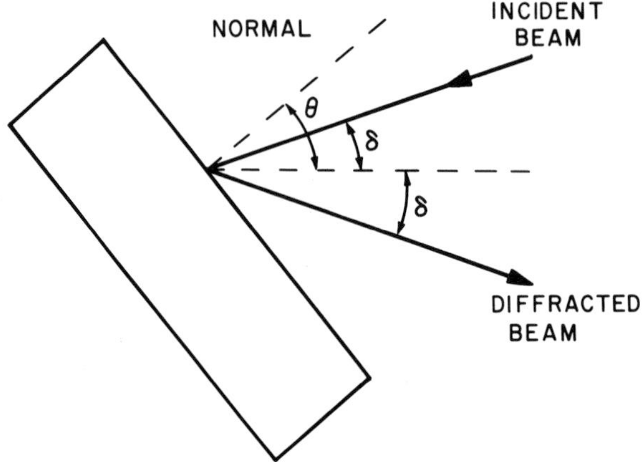

Figure 2.12. Diffraction at the surface of a reflection grating.

the grating and the optical system. The order number is the term used to describe the second property indicated above. By virtue of the theory of diffraction, N is any integer. Thus if a continuum of radiation is incident on a grating, the diffracted energy at a specific angle consists of a series of integrally related wavelengths. For example, if one sets a 100-groove/mm grating at an angle which allows 12 μ first-order radiation to reach the exit slit, one will also detect second-order 6-μ energy, third-order 4-μ energy, fourth-order 3-μ energy, fifth-order 2.4-μ energy, sixth-order 2-μ energy, etc. In order to obtain monochromatic radiation from a grating it is essential to provide a means of eliminating all but one order from this series of integrally related wavelengths.

There are two standard methods for sorting orders in an infrared spectrophotometer. The classical technique is to use a second, prism monochromator, generally placed in front of the grating monochromator. The fore-monochromator allows only a limited spectral band to enter the grating monochromator, so that only a single order remains to be diffracted at any given grating angle. Scanning requires the two monochromators to track together, i.e., to keep in coincident calibration.

The second method for order elimination is the use of filters. Before 1959 this method was limited to far-infrared and near-infrared work. In the former case no prism materials are available, so that inconvenient and often inefficient filtering techniques were used by necessity. In the near-infrared range, simple filters have been available for a long time. Partly in fulfillment of the requirements of military uses of infrared radiation, multilayer interference techniques have been developed to the point where order-sorting filters throughout the entire infrared region are now available. In the majority of applications of filter–grating systems, the gratings are used in first order only, primarily because of the simplicity of filter requirements. First-order operation calls for long-wave-pass filters, i.e., filters which transmit at wavelengths longer than a sharp cutoff and which thoroughly reject radiation of shorter wavelengths or higher orders. The characteristics of a typical long-wave-pass filter are shown in Figure 2-13. Rejection requirements depend both on grating efficiency characteristics and on source emission characteristics. Rejection requirements vary from 99.5% for filters at shorter infrared wavelengths to 99.9999% for filters used at longer infrared wavelengths. Interference filters are relatively insensitive to moisture and are durable in normal instrument environments. In nonrejection regions they transmit between 70 and 90% of the incident energy, which puts their inherent losses in the same class as the reflection losses of prisms.

In general, both filter–grating and fore-prism–grating systems adequately reject unwanted orders for analytical applications. In exceptional cases where the ultimate in low unwanted radiation is required, a fore-prism–grating system, particularly when restricted to lower grating orders, may be expected to provide the edge in this regard. However, the demonstrated advantages of filter–grating systems over fore-prism–grating systems have led to almost universal adoption of these systems for analytical applications where grating performance is desired. Most of the pertinent advantages result from the use of a filter in place of a fore-prism monochromator, and from the versatility of the filter itself.

For example, the optical path length through a filter-grating system is shorter by the path length of a monochromator, which decreases

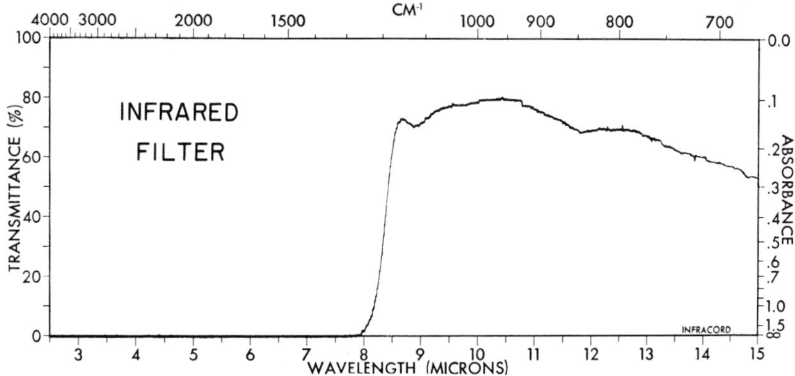

Figure 2-13. Characteristics of a typical long-wave-pass interference filter.

the effect of atmospheric absorption within the instrument. This decreases the need for removing H_2O and CO_2 from the spectrophotometer. The effect of these absorption bands is too frequently overlooked, particularly in high-resolution spectroscopy. Because of the shorter optical path length, the filter–grating instrument is also more compact, thus occupying less laboratory bench space.

In addition, the environmental problems generally associated with infrared materials—moisture sensitivity and the requirement of temperature compensation—are eliminated, and the artificial wavelength range restrictions imposed by the prism materials are lifted. The useful short-wavelength region for a given prism material is determined by the decreasing dispersion, and the long-wavelength limit is determined by absorption.

Filter–grating systems are free from the slit-width effects which are occasionally troublesome in fore-prism–grating systems. These effects arise from the fact that the slit width of the prism monochromator must be large enough to cause its pertinent band of dispersed energy to fill the entrance slit of the grating monochromator. When operating in higher orders of the grating, the angular dispersion is quite high. Under these conditions the slit width of the grating monochromator, and therefore also that of the fore-prism monochromator, is generally quite large in comparison to a normal slit width for a prism monochromator. Particularly if the system is being operated with a high-energy, wide-slit program, there will be a tendency to experience order interference when operating in the higher orders. In a filter–grating system the effectiveness of the filter is independent of the slit width, since the band of energy entering the grating monochromator does not change with slit width.

Angular Dispersion. This fundamental property of any dispersing element is 3 to 30 times as great for a grating as for any prism used in the infrared region. In fact, the only case where the angular dispersion of a grating is less than that of a prism is in the shorter, ultraviolet region. For example, a quartz prism is superior to a grating, with respect to angular dispersion, at wavelengths shorter than about 250 mμ.

Some of the relevant facts about the angular dispersion of a grating may be derived from the following equation, which is a close approximation when the grating is used in the Littrow configuration (δ small in Figure 2-12):

$$d\theta/d\lambda = 2 \tan \theta/\lambda \tag{2-9}$$

From this we may note that the angular dispersion ($d\theta/d\lambda$) depends only on the wavelength λ and the grating angle θ. It is independent of the order number N and the number of grooves per millimeter m, except insofar as these determine the angle at which the wavelength λ is observed. This explains the phenomenon of a sharp change in spectral slit width at a wavelength at which either the grating is changed (in a multigrating system) or the grating order is changed. The changeover wavelength is observed at two different grating angles and the angular dispersion changes proportionally to the two different values of tan θ. For high-resolution work, equation (2-9) indicates the desirability of operating at moderately high grating angles up to the point where the steep aspect of the grating seriously limits the cross section of the incident beam.

Another valuable property of grating spectrophotometers which relates to the angular dispersion properties of gratings is the nearly constant spectral slit width in wavenumbers which accompanies a constant-energy slit program. This contrasts with prism instruments, where spectral slit-width variations, in frequency units, of as much as 10 to 1 will occur. These variations are limited roughly to 2 to 1 for a grating instrument. Since bandwidths and separations are related to energy factors and therefore to frequency differences, the above property of grating instruments, combined with an overall higher resolution, makes grating spectrophotometers exceptionally valuable in the interpretative applications of infrared spectroscopy, such as structure determination.

Grating Efficiency and Polarization. In connection with grating spectrophotometers it is common practice to specify the "blaze" angle and wavelength of the gratings employed. It is, therefore, pertinent to discuss the meanings and implications of these terms to avoid the pitfalls of over- or underemphasis of their significance.

Blazing a grating increases its energy efficiency over a wide range of angles on both sides of the blaze angle, which is the angle at which the diffracted energy is maximized. The blaze angle depends on the shape of the grooves, which is determined by the shape of the ruling diamond and by its angle of contact with the surface being ruled. Figure 2-14 is a diagrammatic representation of the shape of the rulings of a blazed grating. The blaze angle θ_0 is the grating angle between the normal to the broader face of the groove and the grating normal. There also exists a secondary angle of peak efficiency near the "antiblaze," normal to the narrower face of the groove.

Modern blazed gratings are capable of diffracting up to as much as 85% of the incident energy of a given wavelength at angles near the blaze. The efficiency of a grating in the first order holds up quite well over a very wide angular range. For example, a grating blazed at about 27° is usable with no less than half its peak efficiency between approximately 17° and 45°. This angular range becomes much smaller in the higher orders.

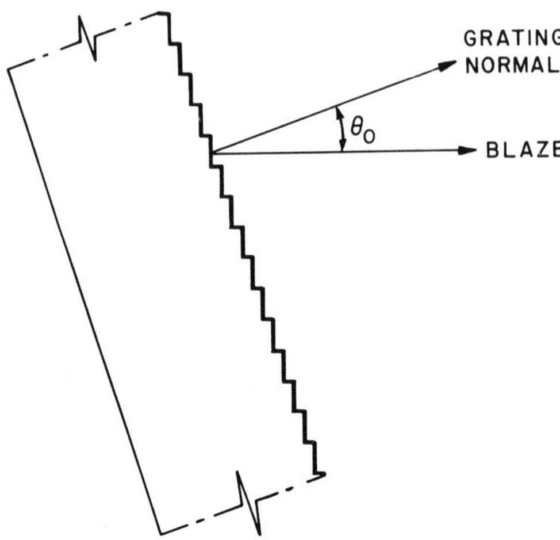

Figure 2-14. Groove shape of a blazed grating.

Although the efficiency curve proves to be asymmetrical when plotted against angle, as well as against wavelength, the curve is nearly symmetric when plotted against frequency. The following are useful guide rules with respect to grating efficiency.

1. A grating operates with 50% or more of its peak efficiency in the first order from $\frac{1}{2}$ its blaze frequency to $1\frac{1}{2}$ times its blaze frequency —i.e., from $\nu_B/2$ to $3\nu_B/2$, or from $2/3\lambda_B$ to $2\lambda_B$. This is a factor of 3 in frequency or wavelength.
2. In higher orders, the width of the 50% or greater relative efficiency range is the same as in first order *on a frequency scale.* Higher orders correspond to higher frequency, which leads to the conclusion that the same frequency interval corresponds to a smaller usable wavelength range $\Delta\lambda$ in higher orders. That is, in higher orders the grating efficiency remains high over smaller wavelength and angular ranges near the blaze.

Actual measurements in the first three orders of a standard infrared grating are shown in Figure 2-15. The close correspondence to the above guide rules may be noted.

The measurements for Figure 2-15 were made with unpolarized light. If polarized light were used, the efficiency curve for light polarized parallel to the grooves would be quite different from the curve for light polarized perpendicular to the grooves. These effects are of significance to the analyst because they show that the energy diffracted by a grating undergoes significant polarization, which varies both in angle and in direction, going through zero polarization in the vicinity of the blaze

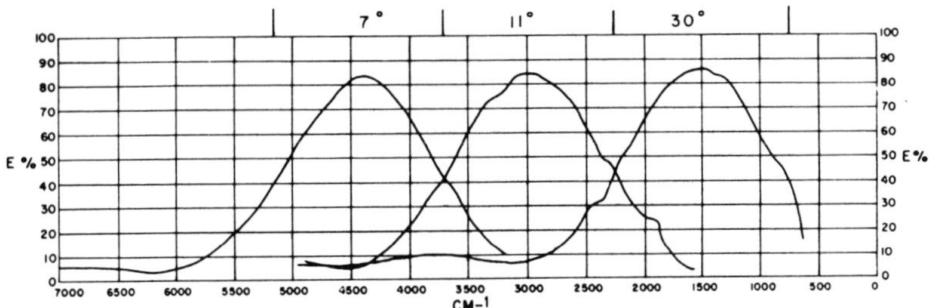

Figure 2-15. Efficiency measurements made in the first three orders of a 100 groove/mm grating blazed at 22°, using unpolarized incident radiation. The angular ranges corresponding to ≥50% peak efficiency are indicated along the top. (Data courtesy Perkin–Elmer Corporation.)

angle. Figure 2-16 is a plot of the degree of the polarization *versus* grating angle based on actual measurements made on the grating that was used to obtain the data for Figure 2-15. The polarization phenomenon must be taken into account in any serious work involving solids which may have selective orientation.

Variable Spike Filters. Multilayer interference filters of the long-wave-pass type as used for order elimination were discussed on p. 37. By variation in the design of the coatings it is possible to make filters which pass only a narrow band of wavelengths. Such a device is called a *spike filter*. Recent developments in coating techniques have led to the availability of variable spike filters. In these, the thickness of the layers is varied as a function of angular position on a circularly shaped disc. If a narrow beam from an infrared source impinges on the disc the

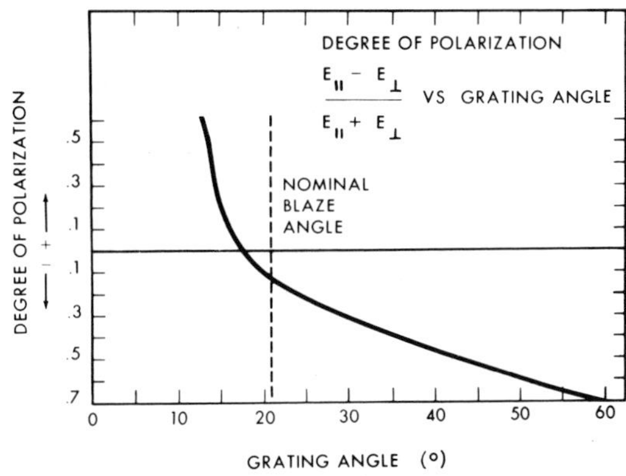

Figure 2-16. The degree of polarization of a blazed grating.

wavelength transmitted by the spike filter is scanned as the disc is rotated. Such a device is not a dispersing element in the literal interpretation, but it may effectively replace a dispersing element when only limited resolution is required. Half bandwidths of roughly 1 to 2% of the peak wavelength are common for spike filters. In general, as higher resolution is required of the device the peak transmittance is lower and the cost is greater.

2.3D. Detectors

The function of the detector is to convert the infrared radiation into an electrical signal. In view of the following two facts, this task should not be taken lightly.

1. The energy of an infrared photon is low. For example, a 3-μ photon has an energy of roughly 1.0 electron volt and a 30-μ photon of 0.1 electron volt. From these energies one may expect that the applicability of infrared solid state detectors, which depend on the activation of some photoelectronic phenomenon within the solid by the photon energy, is limited.
2. Analytical applications almost always require information over a broad spectral range, which necessitates use of a nonselective detector.

Infrared detectors fall into the two broad categories—thermal detectors and photodetectors. The former are relatively nonselective and are therefore the most common in infrared spectrophotometers. The latter encompass mostly solid state devices, which tend to be quite limited in useful wavelength range and have found application in infrared analytical equipment only under unique circumstances.

Thermal Detectors. Thermal detectors essentially respond to the incident radiant power to produce a signal proportional to this power. The power level is about 10^{-9} watt. Obviously the sensitive element, which is blackened to increase its efficiency as an absorbent receiver, must have a very low heat capacity to be responsive to this power. This implies that the target must be extremely thin and small in cross section, which requires that the detector be placed at an image of the slit which has been demagnified in size as much as optical aberrations allow. Most spectrophotometers employ ellipsoidal mirrors as the demagnifying optic. Even with such an aspheric, a reduction of six or seven to one proves to be the point of diminishing returns. Higher magnifications cause such steep rays that the image is made diffuse by aberrations, and more of the energy is lost by reflection at the window of the detector.

The degree to which the blackened target appears black at all wavelengths represents the only limit to the nonselectivity of thermal detectors. A useful though crude picture may be evoked by the statement that an effective black coat has rough spots whose dimensions are larger than the wavelength being absorbed. For wavelengths very long compared to the dimensions of the rough spots the black surface looks smoother and a higher percentage of the radiation is reflected rather than absorbed. This effect of size may be illustrated by analogy to the way in which an automobile tire essentially ignores hairline road cracks but bumps severely in ruts comparable to its own dimensions. Thus, thermal detectors tend to drop in sensitivity in the far infrared. This is rarely a problem in the more commonly used regions of the infrared spectrum.

The thermal detectors used in analytical instrumentation generally fall into one of three categories, depending on the physical basis of the conversion of heat into an electrical signal:

1. thermoelectric;
2. thermal variation of resistance;
3. pneumatic.

Thermocouples. The most commonly used detectors in analytical infrared instruments have been thermocouples. These thermoelectric detectors depend on the Peltier effect, which is the generation of a voltage in a circuit containing two dissimilar metals or semiconductors when the junctions are at different temperatures.

In general construction the sensitive element or receiver is a metal foil, commonly gold or platinum, which has been black-coated with, for example, gold soot deposited in a vacuum. The receiver is spot welded to pins or wires made of the thermoelectric materials used. This serves as the "hot" junction, while the remaining junction is designed to have a relatively large heat capacity and to be shielded from the incident radiation.

The design of a thermocouple represents an interesting balance of design parameters. First, there is a compromise between sensitivity and speed of response. The larger the temperature differential produced by the radiation, the larger the developed electrical signal. This implies minimizing the means of heat transfer from the receiver to its surroundings (i.e., conduction, convection, and radiation). However, a consequence of this isolation is a very slow response to changes in the incident radiation. Nothing can be done about the radiation factor, since one side of the receiver must be black to absorb the radiation with the greatest efficiency. To minimize convection, thermocouples are mounted in an evacuated case. The conduction factor is optimized by choosing the thermoelectric materials that represent the best combination of high Peltier coefficient and low conductivity. Also, the thermocouple lead wires are kept very small in cross section.

Another important compromise is among mechanical life, optical matching, and sensitivity. To maximize the temperature rise of the receiver, it must be very thin and small in area. In fact, a thermocouple is generally sensitive to receiver temperature rises of the order of 10^{-6} C°. If the receiver is too thin, its mechanical life will be too short. If it is too small in area, it will become impossible to concentrate all of the energy which goes through the exit slit onto the receiver. This factor interacts directly with slit-height, slit-width, and aperture ratio limitations in the design of the spectrophotometer. The noise-to-signal sensitivity ratio varies approximately as the square root of the area of the receiver.

Even with the compromises required, the thermocouple has proved to be the most generally useful of all detectors in analytical infrared instruments.

The noise generated by a thermocouple is primarily determined by Johnson noise, which is electrical noise generated by the random thermal fluctuations of the electrons through the resistance R. The rms (square root of the mean square) Johnson noise in a bandwidth Δf appearing across a resistance R at absolute temperature T is

$$(V_n)_{\text{rms}} = (4kTR\Delta f)^{1/2} \tag{2-10a}$$

where k is Boltzmann's constant. At room temperature this becomes

$$(V_n)_{\text{rms}} = 1.3 \times 10^{-10}(R\Delta f)^{1/2}) \text{ [volt]} \tag{2-10b}$$

The bandwidth Δf is inversely proportional to the response time of the system. Thus, as one increases the response time, the bandwidth and therefore the noise decrease.

In infrared spectrophotometers the Johnson noise generated in the detector is the primary limiting factor in the sensitivity. The bandwidth under normal scanning conditions is generally 1 to 2 Hz. For work with narrow slit schedules for high-resolution spectra the response time of the systems is lengthened by a factor of up to 30, thereby decreasing the bandwidth by a factor of 30 and the noise by a factor of more than 5. The resistance of commercially available thermocouples is about 2 to 100 ohms. To a first approximation, the signal sensitivity of these also varies as the square root of the resistance, so that the available signal-to-noise ratio of the different available thermocouples is not greatly different.

The response time of the thermocouple determines the chopping frequency employed in the instrument. At too fast a chopping speed the thermocouple receiver is unable to respond well to the changes in signal. Again, a compromise must be sought. A point of diminishing returns will also be found at too slow a chopping speed, for which the gain in sensitivity is small while the increase in scan time, growth in inherent amplifier noise, and other complications more than counteract the increased sensitivity. For most thermocouples chopping frequencies in the range of 10 to 13 Hz are utilized.

Bolometers. A bolometer is a device which depends on its change in resistance when heated by the incident radiation to produce a corresponding change in an electrical voltage. The sensitive resistance element or receiver is a thin film, mounted so as to isolate it thermally from its surroundings as far as possible. The film may be either a metal or a semiconductor, prepared by vacuum condensation, rolling, or sputtering. Electrically it is usually one element of a bridge circuit such as a Wheatstone bridge. This bridge circuit and the need for a power supply stable enough to afford sensitivity to minute temperature and resistance changes represent complicating factors in the use of bolometers as compared to thermocouples. They account to a great extent for the less frequent utilization of bolometers in analytical instrumentation.

Most of the compromise factors discussed in reference to thermocouples apply equally to bolometers. In addition, the previously discussed noise considerations apply equally well to metal bolometers whose primary source of noise is Johnson noise. The resistance of commercially available metal bolometers is 30–100 ohms. Response times are comparable to those of commonly available thermocouples.

Semiconductor bolometers have a much higher impedance than metal bolometers and are also characterized by a much larger temperature coefficient of resistance. Another unique characteristic of semiconducting bolometers is that they cannot be represented by a single time constant; instead they require two different time constants to describe their response function. In addition, their sensitivity varies as $A^{0.7}$, where A is the area of the receiver. For most other thermal detectors the variation is proportional to $A^{0.5}$.

Sensitivity is usually defined as the power which must fall on the detector to generate an electrical signal equal to the noise generated by the detector itself. The result of the above-mentioned difference in area functions is that the noise level in semiconducting bolometers varies more rapidly with area than in metal bolometers and thermocouples. For the receiver areas required in infrared spectrophotometers, the semiconducting bolometers have not found general application.

Pneumatic Detectors. The third type of detector which has been used in infrared spectrophotometers is the pneumatic detector, which is an extremely sensitive gas thermometer. The radiation which reaches the sensitive element or receiver heats the enclosed gas. The resulting small rise in pressure is converted to an electrical signal by one of several possible methods. The most commonly applied version of a pneumatic detector is the Golay cell, whose receiver is an aluminum-coated plastic. In this case the receiver does not necessarily depend on a black coating to absorb the radiation. Instead, in most versions of the Golay detector, the receiver characteristics are determined by electromagnetic theory to provide a flat sensitivity over the entire electromagnetic spectrum from far ultraviolet through radio frequencies. Golay cells are advantageous when relatively larger receivers are required, as in far-infrared or large grating spectrophotometers. The standard Golay cell has a receiver diameter of 3 mm compared to a standard thermocouple rectangular target size of 2 by 0.2 mm.

Among the disadvantages of the pneumatic type of detector is its physical size. Thermocouples and bolometers are mounted in a thin stalk, which permits them to be placed within the beam, utilizing an ellipsoidal mirror to reflect and to concentrate the energy on the target with very little of the energy being lost in the shadow of the stalk. However, the size of the Golay detector precludes such an arrangement. One commonly used solution to this dilemma is a 45° off-axis ellipsoidal mirror, a relatively expensive optical element.

Another problem unique to this type of detector is its susceptibility to destruction on exposure to high levels of radiation, such as normal room radiation. In general, problems relative to durability and sensitivity to microphonics tend to be somewhat more severe for Golay detectors than for other thermal detectors.

Photodetectors. The second major class of infrared detectors—photodetectors—has found negligible application in analytical instrumentation except for the use of PbS photoconductive detectors in the near infrared. The principal reasons for this are:

1. the previously mentioned limited spectral regions of high sensitivity for each type,
2. the fact that low-temperature operation, usually inconvenient, is necessary with longer-wavelength photodetectors and is desirable even for shorter-wavelength photodetectors.

Increasing convenience in suitable cryogenic devices as well as further developments of photodetectors may change the above indicated situation within the next few years. Only time will tell.

2.3E. Amplifiers and Recorders

Beyond the preamplifier, the amplifier requirements are essentially independent of the detector. Because the preamplifier requirements for thermal detectors are unique and more stringent than for photodetectors, only the case of thermal detectors will be considered here.

The unique requirements of the preamplifier with respect to thermal detectors have their origin in the generally low-impedance and low-signal levels connected with these. These rms signal levels, of about 10^{-9} volt, require the utmost care in handling to avoid being swamped by magnetic or electrical pickup generated within or outside the spectrophotometer. At these signal levels, magnetic shielding by common soft iron or steel is ineffective because these materials have very low permeability at the low magnetic fields capable of inducing spurious electrical signals of this magnitude. Instead, special low-field high-permeability materials such as "Mu-metal" and "Hy-mu 80" must be utilized. To be effective these materials must be dead soft annealed. They are subject to loss of their essential properties by minor mechanical stressing, and therefore the magnetic shields must be handled with care. In addition, it is essential for all wires in the low-level circuit to be tightly twisted to avoid presenting loops to stray magnetic fields, a situation which could also generate spurious voltages.

Analogous protection against stray electrical pickup necessitates extraordinary precautions. Chief among these are the grounding requirements. First of all, the elements in the low-level circuit through the preamplifier must be securely tied to the instrument ground wire, which itself should be connected to a reliable, solid electrical ground such as a cold water pipe. In addition, it is important that the entire circuit be tied to this ground at only one point. The penalty for improper grounding is the existence of ground "loop currents" which, although minute, may generate signals in the low-level circuit, leading to erratic instrument operation.

The low-level, low-impedance signals generated by thermal detectors are not suitable for direct coupling to a vacuum tube amplifier. The input grid circuit of a typical vacuum tube works best with an impedance of the order of 100,000 times the impedance of a thermal detector. Furthermore, the noise level inherent in a lower-noise vacuum tube is equivalent to an input signal of about 0.5 μv, which is roughly 1000 times the minimum detectable Johnson-noise-limited signal of the detector itself. These mismatched conditions are bridged by a transformer especially designed for high-efficiency amplification of the signals involved. This unit must be protected by extraordinary shielding, both magnetic and electrical, and potted so as to minimize susceptibility to microphonics. The first stage of amplification beyond the transformer is still at a sufficiently low level to necessitate precautionary steps and careful selection of a low-noise vacuum tube. In a well designed and constructed preamplifier the noise contributed by the input vacuum tube is no greater than the noise level from the detector after amplification by the transformer. In a properly operating system there is no significant noise contribution from the circuit beyond the input stage.

Transistors generally have too high an equivalent noise input for direct coupling to a thermal detector and they have too low an input impedance for use of the type of transformer described above. However, with the development of the so-called field-effect transistor this picture has changed to some degree. These are lower-noise, higher-input-impedance devices than normal transistors. Useful solid state preamplifiers have been designed for infrared spectrophotometers with field-effect transistors, but, in general, they have a lower signal-to-noise ratio than well designed vacuum tube amplifiers.

The amplifier system subsequent to the input stage generally follows a common pattern. The low-frequency signal (e.g., 13 Hz) is amplified to a level of many volts by several stages of more or less standard design, with a gain potentiometer inserted in one of these stages. The resulting high-level, low-frequency signal is then rectified to direct current by a rectifier synchronized with the optical chopper. Synchronous rectification serves to minimize the effect of spurious signals. Such signals tend to be either random in frequency or of a frequency different from the synchronous frequency. Therefore, they have little effect on the synchronously rectified signal. In addition, for a double-beam optical null system synchronous rectification provides a dc signal whose polarity indicates which of the two beams is more intense, and whose magnitude indicates the degree of unbalance between the two beams. In the case of a single-beam system the polarity is fixed, while the magnitude of the rectified voltage is directly indicative of the spectral transmittance through the sample. The latter fact makes it clear that highly linear amplifiers must be used to make accurate transmittance measurements in single-beam systems.

Regardless of the actual method of synchronous rectification, the rectifier derives its synchronizing signal by some means of coupling to the shaft of the optical chopper. In one commonly used method a generator powered by a rotating permanent magnet is mounted on this shaft. The generated signal serves as the reference switching signal across a bridge circuit of rectifying diodes. Another method is the utilization of cam-operated switches very similar to automobile distributor breaker switches. The feature unique to the instrument breaker switches is the coating of the switch contacts with a material, such as gold, which exhibits low-contact resistance to low-signal voltages. Still another method of rectification involves the use of magnetized reed switches which are opened and closed by a rotating permanent magnet attached to the chopper shaft.

The subsequent use of the rectified signal depends on the photometric system employed. In single-beam systems it is fed directly to a potentiometric recorder. However, in a double-beam optical null system the rectified signal is remodulated, this time at the line frequency (e.g., 60 Hz). The object of this is to obtain an ac signal which can be amplified to sufficient power to drive a servomotor, which then positions the optical attenuator to establish a null signal. Line-frequency remodulation is required because of the commercial unavailability of servomotors which will operate at frequencies of 13 Hz. The phase of the remodulated 60 Hz signal is determined by the polarity of the rectified signal and is, therefore, also sensitive to the direction of unbalance. The servomotor is a phase-sensitive device which derives a reference voltage from the power line. The phase of a voltage in the separate signal winding of the servomotor determines the direction of rotation so as to drive the optical attenuator toward a null signal.

In order to record a spectrum the pen of the instrument recorder must indicate the position of the wedge while the abscissa is scanned at a rate dependent on the scanning of the spectrophotometer. The following will briefly describe the most common methods of accomplishing these functions.

The more direct the coupling between the recorder abscissa and ordinate, and the instrument scan and wedge positions, the less the versatility of the instrument. This is one of the typical characteristics of the less expensive spectrophotometers, which do have relatively limited versatility. In the most direct case, for the abscissa drive, the recorder drum may be mounted directly on the wavelength or frequency drive shaft. This will have attached to it either the wavelength (or frequency) cam or the drive pulleys which actuate the trigonometric function generating mechanisms used in some grating instruments. If a flat-bed or a strip-chart recorder is used instead of a drum, its abscissa controlling function may be driven by a metal tape attached to the spectral drive shaft. Similarly, for the ordinate the pen carriage may be attached to a cable which is driven by the same shaft that positions the optical attenuator.

We may now explore some of the functions and advantages of less direct coupling. For the abscissa, use of a gear-box coupling with interchangeable gears allows for abscissa scale changes. One may then expand the scale by a large factor to suitably record high-resolution spectra. Conversely, the spectrum may be condensed for a more convenient form of storing or filing. For the ordinate the indirect coupling is usually electrical; that is, a potentiometer is mounted on the attenuator drive shaft. Its output serves as the input master signal for a separate pen servocircuit with a potentiometer on the pen-carriage drive shaft serving as the slave potentiometer. This type of arrangement also allows for convenient scale expansion or compression.

The ordinate speed of a recorder, which limits how rapidly information may be recorded, is ultimately determined by the chopping speed of the spectrophotometer. In general, if the sample beam is blocked, most recorders are capable of traveling full scale in 1 to 2 seconds. This is sometimes referred to as the "slewing" time. It should be noted, however, that a properly operating null system never runs under the above conditions, since the changes in transmission are so slow that the signal is never far from null. Nevertheless, the full-scale slewing pen speed is a good indication of the speed of response of the system.

2.3F. Polarizers

Because of the close relationship of polarizer accessories to fundamental components of spectrophotometers, these devices, which are gaining in utilization, will be discussed here.

The relationship referred to above is two-fold in nature. First, as has been pointed out in the section on dispersing elements (Section 2.3C), both prisms and gratings produce significant degrees of polarization which in the latter case vary drastically with wavelength. For analyses of solids these effects may cause distortion of the infrared spectrum making the use of polarizers desirable for insuring the accuracy of results as well as essential for structural analysis. The second aspect relating polarizers to basic components of spectrophotometers is that the physical causes of polarization effects generated by dispersing elements are utilized for making polarizer accessories.

Available infrared polarizers are either the pile-of-plates type or the wire-grid type. The operation of the pile-of-plates type depends on the phenomenon described on p. 35. That is, at angles of incidence different from 0° the reflection coefficient differs for the two components of polarization. The thin plates, commonly made of AgCl, are placed at a steep angle relative to the beam such that one component is almost completely reflected out of the transmitted beam while the other component is only partially reflected.

The so-called wire-grid polarizer depends on the polarization properties of gratings which have been discussed on p. 40 ff. In practice, the wire grid is the replica of a fine-lined grating formed on an AgCl or polyethylene substrate. The "wire grid" consists of metal vapor deposited on the tips of the grooves. One polarization component is diffracted out of the beam while the other is directly transmitted with high efficiency. The relative intensities of the two components is generally of the order of 500 to 1 as compared to the order of 50 to 1 for the pile-of-plates polarizer. This is not the only advantage of the wire-grid polarizer over the pile-of-plates polarizer. The former are also more compact allowing simultaneous use of other sampling accessories. They are more rugged and they cause less distortion of the beam on insertion.

2.4. SPECIAL OPERATING FEATURES

In the previous discussion we have described the fundamental components of infrared spectrophotometers. In actual practice many instruments, particularly in the more versatile, high-performance category, incorporate many features which enable the analyst to take better advantage of the performance inherent in the instrument. These features may be classified as follows:

1. variation of basic system parameters;
2. variation of recorder parameters;
3. optimization of scan time;
4. compensation for nonprogrammed energy losses.

2.4A. Variation of Basic System Parameters

The features in this category allow the analyst to manage the fundamental variables in a way that is most advantageous to the problem at hand. The interdependence of fundamental operating variables of a spectrophotometer was discussed in Section 2.2. The controls which more directly affect resolution, photometric accuracy, and scanning speed follow.

Slit Program. This directly affects the resolution. If one is working over a limited spectral region a constant, nonprogrammed slit width may be used. Versatile instruments provide both a wide range of constant-energy slit programs and adjustable nonprogrammed slits which may be made both narrower and wider than normal survey slit widths.

Response Speed (Pen Speed). This determines the rate at which the system can respond accurately to, and record, the scanned information. The slower the response or pen speed the narrower the effective bandwidth of the system. As indicated by equation (2-5), slowing the pen speed or response time serves to reduce the recorder noise level, thereby enhancing photometric accuracy; that is, the quieter the recorded spectrum, the more accurately one may determine the peaks of the absorption bands. In general, the response control adjusts electronic time constants within the system, while the pen-system speed control mechanically changes the rate of response of the pen system. Some instruments incorporate both controls, but many have either one or the other. This type of control is generally incorporated in only the more versatile instruments.

Scanning Speed. The function of this control is self-explanatory. In the more versatile instruments the scanning speed may be adjusted continuously over factors of several hundred. In the less versatile instruments there is generally a choice of two fixed scanning speeds, in a ratio of three or four to one. It should be noted in the latter case that only the slower available speed may be relied on for providing the full photometric accuracy of which the instrument is capable. The faster speed is provided for fast survey scans, for quick identification applications where photometric accuracy is not critical.

So far the direct effect of these controls has been discussed. However, as described at the beginning of this chapter, there is considerable interaction among all three basic variables. For example, a narrow slit program, in addition to providing higher resolution, is also useful in order to record more nearly the correct depths of narrow absorption bands. However, the narrower slit program requires a slower response time to reduce the noise level resulting from the lower energy and, in addition, the scanning speed must be made slower in proportion to the longer response time. To consider another possible case, assume that speed is of the essence. The compromise decision must be made between photometric accuracy and resolution. The extreme possibilities are either a high-resolution, high-noise-level spectrum, or a low-resolution, "quiet" spectrum. Although in many laboratories a fixed set of conditions such as are offered in less versatile instruments will suffice for the great majority of work, instruments of higher versatility are required to enable the analyst to obtain the detailed information that infrared analyses are capable of providing.

2.4B. Variation of Recording Parameters

The second category of operating features mentioned above is the variation of recording parameters. By this is meant the adjustment of abscissa and ordinate scale factors. In the case of the abscissa on the more versatile instruments the scale factor is adjusted through changes in the gearing to the recorder abscissa drive. For the instruments of limited versatility there is generally direct coupling between the main cam or drive shaft and the recorder abscissa. In this case the only way of varying the abscissa factor is to utilize the generally available pen position readout units, which make possible recording the instrument pen position on an external recorder whose paper feed rate determines the abscissa scale factor. Since versatile instruments provide charts much larger than a standard $8\frac{1}{2} \times 11$ in., a common accessory for these is a slave recorder which traces the spectrum on prepunched standard notebook paper for convenient filing. Some of the less expensive instruments normally record on notebook-size charts, which greatly simplifies filing and storage problems.

Ordinate scale expansion is an important feature in versatile instruments. It affords the possibility of ordinate scale magnification frequently needed to verify the relative magnitude of side bands or of minor absorption bands in normal or differential analyses. In such cases it obviates the requirement for adjusting the cell lengths or sample concentration to optimize the depth of a particular band. When utilizing scale expansion the analyst must take into account the fact that the noise, as well as the signal is magnified. Thus it is sometimes necessary to lower the noise level by increasing the response time or increasing the slit width in order to take full advantage of ordinate scale expansion.

Another type of ordinate scale variation is absorbance presentation. For this purpose the normal transmittance drive shaft is coupled to a set of logarithmic gears or a logarithmic potentiometer which translates the information directly to an absorbance scale.

2.4C. Optimization of Scan Time

The third class of features provided in some infrared spectrophotometers pertains to the optimization of the scanning time. These features decrease the running time of a spectrum below the time it would take if the scan rate were constant in abscissa units per unit time (e.g., μ/min or cm^{-1}/min). The importance of these features in a given laboratory depends on the demands on the instrument running time and on the operator's time.

The more common of the two techniques for minimizing the time for scanning a spectral region is termed "programmed scan." The programming takes into account that for a constant-energy slit program the spectral slit width may vary by a significant factor over the spectral region concerned. One may consider a spectral slit width as one "bit" of information. Now if the rate of scan is varied inversely as the spectral slit width, the result is to scan a constant number of bits of information per unit time. In other words, where the spectral slit width is relatively large there are correspondingly fewer bits of information in a given abscissa interval and this interval may be passed over relatively rapidly. Conversely, where the spectral slit width is relatively small (higher resolution) there is more information in a given abscissa interval and more time must be taken to record it properly.

As a general rule, there is more to gain from the use of programmed scan in prism instruments than in grating instruments. The reason for this is the larger variation in resolution inherent in prism instruments. In grating instruments with linear frequency abscissa presentation, the variation in spectral slit width in frequency units is generally no more than 2:1.

Another means of minimizing scan time in an automatic way is speed suppression. This enables one to scan faster in regions between absorption bands. As soon as an absorption band edge is reached and a significant deviation from optical null results from the overly fast scan, an electronic system automatically slows down the scan rate to attain a more nearly optimum null condition. In spectra with very few bands the savings in time with automatic speed suppression may be as much as 75%. It is profitable to operate with the highest suppression control setting, just short of causing speed suppression on noise peaks; too high a suppression setting may actually result in requiring a longer total scan time than no suppression at all.

2.4D. Compensation for Nonprogrammed Energy Losses

The last category of features to be considered makes possible automatic compensation for nonprogrammed energy losses such as occur in regions of solvent absorption in differential analyses. Whenever there is an absorption band in the reference beam which is not accounted for in the constant-energy slit program in the instrument, response will tend to become sluggish in proportion to the deficiency in reference-beam energy. Under these conditions, band depths and positions will be inaccurate, and some bands may be missed completely.

Either of two methods may be used to compensate for the loss in reference-beam energy—automatic gain control, or automatic slit control. In the former case, the gain of the system is automatically increased to compensate for the energy loss by means of a feedback circuit which is sensitive to reference-beam energy only. In the latter case, a similar feedback circuit is made to operate the slit servosystem to widen the slits, thereby regaining the lost energy. Generally these systems can compensate for losses in reference-beam energy of as much as 90–95%. Figure 2-17 demonstrates the application of an automatic slit-control system in a differential analysis involving the same sample in both beams, with a slightly larger absorption path in the sample beam.

The two systems are not identical in function or effect. In the case of automatic gain control the normal slit schedule is unaffected, but there is an increase in noise level in the region of a reference-beam absorption, corresponding to the increased system gain. However, the noise can be kept to a tolerable increase by selecting suitable operating conditions. With the automatic slit-control system, the slits are opened beyond the programmed width to maintain constant energy. The advantage of this is to maintain a constant noise level. The disadvantage is unexpected, uncharted loss in resolution. For a solvent band which absorbs 90% of the reference-beam energy, the spectral slit width will vary over the band by more than a factor of three.

2.5 AVAILABLE INSTRUMENTS AND THEIR SPECIFICATIONS

A large and ever-increasing selection of infrared spectrophotometers is available to the analyst. This section will summarize some of the distinguishing factors in these instruments and provide some basis for interpreting manufacturers' specifications. Final selection of a particular instrument depends not only on the best match between the specifications and the requirements of the particular laboratory, but also on many factors that lie beyond the scope of this chapter. These include the personal confidence of the analyst in the manufacturer; the availability of service, spare parts, and supplies; delivery time; cost;

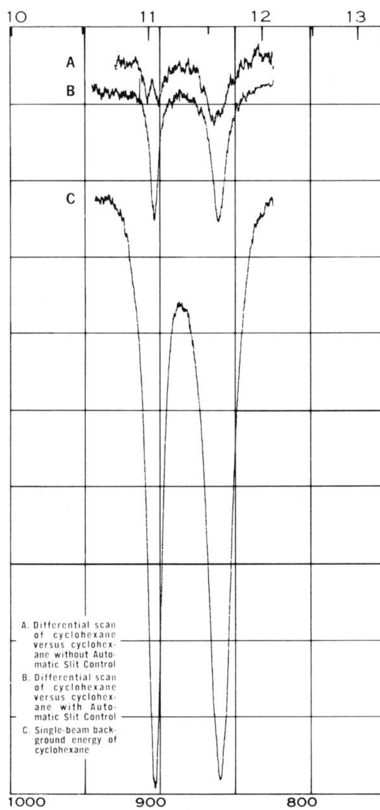

Figure 2-17. Automatic slit control applied to a differential analysis. (Courtesy of Beckman Instruments.)

instrument size; minimum sample size required; operator convenience factors; reliability and quality control; and selection of accessories.

Spectrophotometers presently available fall into two basic categories, with some overlap between them: the more versatile, high-performance instruments, and the less expensive instruments of limited versatility. In each category both prism and grating versions are available. Some of the high-performance instruments provide for interchangeability of monochromators in order to cover a wide range of overlapping spectral regions. In cases where interchanges are not feasible there are frequently different versions of the same type of instrument, covering different wavelength regions.

For example, instruments of the less versatile type are available with NaCl, KBr, or CsBr prisms, and with gratings covering various spectral regions from 0.83 to 50 μ (12,000 to 200 cm^{-1}). More versatile instruments are available with NaCl, KBr, CsBr, KRS-5, CsI, LiF, and CaF_2 prisms, and with gratings covering regions from 1 to 300 μ

(100,000 to 33 cm^{-1}). Both filter–grating and fore-prism–grating instruments are manufactured.

As previously indicated, most of the spectrophotometers used for analytical purposes are double-beam instruments, and most of these are of the optical null type. A few are in the ratio-recording category. Almost all utilize thermal detectors, either thermocouples or metal bolometers, although some incorporate a Golay pneumatic detector over part or all of their spectral range. Sources employed by commercial instruments run the gamut of all the sources described previously.

In reference to specifications, it is possible to discuss only the generally accepted interpretation of published specifications. In view of the fact that there are no established, industry-wide, national, or international standards, the manufacturer of a particular instrument should be consulted on his interpretation of any specification critical to the selection of a particular instrument. However, there are some generally accepted customs, knowledge of which may prove useful to the reader.

Resolution. Two bands are considered resolved when both may be consistently measured with an absorption-peak-to-valley depth that exceeds the peak-to-peak noise level—i.e., each band will be distinguishable from noise peaks. Under this definition a side band that appears as a shoulder on another band is not considered resolved.

Resolution performance of an instrument is stated in many different ways, which depend to some extent on the type of instrument. For the less versatile instruments, resolution may be given in terms of the band depth of a side band of a readily available test material under standard scanning conditions. A commonly used material is polystyrene film, usually 0.07 mm thick. In a rock-salt instrument the depth of the 3.51 μ or the 3.30 μ sideband shown in Figure 2-18 is frequently used as the reference. The validity of this test is based on the fact that the physical slit width of a rock-salt instrument is relatively narrow in this spectral

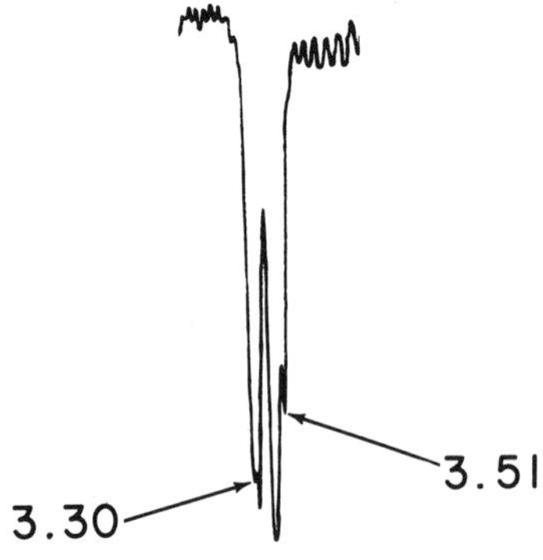

Figure 2-18. Resolution of a polystyrene film sample on an NaCl prism instrument.

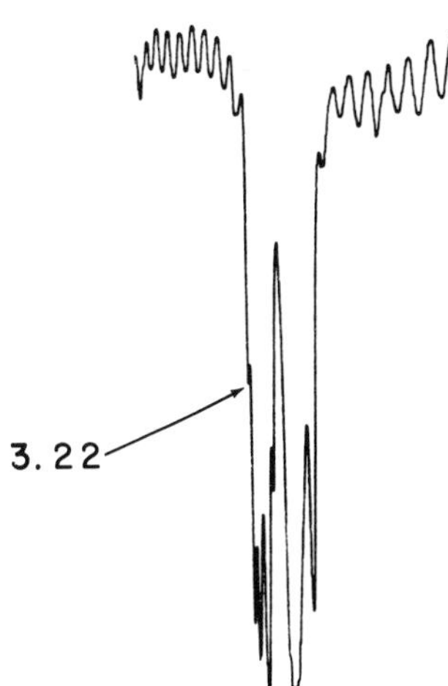

Figure 2-19. Resolution of a polystyrene film sample on a grating instrument.

region, and consequently the depth of these bands is sensitive to many instrument factors. Figure 2-19 shows the same spectral region scanned on the same type of instrument with a grating as the dispersing element, instead of a rock-salt prism. The increased resolution supplied by the grating shows up dramatically in this region. For the grating spectrum, the depth of the first band of the quintuplet at 3.22 μ is frequently used as an indication of resolution under survey scan conditions.

Also, for the less versatile instruments the resolution for a particular wavelength may be quoted. Experience serves to indicate the implied resolution throughout the region. Such extrapolation is easier for a grating instrument because, it may be recalled, the resolution of grating instruments varies much less than that of prism instruments. In some cases a plot of spectral slit width *versus* wavelength or wavenumber has been published for the instrument in the scientific literature or in manufacturers' brochures. This is also true of the more versatile instruments in reference to their standard survey conditions.

For the more versatile instruments the "ultimate" resolution in one or more spectral regions is frequently quoted. To duplicate this type of performance a good deal of patience on the part of the analyst is required. For example, most of the bands used for reference are in vapors, and sampling conditions must be carefully adjusted to provide an adequate path length and concentration for well-defined bands, but the pressure must be quite low to avoid pressure-broadening of these bands. In addition, the instrument must be scanned very slowly, with slow response speed, large abscissa expansion, and a narrow slit program. Considerable experimentation is required to determine the proper conditions. Figure 2-20 shows the NH_3 spectrum in the 967 cm^{-1}

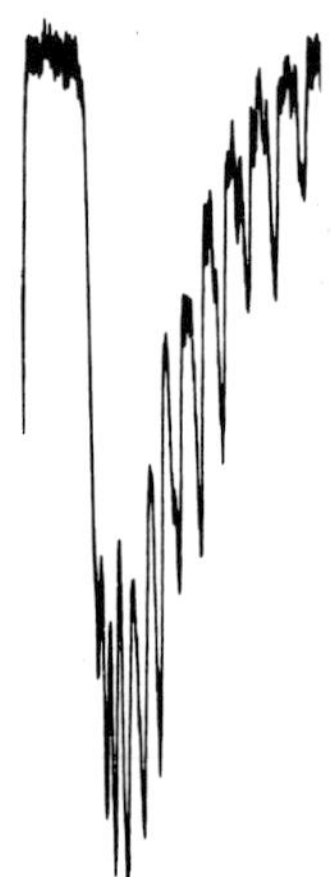

Figure 2-20. High-resolution NH_3 vapor spectrum.

region, which is frequently used as an indication of ultimate resolution. The first resolved band is 0.24 cm^{-1} from its neighbor, which in turn is 0.39 cm^{-1} from its neighbor.

Scattered Radiation. This is sometimes referred to as "unwanted radiation," a more inclusive term which is really more accurate, since the object of this specification is to indicate the purity of the spectrum. Unwanted radiation of any origin affects the depth of the absorption bands and therefore the quantitative accuracy of measurements. This becomes apparent if one considers a totally absorbing band at a wavelength λ_1 when the spectrophotometer is set to λ_1. If some energy of a second wavelength λ_2 not absorbed by the sample also reaches the detector, the totally absorbing band at λ_1 will falsely show some transmission. Instead of transmitting zero energy, it will show a transmittance level equal to the amount of unwanted energy, originating from all other spectral regions, which exists at the wavelength setting λ_1.

In a prism instrument, unwanted radiation consists generally of a continuum of energy which has been scattered within the monochromator and is reflected through the exit slit to the detector. The scattered energy obviously is concentrated near the peak of the blackbody curve, and makes itself evident at longer wavelengths, where the slit width is relatively large for a given slit program. Under these conditions, more energy enters the monochromator to be scattered, and a larger amount of the scattered energy can find its way out through the wider exit slit. Thus the scattered energy becomes a greater percentage of the desired monochromatic energy. In practice, the level of scattered radiation is usually trivial except in the longer-wavelength region. Therefore, the maximum scattered radiation level at a particular wavelength near the longer-wavelength limit is generally specified for prism instruments.

General scattered radiation also may be significant for grating instruments. There is the additional possibility of overlapping grating orders adding to the unwanted radiation. If one is working in the first order, the overlapping orders are all at shorter wavelengths. If the operation is in the higher orders, overlapping orders may be at either shorter or longer wavelengths than the desired one.

There are two fundamental ways of measuring the level of unwanted radiation; both have limitations which should be understood. One method is to trace the cutoff edge of appropriate optical materials, indicating the deviations from zero transmittance in the cutoff region. A typical series of such cutoff curves is shown in Figure 2-21. A second method of measurement is to observe the absorbance depth of bands which are known to be totally absorbing or almost so. The most common pitfall here is to make such a measurement with a thin film. For example, if an 0.07-mm film of polystyrene is used, the 13.1 and 14.2 μ bands should reach zero transmittance and therefore be useful for indicating scattered radiation level. However, the film will heat up on exposure to the source by as much as 40–50C°. At the wavelengths involved, an object above ambient temperature will reradiate a detectable amount of energy (see Section 2.3A) and will give a false reading for the band bottom. Consequently these bands may spuriously indicate a significant "scattered radiation level." One interesting experiment is to blow cold air on the film with the instrument set on the bottom of one of these bands. In fact, if the film is cooled below room temperature, it is even possible for the detector to radiate enough energy to the cold sample to indicate a transmittance below zero. Use of a liquid sample for this purpose tends to minimize the above indicated problem.

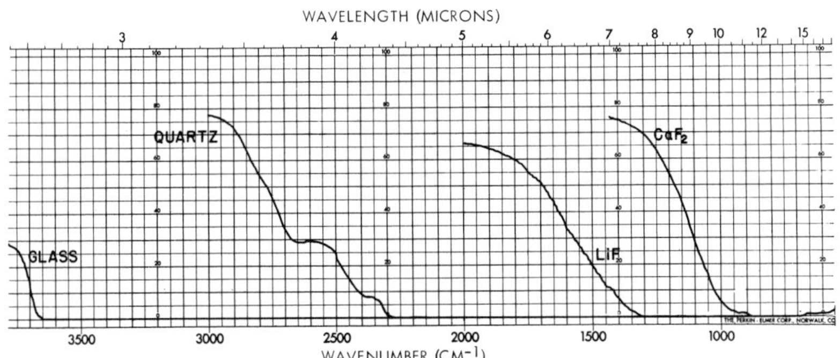

Figure 2-21. Typical cutoff curves used for unwanted radiation measurements.

However, it is essential always to select a sample which is not overly rich in absorption bands, since such a sample would absorb a great deal of the energy which can be scattered and will yield a fallaciously favorable result.

Abscissa Accuracy and Reproducibility. These specifications determine the accuracy and precision with which the spectral position of absorption bands may be located. The abscissa accuracy, given as $\pm n$ (μ or cm^{-1}), indicates the maximum deviation between the measured and the true position of a band. Thus it is a measure of the calibration accuracy of the instrument. In the case of the more versatile instruments the abscissa reference may be a counter or dial reading. For the less versatile instruments the abscissa reference is generally the preprinted recorder paper located relative to a fiduciary mark on the recorder.

The abscissa reproducibility, in contrast to the accuracy, indicates the repeatability of band position from spectrum to spectrum, in general even over a long period of time. Since this specifically indicates the amount by which any spectrum may deviate from any other spectrum along the abscissa, there is no direct relationship to an abscissa standard. Therefore, the reproducibility is stated in terms of a number M (μ or cm^{-1}) without the $\pm$ prefix. The inclusion of the $\pm$ prefix for the reproducibility specification is, however, not uncommon in the literature, and requires interpretation because of the resulting ambiguity.

The reproducibility specification indicates the ultimate abscissa precision attainable with the instrument; i.e., when abscissa precision is an important factor, the analyst may use wavelength or wavenumber standards to formulate a working curve for the abscissa calibration of the instrument. However, this working curve can be valid only within the abscissa reproducibility of the instrument. Table 2-II lists many of the commonly used calibration standards. More extensive lists are available in the literature.

As previously mentioned, the calibration accuracy specification on high-performance instruments applies to the direct readout element of the monochromator, which is usually a counter or a dial. The large-size-chart abscissa scales generally used with these instruments limit the accuracy on the chart itself more severely due to the humidity sensitivity of even high-grade recorder paper, which affects the dimensional stability. Sprocket-driven strip-chart recorders tend to be self-compensating for this effect. For less versatile instruments, the recorder paper itself, usually with a smaller abscissa scale, is used as the abscissa reference.

Ordinate Accuracy and Reproducibility. These specifications determine the ordinate accuracy and precision of the transmission measurements which directly affect the quantitative results. Analogous to the case of the abscissa, the accuracy specification relates to the accuracy of the built-in ordinate calibration, while the reproducibility indicates the precision attainable by utilizing a working curve.

In general the ordinate accuracy specification applies to the linearity of the optical attenuating device and its associated pen drive system. Two of the devices used to test this are optical chopper wheels and electrical potentiometers. In the electrical potentiometer case, the sample beam is replaced by a test signal of proper phase, and the recorder pen and/or attenuator position are checked against a highly

TABLE 2–II. Calibration Standards

(All lines refer to vapor absorption bands unless otherwise noted)

λ (μ) (in air)	ν (cm^{-1}) (*in vacuo*)	Material
1.0140	9859.4	Hg (emission)
1.1287	8857.0	Hg (emission)
1.3673	7311.5	Hg (emission)
1.5296	6535.9	Hg (emission)
1.7073	5855.6	Hg (emission)
2.3253	4299.3	Hg (emission)
2.605	3837.9	H_2O
2.913	3432.0	NH_3
3.302	3027.1	Polystyrene film
3.420	2924.0	Polystyrene film
3.507	2850.7	Polystyrene film
4.254	2349.9	CO_2
4.258	2347.6	CO_2
5.142	1944.0	Polystyrene film
5.146	1942.6	H_2O
5.348	1869.4	H_2O
5.421	1844.2	H_2O
5.549	1801.6	Polystyrene film
5.577	1792.6	H_2O
5.640	1772.6	H_2O
5.708	1751.4	H_2O
5.763	1734.6	H_2O
5.988	1669.4	H_2O
6.074	1646.0	H_2O
6.184	1616.7	H_2O
6.211	1609.6	Indene liquid
6.243	1601.4	Polystyrene film
6.315	1583.1	Polystyrene film
6.342	1576.2	H_2O
6.414	1558.5	H_2O
6.436	1553.3	Indene liquid
6.824	1464.9	H_2O
6.958	1436.7	H_2O
7.044	1419.3	H_2O

TABLE 2–II (continued)

λ (μ) (in air)	ν (cm^{-1}) (*in vacuo*)	Material
7.176	1393.2	Indene liquid
7.344	1361.3	Indene liquid
8.244	1212.7	NH_3
8.366	1195.0	NH_3
8.493	1177.1	NH_3
8.626	1158.9	NH_3
8.661	1154.3	Polystyrene film
8.765	1140.6	NH_3
9.060	1103.4	NH_3
9.217	1084.6	NH_3
9.292	1075.9	NH_3
9.725	1028.0	Polystyrene film
9.814	1018.6	Indene liquid
10.072	992.6	NH_3
10.503	951.8	NH_3
11.007	908.2	NH_3
11.026	906.7	Polystyrene film
11.607	861.3	Indene liquid
12.380	807.5	NH_3
13.693	730.1	Indene liquid

The following bands are not established standards but are commonly used for calibrating infrared spectrophotometers in the longer-wavelength regions

λ (μ) (in air)	ν (cm^{-1}) (*in vacuo*)	Material
13.680	730.8	Polyethylene film
13.890	719.7	Polyethylene film
14.986	667.1	CO_2
16.178	617.9	H_2O
17.400	574.5	1,2,4-Trichlorobenzene liquid
18.160	550.5	1,2,4-Trichlorobenzene liquid
19.008	525.9	H_2O
19.907	502.2	H_2O
21.161	472.4	H_2O
21.790	458.8	1,2,4-Trichlorobenzene liquid
21.860	457.3	H_2O
21.872	457.1	H_2O
22.617	442.0	H_2O
22.760	439.2	1,2,4-Trichlorobenzene liquid
23.860	419.0	H_2O
25.140	397.7	H_2O
26.620	375.5	H_2O
29.830	335.1	H_2O
30.510	327.7	H_2O
33.010	302.8	H_2O

linear potentiometer, which varies the test signal. If optical chopper wheels are used, they are mounted in the sample space, in the sample beam. Their ratio of open to closed areas is precisely computed. They are made to rotate at a frequency high compared to the normal 10–13 Hz basic chopping rate of the instrument, to achieve a suitable averaging effect of the on and off pulses transmitted by the chopper wheels.

It should be noted that the above techniques, as well as other means of checking instrument linearity, serve only to indicate the limitations of the ordinate measuring device of the instrument. The actual photometric accuracy which can be realized in a particular problem is affected by a large number of factors. For example, high photometric accuracy depends on a selection of suitable operating conditions for the spectrum as previously discussed in this chapter (see Sections 2.2B and 2.4A). In addition, many precautions must be taken with respect to sampling techniques, as will be described in Chapter 7.

Over and above the fundamental specifications discussed here, there is an extensive list of other specifications which accompany many spectrophotometers. These cover such matters as range of variables, chart presentation, special features, operating controls, sources, detectors, and fundamental systems data.

2.6. THE COMPUTER REVOLUTION

It has been the objective of this chapter to consider prevalent infrared photometric systems. The growing application of digital data-recording (DDR) techniques and computer processing of spectral data has produced results pertinent to subjects discussed in this chapter; these techniques are certain to have a revolutionary effect on the field.

With further development of DDR techniques, the instrument analog recorder, which has provided the basic record of spectral data, may be relegated to the role of furnishing the quick visual check for the data being recorded in digital form on punched or magnetic tape.

Under these conditions, the DDR and the computer become an integral part of the instrument system. The computer can then take over some of the previously described functions of the instrument. For example, the computer can accept relatively "noisy" data and smooth the data by polynomial techniques as well as averaging of multiple runs. Furthermore, the computer will be able to correct routinely for systematic instrument errors, such as calibration, and for inherent factors, which will enable plotting of true band shapes for example.

REFERENCES

1. N. L. Alpert, "Infrared Filter Grating Spectrophotometers—Design and Properties," *Appl. Opt.* **1**: 437 (1962).

2. R. P. Bauman, *Absorption Spectroscopy*, Wiley, New York (1962).
3. S. S. Ballard and J. S. Browder, "Thermal Expansion and Other Physical Properties of the Newer Infrared-Transmitting Optical Materials," *Appl. Opt.* **5**: 1873 (1966).
4. G. R. Bird and M. Parrish, Jr., "The Wire Grid as a Near-Infrared Polarizer," *J. Opt. Soc. Am.* **50**: 886 (1960).
5. G. R. Bird and W. A. Shurcliff, "Pile-of-Plates Polarizers for the Infrared: Improvement in Analysis and Design," *J. Opt. Soc. Am.* **49**: 235 (1959).
6. G. K. T. Conn and D. G. Avery, *Infrared Methods, Principles and Applications*, Academic Press, New York (1960).
7. M. J. E. Golay, "Comparison of Various Infrared Spectrometric Systems," *J. Opt. Soc. Am.* **46**: 422 (1956).
8. D. Gray (Coordinating Editor), *American Institute of Physics Handbook*, 2nd Ed., McGraw-Hill, New York (1963).
9. H. L. Hackforth, *Infrared Radiation*, McGraw-Hill, New York (1960).
10. L. W. Herscher, H. D. Ruhl, and N. Wright, "Improved Optical Null Infrared Spectrometer," *J. Opt. Soc. Am.* **48**: 36 (1958).
11. International Union of Pure and Applied Chemistry, Commission on Molecular Structure and Spectroscopy (Editors), *Tables of Wavenumbers for the Calibration of Infrared Spectrometers*, Butterworths, Washington, D.C. (1961).
12. J. Overend, A. C. Gilbey, J. W. Russell, C. W. Brown, J. Beutel, C. W. Bjork, and H. G. Paulat, "A Littrow–McCubbin High Resolution Infrared Spectrometer," *Appl. Opt.* **6**: 457 (1967).
13. W. J. Potts, Jr. and A. Lee Smith, "Optimizing the Operating Parameters of Infrared Spectrometers," *Appl. Opt.* **6**: 257 (1967).
14. R. A. Sawyer, *Experimental Spectroscopy*, 3rd ed., Dover, New York (1963).
15. R. A. Smith, F. E. Jones, and R. P. Chasmar, *The Detection and Measurement of Infrared Radiation*, Oxford University Press, London (1957).
16. J. Strong, *Concepts of Classical Optics*, Freeman, San Francisco (1958).
17. J. Strong, "Resolving Power Limitations of Grating and Prism Spectrometers," *J. Opt. Soc. Am.* **39**: 320 (1949).
18. J. U. White, N. L. Alpert, A. G. DeBell, and R. M. Chapman, "Infrared Grating Spectrophotometer," *J. Opt. Soc. Am.* **47**: 358 (1957).
19. J. U. White and M. N. Liston, "Construction of a Double Beam Recording Infrared Spectrophotometer," and "Amplification and Electrical System for a Double Beam Recording Infrared Spectrophotometer," *J. Opt. Soc. Am.* **40**: 29, 36 (1950).
20. V. Z. Williams, "Infrared Instrumentation and Techniques," *Rev. Sci. Inst.* **19**: 135 (1948).

CHAPTER 3

Accessories

To obtain the desired spectra of samples, the spectroscopist will require a variety of accessories in addition to the instrument. The purpose of a number of these accessories is to hold the sample; that of others is to permit the use of additional techniques; and that of the remainder is to extend the usefulness of the instrument. This chapter will very briefly discuss the construction and use of the most common types of commercially available accessories. For a more detailed description, the spectroscopist should refer to the literature of the various instrument and accessory manufacturers.

3.1. CRYSTALS

Many accessories require windows or lenses of an infrared transmitting material. These materials are listed in Table 2-I of Chapter 2. The most commonly used crystals are sodium chloride (NaCl), potassium bromide (KBr), cesium bromide (CsBr), and cesium iodide (CsI). As these crystals are all water soluble, hygroscopic, and easily damaged, they require special care and handling. The wider the infrared transmission range, the more hygroscopic and easily damaged they are.

Handling. Crystals should be touched only on the edges, never on the faces, and frequently it is necessary to wear rubber gloves or finger cots when handling them. The latter is always required during the polishing operations described below. It is good practice to hold the crystals or the accessories in which they are mounted with the tips of the fingers and as far from the palm as possible; otherwise they may become badly fogged. Fogging does not necessarily make them unusable, but will decrease the amount of energy they transmit. Crystals and accessories in which they are mounted should always be stored in a dessicator when not in use.

Cleaving. If it is desired to change the size or shape of a crystal, it can easily be cleaved. This can be done by laying the crystal on a flat surface, holding a new clean razor blade on end, perpendicular to the crystal at the spot where it is to be cleaved, and giving the blade a sharp rap with the handle of a long wooden-handled spatula or a screwdriver. In cleaving a large crystal into several smaller pieces, the crystal should be cleaved into equal-sized pieces whenever possible. The thickness of the crystal can be reduced by placing it on edge and proceeding in the same manner.

Grinding. Crystals with badly marred faces can be sanded smooth and flat on 600 grit sandpaper placed on a flat surface. The grinding should be done with a random rotary motion. Care must be taken in exerting pressure on the crystal, or the edges may be ground faster than the center, thus producing a curved surface. After sanding, the crystal is transferred to a piece of very fine emery cloth and ground further. This operation will level off the hills and valleys from the sanding and make the crystal easier to polish.

Polishing. After grinding is complete, or if the crystal is not badly marred, it is polished. This is usually done on a cloth lap, but in some cases, especially cesium iodide, a pitch lap is preferable. The cloth can be any soft, lint- and size-free material. It is stretched over a piece of plate glass so tightly that wrinkles are not raised when the hand is rubbed over it. The polishing compound is usually fine jeweler's rouge or barnsite (for water insoluble crystals, other oxides may give faster polishing). The lubricant used is isopropyl alcohol—90% for sodium chloride and anhydrous for all others.

A small quantity of the polishing compound is placed on the lap and the whole well soaked with alcohol. (As mentioned earlier, finger cots or rubber gloves should be worn during the polishing operations.) The crystal is placed on the lap and removed with a figure-eight motion under firm pressure. Again the pressure must be exerted with care to guard against polishing the edges faster than the center. When the polishing is complete, the crystal is transferred to a second cloth and dried with short straight strokes; it should never be allowed to air dry. If the polished surface is satisfactory, the crystal can be turned over and polished on the other side in the same manner. If the surface is not satisfactory, it can be repolished.

Directions for construction of a pitch lap can be found in a good book on optical polishing.

3.2. GAS CELLS

Standard commercially available gas cells can be divided into two main classes. Glass-body cells and metal-body cells. Both classes of cells have windows of infrared-transmitting crystals, either cemented to the body or held in place by retainer rings. Not only must care be exercised in the handling of these cells (see Section 3.1), but the samples should be dry before they are put into the cell. Cells should be cleaned by evacuating the sample followed by flushing with clean dry air.

Glass-Body Cells. Glass-body cells are available in three standard path lengths; 2 cm, 5 cm, and 10 cm. Until recently, the windows have been cemented to the ends of the glass cylinder; now the windows may be held in place with metal retainer rings. The latter arrangement not only makes it easier to remove the windows for polishing, but eliminates the possibility of attack on the seal by the sample. The cell can be evacuated, filled, and flushed through stopcocks sealed to the glass body.

These cells can be evacuated to less than 1 mm Hg pressure, but cannot be filled to more than one atmosphere pressure. Excess pressure inside the cells will tend to break the seals holding the windows to the body and cause leaks. The cells also can be converted for use at elevated temperatures by wrapping them with heating tape. When heating the cell, do not increase the pressure above one atmosphere.

Metal-Body Cells. It is possible to get cells similar to those above with metal instead of glass bodies. Usually, however, the term "metal body" implies cells of longer path length, or high-pressure cells. Windows are held in place with retainer rings.

Long-path-length gas cells have internal optics to reflect the radiation back and forth through the same volume of sample. This type of cell is called a "folded path" cell. (Figure 3-1.) In many of these cells, the path length can be varied in discrete steps by changing the angles of one or more mirrors. Common maximum path lengths of this type of cell are 1 m, 10 m, and 40 m.

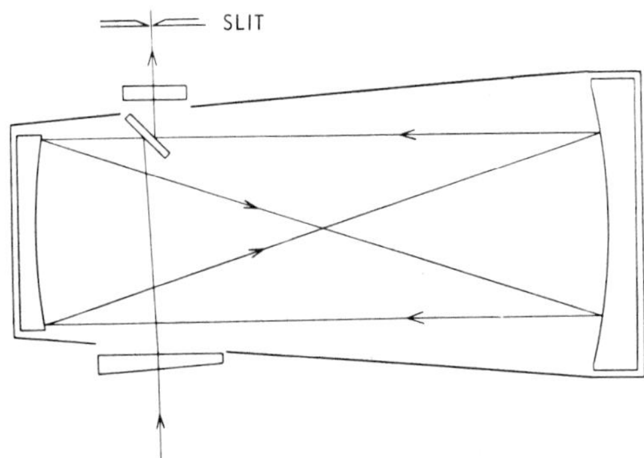

Figure 3-1. Optical path of a long-path-length gas cell.

Metal-body cells are equipped with either valves or hose nipples to allow evacuation, filling, cleaning, or flushing. The hose nipples can be replaced by valves, if the operator so desires.

Care must be used in deciding what samples to run in these cells, or corrosion of the metal bodies, mirror mounts, or mirror surfaces may result. Another frequently encountered problem is the adsorption of sample by the cell walls. The current practice of lining the cell with a resin or plastic has reduced but not eliminated this problem.

Most metal-body cells can be filled to a pressure well above one atmosphere. Before filling the cell it is recommended that the user check the manufacturer's specifications, especially concerning maximum pressures. This latter limit may vary with the type of crystal used for the windows.

Metal-body cells can also be operated at elevated temperatures by either wrapping them with heating tape, or by shining a heat lamp on them. As in the case of glass-body cells, care must be taken not to build up the pressure beyond the allowable maximum.

3.3. LIQUID CELLS

Liquid cells can also be divided into two main classes: sealed cells and demountable cells. Other cells that are available will fit into one of these two classes, except for a recent development called a demountable sealed cell.

Sealed Cells. Sealed cells are those which can be filled and cleaned without changing the path length of the cell. They were originally made by cementing a lead spacer between two windows, using lead amalgam, thus giving a path length of about the thickness of the spacer.

The construction of such a cell is shown in the Figure 3-2. The lead spacer and lead gasket must be cleaned in dilute acetic acid, rinsed, dried, and amalgamated just before the cell is assembled. The amalgamation is done either by dipping the lead into a pool of mercury and wiping off the excess, or by spreading a drop of mercury over the lead surface with a glass rod. Once the lead has been amalgamated, the cell must be assembled immediately or the amalgam will set and will not act as an adhesive. It is possible for the user to rebuild these cells in his laboratory, but unless he has had previous experience, it is not an easy task. The beginner should check with the manufacturer for details before attempting it the first time.

These cells are filled by inserting a syringe partially filled with sample into the lower port of the cell, and very gently pushing the sample into the cell. If too much force is used, pressure is built up inside the cell, and spreading of the windows may result. This spreading

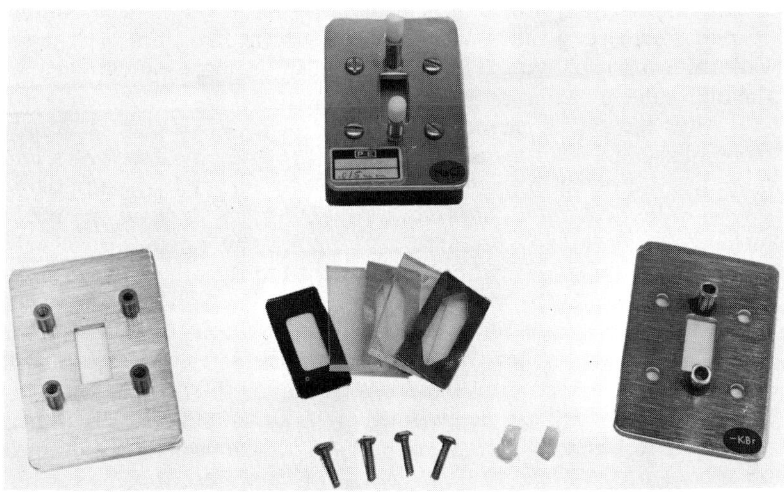

Figure 3-2. A sealed cell and its component parts.

will change the path length, and may cause leaks. As the sample is pushed into the cell, it can be seen rising between the windows, and when it can be seen in the top port the cell is filled. Overfilling is to be avoided, as this may also damage the cell. When the cell is filled, a stopper is inserted in the top port, the syringe is removed, and a second stopper is inserted in the bottom port.

Another technique that may be used for filling these cells is to place a syringe containing the sample in the upper port, and a clean syringe in the bottom port. Sample is now sucked into the cell by pulling out the piston of the lower syringe. After the cell is filled, the syringes are again replaced by stoppers as above. This technique avoids the possibility of pressure build-up inside the cells, but as liquid is being dropped through space the operator must be very careful to keep the sample free of bubbles. The primary use of this technique is the filling of thin cells.

In cleaning the cell, a clean syringe is inserted in the top port, and (again with very gentle pressure) as much sample as possible is forced through the port used for filling. The cell is now flushed five or six times, each time with a syringe full of solvent. (Flushing is done in the same direction as the cell is emptied). This normally suffices to remove all the sample from the cell; inspection of the cell will show whether this step needs repeating. The best solvent to use is determined by the sample to be removed from the cell. The solvent should have a low boiling point and not leave any residue (including water) on drying;

if either of these requirements is not met by the solvent, the cell should be reflushed several times with a second solvent that does meet them. Chloroform is usually used for this final flushing, and frequently will be the only solvent required.

After the cell is cleaned, it must be dried. This step is usually accomplished by inserting a clean, dry syringe into one of the ports, and pumping the piston until no solvent can be detected (visually or by odor). Other techniques are to blow out the cell with clean, dry air or nitrogen (using very low pressures), or to pull a vacuum on the cell with a pump. (A drying tower should be attached to the other port to clean and dry the air that is pulled through the cell.) Solvent that is not removed will probably show up in the next sample, so the operator should know where the strongest band in the solvent spectrum will appear (for chloroform, this band is at 13.2 μ, or 760 cm^{-1}).

This type of cell can be obtained in various standard path lengths from 0.015 to 1 mm. Nonstandard path lengths can usually be obtained on special order. Cells of 0.025-mm path length are those most frequently used. However, for work with solutions of weak absorbers (nonpolar materials), or for differential analysis, a path length of 0.050–0.100 mm or greater may be better. Matched pairs can also be obtained for a premium price.

Cells of the sealed type can also be obtained as micro- or ultramicro cells. The micro- and ultramicrocells can be purchased in the same path lengths as the standard cells, but because of differences in design they require less volume for filling. Micro- and ultramicrocells frequently require beam condensers.

Cavity cells fit the sealed-cell definition, but are made according to a different principle. These cells consist of a single block of window material into which a rectangular cavity has been bored. They can be filled, cleaned, and flushed with a hypodermic needle and syringe. A conical hole in the block above the cavity will accept a plastic stopper; The stopper prevents evaporation of the sample. If these cells are damaged or broken, they cannot be repaired.

The above types of cells are primarily used for quantitative analysis and to hold volatile liquids (a rule-of-thumb criterion for the latter is a boiling point of 80°C or less). As they are difficult to clean and repair, they are not usually used for qualitative work on higher-boiling samples, unless a specific cell path is required.

Other types of cells which can be put into this category are variable-space cells and wedge cells. These cells are used for compensation purposes and not for holding samples. The variable-space cell consists of two barrels, one inside the other, each of which contains a window. The inner barrel is attached to a cap threaded onto the outer barrel. Turning this cap changes the spacing (path length) between the two windows and allows the operator to match it to the path length of the sample cell. They are filled and cleaned in the same way as the other sealed types. Wedge cells are constructed in a wedge shape, and mounted in such a way that they can be moved in a direction perpendicular to the beam axis. This movement will change the path length of the cell in the beam.

Demountable Cells. Demountable cells differ from sealed cells in that they must be disassembled for filling and cleaning. The very short path lengths used in infrared work make it impossible to reassemble these cells with a reproducible path length, thus making them useless for quantitative work. As they are not sealed, they will not hold volatile materials for a sufficient length of time to obtain a spectrum. Their primary use is for high-boiling liquids, especially viscous liquids, in qualitative work. They can also be used for solids.

Demountable cells are similar in construction to ordinary sealed cells, but have no filling ports or drilled windows. The cell is frequently assembled without a spacer; the sample path length is then listed as a capillary film. When using such a cell with a spacer, the operator must be careful not to get gas bubbles in the sample. The best techniques to use in assembling these cells are to slide, rather than drop, the top window into place, or to contact one edge and then lower the other edge until the top window is in place.

The cell is held together by knurled nuts on threaded posts or by screws in threaded studs. When tightening the nuts or screws, care must be exercised to keep the windows parallel, otherwise the sample will tend to run to one side, and this may leave an air space in the beam. The operator should never exert excessive pressure on the cell by turning the screws or nuts too tight, as this usually breaks the windows. If the sample is too thick, it is better to use a thinner spacer, or to squeeze some of the sample out of the cell by rotating the windows under pressure with the fingers, than to further tighten the nuts or screws.

Whenever the cell has been used, it must be taken apart and cleaned. If the windows were not damaged during use, they should be polished as in Section 3.1 (*polishing*); if they are damaged, they should be sanded prior to polishing—Section 3.1 (*grinding*). The metal plates, rubber gaskets and spacers should also be thoroughly washed and dried to insure that no leftover sample remains for possible contamination of future samples.

Demountable cells are available in micro- and macrosize. Microsize demountable cells require a beam condenser.

Demountable Sealed Cells. These cells combine some of the advantages of both the sealed-cell type and the demountable-cell type. They can be assembled with the ease of demountable cells, and can be filled and cleaned without changing the path length (as in sealed cells).

The spacers are made of plastic (teflon) instead of amalgamated lead, thus eliminating one of the major difficulties in the assembly of sealed cells. However, they are not as tight as sealed cells, and will not hold volatile liquids for prolonged periods.

3.4. KBr-PELLET ACCESSORIES

KBr pellets are made in dies of various types, most of which require laboratory presses capable of delivering at least ten tons of pressure. Frequently, the pellets will require a special holder to enable the operator to mount them in the spectrophotometer. In other cases, the pellet is made directly in the holder, and the holder will require a special device for mounting in the spectrophotometer. This latter is especially true of microsize pellets, which usually are mounted in a beam condenser rather than directly on the instrument.

Most dies produce circular pellets, although one manufacturer makes a die that produces rectangular pellets. Plastic-on-paper forms can be obtained that will fit into standard circular dies, and convert the die to produce rectangular rather than circular pellets. In this case, the pellet is made in the form and the form plus the pellet are transferred to a standard holder.

As the die or pellet holder frequently requires certain features for optimum operation with a given spectrophotometer, it is an excellent idea to check the literature put out by the various manufacturers before purchasing these accessories. It should be borne in mind that as macro- and microdies may not be interconvertible, the anticipated sample sizes available should also be considered.

A unique entry in this area of accessories is the *MiniPress* of Wilks Scientific Company. This accessory makes the pellets without a press and directly in the holder; in fact, the press is the holder. A special adapter is required to mount the holder in the spectrophotometer. The MiniPress has the disadvantage of vignetting the beam.

In using these accessories, the operator should follow the manufacturer's operating instructions to be certain that he does not damage the various parts. He should also remember that KBr is corrosive, and therefore he should thoroughly clean the die after each pellet has been pressed. If the die will not be used for a period of time, it should be stored in a dessicator. Care must also be exercised in handling the polished faces of the rams, as scratches, pits or dents in these faces will usually result in poor pellets.

3.5. BEAM CONDENSERS

Beam condensers are used to concentrate the radiation beam into a smaller cross section. This is accomplished with either lenses or mirrors. Beam condensers using lenses will become opaque at the frequency of the lens-material cutoff, and may not be usable over the range of the instrument. Beam condensers with lenses should be handled with care to prevent fogging of the lenses, and stored in a dessicator when not in use.

As in the case with KBr-pellet accessories, the manufacturer's literature should be checked before purchasing a beam condenser, to insure that the accessory is compatible with the spectrophotometer. It is possible to obtain beam condensers with varying condensing factors, and this feature should also be checked to insure that the beam size is reduced to as close to the desired area as possible.

The operator should remember that the use of a beam condenser will increase the heat problem. This may make it difficult to prevent evaporation of volatile samples unless steps are taken to dissipate the excess heat. If heat is a problem, the operator should look into either using a filter ahead of the sample, or building a heat sink around the sample. In addition to the volatility problem, some samples are also degraded by the heat.

3.6. POLARIZERS

Polarizers are another type of accessory that may have special requirements for compatibility with the spectrophotometer. Again the manufacturer's literature should be checked to be certain that the polarizer will be usable on the instrument, and has the desired frequency range.

The operator should be aware of the fact that in a grating instrument, the gratings produce a varying degree of polarization. This same problem is also present, but to a lesser degree, in prism instruments. Frequently it is more useful to set the polarizer at a 45° angle to compensate for the grating polarization, and then to rotate the sample. Rotatable cells are available for this purpose.

3.7. SPECULAR REFLECTANCE

Specular-reflectance accessories are another class of devices that are built for a specific type of spectrophotometer. In such an accessory, a transfer optical system is used to impinge the radiation on the sample and refocus the reflected radiation on the slit. In current practice the radiation is also focused on the sample, as opposed to the previous practice of spreading it out over a large area.

Micro- as well as macro- reflectance accessories are available. Macroaccessories can be used for micro- (small-spot) work by placing a mask over the sample to reduce the irradiated area. The finish on the mask may have a higher reflectivity than the sample and become a problem with low-reflectivity samples. Graphite sheeting is a good material for masks.

By moving the sample across the opening in the mask or over the illuminated area in a micro-accessory, it is possible to obtain a reflection profile of the sample. If much of this work is to be done, a platform with micrometer screws for positioning the sample can be substituted for the sample holder. This technique is especially useful in the semiconductor industry, where obtaining an epitaxial layer-thickness profile may be required.

These accessories use mirrors as transfer optics; they may also use condensing lenses for illuminated-area reduction. Except for those using condensing lenses, such accessories can be used over the entire range of the spectrophotometer. If the aluminum coating on the mirrors is of sufficient quality, they may also be used on ultraviolet- and visible-range spectrophotometers.

Normally accessories of this type are used in pairs; one unit in the sample beam and the other in the reference beam. When they are used in pairs, energy losses in the two beams are equalized. This also allows the operator to compare a sample to a reference without making two runs.

3.8. ATR

Attenuated total reflection (ATR) is a principle known since the days of Sir Isaac Newton, but which has only recently been put into practical use in infrared work. If radiation traveling through a crystal of high refractive index impinges on a flat interface where the other medium is of significantly lower refractive index, and the angle of incidence is greater than a critical angle, the radiation will be reflected at the interface without loss of energy. Even though there is total reflection, the radiation will travel outside of the crystal for a very short distance. The papers of Fahrenfort [1] and Harrick [2] give a good presentation of this phenomenon.

ATR units were originally built for a single reflection, but were later modified for several reflections. Currently these units are available for multireflections, and are usually called FMIR (frustrated multiple internal reflection).

As in the case with specular reflection accessories, the ATR–FMIR accessories have transfer optics to direct the radiation onto the crystal and refocus the radiation transmitted by the crystal onto the slit. Usually these optics also reduce the size of the beam before it hits the crystal, to enable the use of smaller crystal and sample sizes.

In ATR units the crystals are usually hemicylinders with a different radius for each material. The radius is chosen to give collimation of the radiation on entry into the crystal and focusing on exit from the crystal. FMIR crystals—normally flat bars—are cut for specific angles at the entrance and exit edges.

ATR units rotate the crystal and some of the optics to allow the unit to be used at different angles. FMIR units can only be used at the specific angle for which the crystal has been cut, but it is possible to use crystals cut for different angles with the same FMIR unit. The angle is an important item, as the distance the radiation travels out of the crystal and, therefore, the depth of penetration into a sample placed at this point, varies with the angle; the higher the angle, the less the penetration.

Crystals usually are made of KRS-5—a double salt of thallium bromide and thallium iodide. Other types of crystal are available, the more common ones being germanium and silver chloride. KRS-5 crystals should be handled with caution as they are toxic. Whenever they are touched, the operator should wash his hands thoroughly and as soon as is convenient. These crystals are also very soft and easily bent, scratched, or deformed. Silver chloride crystals are very corrosive when in contact with the holder and a wet sample, even when the holder is silver-plated.

These accessories can be used for both solids and liquids, but are usually used for solids. Solid samples are held against the crystal with a back plate under pressure from a screw clamp. Usually, an elastomeric gasket is placed between the sample and the back plate to give a more even distribution of pressure. For liquid samples, the crystal and holder are fitted with a back plate containing two ports, giving a system similar to a demountable sealed cell.

The transfer optics for both these and the specular reflection accessories must be aligned the first time the accessory is used. Normally, subsequent use will require only small adjustments, as long as the accessory is used in the same instrument and there has been no realignment of the instrument optics.

3.9. MISCELLANY

There are other varieties of accessories available, mostly for special techniques. In the latter category are heatable cells (and the necessary controls), cryostats, horizontal cells, high-pressure cells, etc. One item, however, deserves more than just mention, as it is a very useful and versatile accessory: this is the beam attenuator. There are several types of attenuators available commercially that are continuously variable. (It is also possible to use wire screens for the variation.) The attenuator is used in the reference beam in combination with many of the accessories listed above to compensate for large energy losses in the sample beam.

In conclusion, for a more detailed description of these accessories, the spectroscopist should contact the various instrument and accessory manufacturers for their literature. He should also ascertain whether the accessory being considered will do the required job and be compatible with the spectrophotometer to be used.

REFERENCES

1. J. Fahrenfort, *Spectrochim. Acta* **17**: 698 (1961).
2. N. J. Harrick, *Internal Reflection Spectroscopy*, Interscience, New York (1967).

CHAPTER 4

Theoretical Considerations in Infrared Spectroscopy

A complete presentation of the theoretical foundations of infrared spectroscopy would not only fill several volumes, but would require that the reader bring to the study an extensive mathematical background to enable him to understand and appreciate the concepts involved. Much of the material that would have to be included in such a thorough presentation is readily available in the current literature of spectroscopy, and the author has therefore chosen simply to introduce most of these concepts so as to provide the reader with the necessary background and perhaps furnish the impetus for further study in more advanced monographs.

It seems logical that we should begin by discussing diatomic molecules, since they lend themselves to more thorough analysis than polyatomic molecules, and the concepts encountered in their study are less abstract. First, however, we have to outline the notation that will be employed throughout this discussion.

4.1. NOTATION

In the course of the discussion, we shall refer to the rotational constant B and the quantum numbers v and J. Accepted notation employs the following convention concerning the three symbols. The energy state of the molecule prior to the absorption of infrared radiation is described by the double-primed quantities, B'', v'', and J''; the energy state of the molecule after infrared absorption has occurred will be designated by the single-primed symbols B', v', and J'; and the general expression for the energy, i.e., the system used to indicate that neither state is specified, will be simply B, v, and J.

We shall also have occasion to refer to several functions of these three quantities. Such functions will be designated by a symbol followed by the variable in parentheses; for example, in the expression

$F(J)$ the symbol F denotes the function and the term (J) indicates that J is the variable.

The following symbols retain the same definition throughout the text and are presented here so that the student may become familiar with them.

- ν is the frequency in cycles per second, or sec^{-1}
- ν_{osc} is the frequency of vibration which the molecule would have if it were a perfect harmonic oscillator
- $\tilde{\nu}_0$ is the wavenumber of the band origin in cycles per centimeter, or cm^{-1}
- $\tilde{\nu}_R$ is the wavenumber of a line in the R-branch of the band in cycles per centimeter, or cm^{-1}
- $\tilde{\nu}_P$ is the wavenumber of a line in the P-branch of the band in cm^{-1}
- $\tilde{\nu}_Q$ is the wavenumber of a line in the Q-branch of the band in cm^{-1}.

Although there is a real distinction between $\tilde{\nu}$, the wavenumber in cycles per centimeter, and ν, the frequency in cycles per second, most of the current literature uses the symbol ν indiscriminately for both. In order to conform to general usage, we shall omit the bar in the former symbol when presenting data from the literature in parts of this and other chapters. However, we feel that in the interests of maximum clarity, when these methods of describing frequency are presented to the student for the first time, the dual notation should be retained.

4.2. DIATOMIC MOLECULES

4.2A. Absorption of Infrared Radiation

Molecules can absorb electromagnetic radiation of many wavelengths, but in this discussion we shall restrict ourselves to describing absorption at wavelengths between 2 and 50 μ. Radiation at these wavelengths will be loosely termed *infrared radiation* in this text. Molecular absorption of radiation is selective, i.e., only radiation of certain wavelengths will be absorbed by any given molecule, and furthermore, not all molecules are capable of absorbing discrete infrared radiation. For reasons that we will describe in detail later on, only those diatomic molecules having a permanent dipole moment will absorb radiation.

The model chosen to describe the absorption process pictures a molecule in which the individual atoms, held together by chemical bonds, are in vibratory motion along these bonds, while the molecule as a whole is rotating. Thus we are concerned with both the vibrational and the rotational energy of the molecule. We describe a diatomic molecule such as HCl as being in a state of vibratory motion brought about by the alternate stretching and contracting of the chemical bond as the hydrogen and chlorine atoms move toward and away from each other. This vibratory motion is superimposed on a rotation of the molecule about an axis perpendicular to the chemical bond. When infrared radiation of the proper frequency impinges on the molecule and is absorbed, the vibration and/or rotation is changed. Thus, if the vibrational energy of the molecule is affected, the atoms will vibrate with a greater amplitude and the chemical bond will be elongated slightly, and if the rotational energy of the molecule is affected, it will rotate at a higher frequency than before the radiation was absorbed.

Let us consider the longest wavelength (lowest energy) of infrared radiation absorbed by a diatomic molecule which results only in a change of the molecule's vibratory motion. The wavelength of this absorbed radiation locates an absorption band in the infrared spectrum of the molecule. This longest-wavelength absorption band is called the *fundamental band*, since it represents the lowest vibrational energy change of the molecule. Diatomic molecules have only one fundamental absorption band, although multiples of this fundamental, called *overtones*, can occur. If the fundamental absorption band is at a frequency ν_1, the first overtone is approximately at the frequency $2\nu_1$.

If the absorbed infrared radiation excites both vibrational and rotational changes in the molecule, instead of a single frequency of infrared radiation ν_1, a band of frequencies centered about ν_1 will be observed for the fundamental band. For example, the fundamental absorption band of HCl is shown in Figure 4-1. The band center is at 2886 cm^{-1} and the band envelope has a width of about 150 cm^{-1}. If the band is examined at low gas pressures and under high resolution, a series of maxima are seen, as shown in Figure 4-1B (located

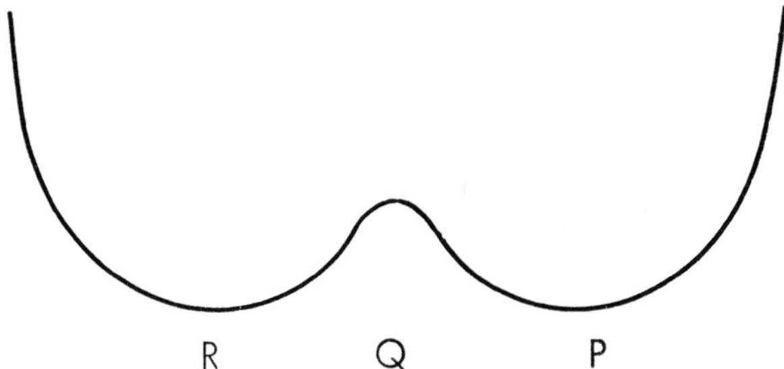

Figure 4-1A. The fundamental absorption band for HCl under medium resolution.

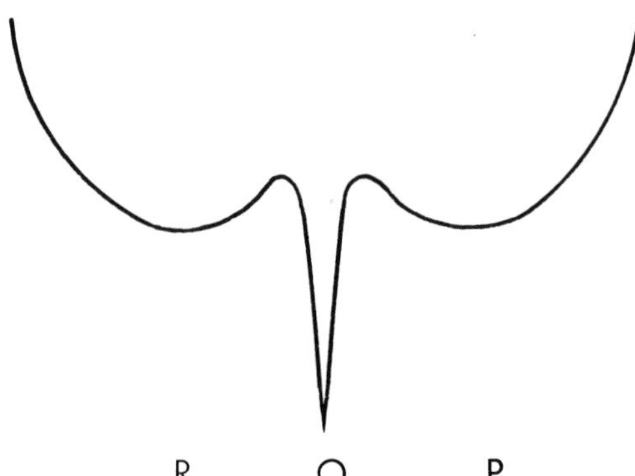

Figure 4-2. The P-, Q-, and R-branches of an absorption band.

in pocket on inside back cover). If the pressure of the gas is high or the resolution of the spectrophotometer is low, a broad band envelope is obtained, as shown in Figure 4-1A. Here, in place of the many individual peaks of the high-resolution curve, we have two maxima, with a minimum at the band center. The maximum on the high-frequency side is called the R-branch, the central minimum the Q-branch, and the low-frequency maximum the P-branch. For some molecules, the Q-branch will appear as a maximum, as illustrated in Figure 4-2. Polyatomic molecules may have very poorly defined maxima for which it is difficult to assign the correct symbols to the branches.

The high-resolution spectrum of HCl shown in Figure 4-1B can be the subject of a fairly exact mathematical analysis. The first step is to determine how the spectrum can be related to the changes in vibrational and rotational energy that the molecule has undergone as a result of the absorption of infrared radiation. An interesting paper on the line shape of the vibrational–rotational band of HCl perturbed by an Argon matrix has been published by Illinger and Trindle [24].

4.2B. Relationship Between Band Contours and Vibrational–Rotational Energy

Let us first consider the possible changes in the vibrational energy of a molecule. This energy is quantized, that is, the molecule can vibrate only at certain discrete energies. An examination of a large number of HCl molecules at 25°C would reveal that most of the atoms are vibrating at the lowest possible energy. Since only discrete values of the energy are permitted, energy levels can be indicated by integral values of a quantum number v. Thus, the lowest level is assigned a value of $v = 0$; the next higher level is $v = 1$.

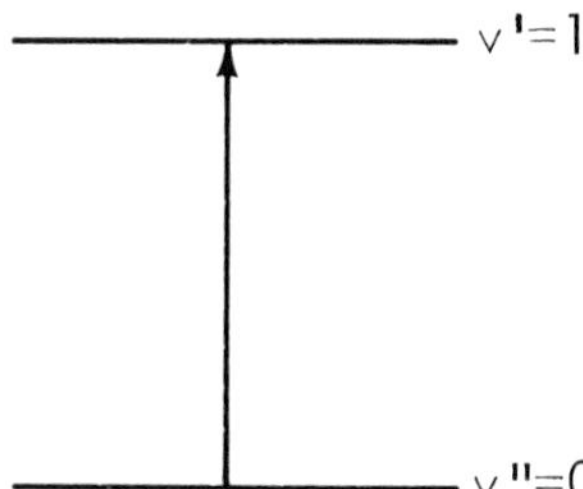

Figure 4-3. A vibrational energy transition from $v'' = 0$ to $v' = 1$.

The fundamental vibrational energy change for an HCl molecule is a jump from $v'' = 0$ to $v' = 1$, shown schematically in Figure 4-3, where the arrow represents the transition. This fundamental transition corresponds to the fundamental vibrational absorption band. As explained earlier, when such a band is observed, it is found to have a finite width as a result of simultaneous changes in the rotational and vibrational energy of the molecule.

Thus, to represent a band or indicate how energy transitions occur when radiation is absorbed and an infrared band formed, one must diagram rotational as well as vibrational energy levels.

Rotational energy levels are also quantized and are indicated by the quantum number J.

The magnitude of rotational energy changes in molecules is less than that of changes in vibrational energy, and therefore rotational

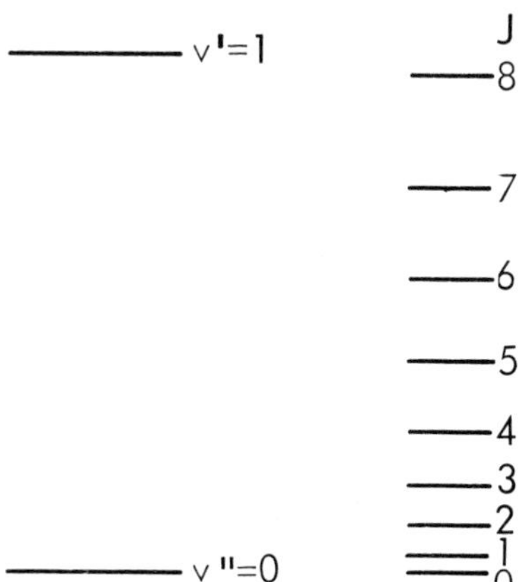

Figure 4-4. Vibrational and rotational energy levels.

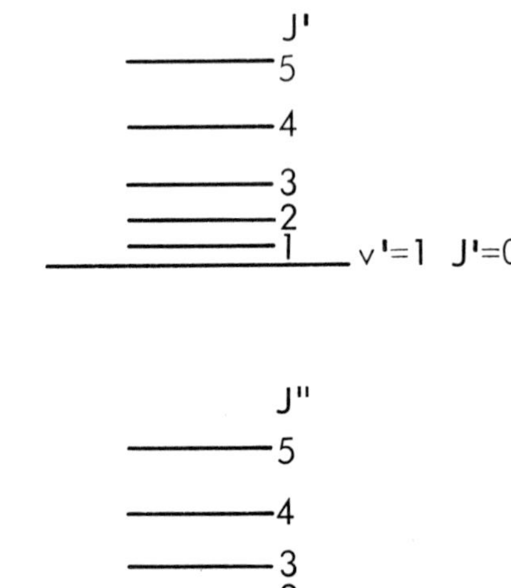

Figure 4-5. A vibrational–rotational energy level diagram.

energy levels can be represented as being more closely spaced than vibrational levels. A schematic representation of rotational levels is shown in Figure 4-4. We can combine the schematics of rotational and vibrational energy levels that exist for a molecule and represent these levels as shown in Figure 4-5 (where the magnitude of the rotational levels has been exaggerated).

In the diagram a prime mark is used on those rotational quantum numbers that are associated with the vibrational energy level $v' = 1$, i.e., the energy level after absorption. Such a diagram implies that an energy level can be made up from the sum of a rotational and a vibrational energy, and the two kinds of energy can be added together to give a fixed level for the molecule, which can be termed a *vibrational–rotational level*. In the illustration, we can thus consider an energy level as existing with, for example, a vibrational quantum number equal to 1 and a rotational quantum number equal to 2, or we can conceive of other levels having any combination of vibrational and rotational quantum numbers. This type of diagram can be used to illustrate the energy transitions which occur when infrared radiation is absorbed and an infrared band formed.

If we consider a large number of molecules of one compound, we encounter a variety of rotational and vibrational energies. Most of the molecules, however, will be in the lowest vibrational level $v'' = 0$. While most of the molecules are in this lowest level, they can have various rotational energies. For example, at room temperature a large number will be distributed from $J'' = 0$ to $J'' = 20$, with higher values of J'' for higher temperatures. The exact distribution depends on a number of factors, but for the present discussion we can assume that most of the molecules will be in levels for which J'' is less than 20.

When infrared radiation is absorbed by a molecule, an energy transition can occur for which both the vibrational and rotational quantum numbers change. These changes in the quantum numbers can be written as Δv and ΔJ. For example, if the radiation changes the vibrational quantum number from 0 to 1, we can write

$$\Delta v = v' - v'' = 1 - 0 = +1$$

v' represents a higher energy state than v''.

We can now restate a number of definitions in terms of the vibrational and rotational quantum numbers. For example, energy transitions for which the vibrational quantum number change is

$$\Delta v = +1 \tag{4-1}$$

represent the wavelength of the fundamental absorption band. The changes in rotational quantum number associated with these vibrational energy changes in vibrational–rotational infrared bands are

$$J' - J'' = \Delta J = 0, \pm 1 \tag{4-2}$$

Both the vibrational and the rotational energy transitions that occur in the fundamental transitions combine to make up the fundamental absorption band. When the vibrational quantum number change is

$$\Delta v = +2 \tag{4-3}$$

the rotational quantum number changes are again

$$\Delta J = 0, \pm 1 \tag{4-4}$$

These transitions correspond to the first overtone, and represent the first overtone band.

We now have sufficient information to draw the fundamental absorption band for HCl. The changes in quantum number which occur when the HCl band is observed are $\Delta v = +1$ and $\Delta J = \pm 1$. Energy transitions for which $\Delta J = 0$ are called Q-branch transitions and usually do not occur for diatomic molecules such as HCl. (There are several exceptions, such as NO, for which $\Delta J = 0$ transitions can occur. This is explained in terms of the odd electron present in NO.) The transitions where $\Delta J = +1$ are R-branch and those where $\Delta J = -1$ are P-branch transitions. Thus, the concept of P-, Q-, and R-branches described earlier is related to the change in

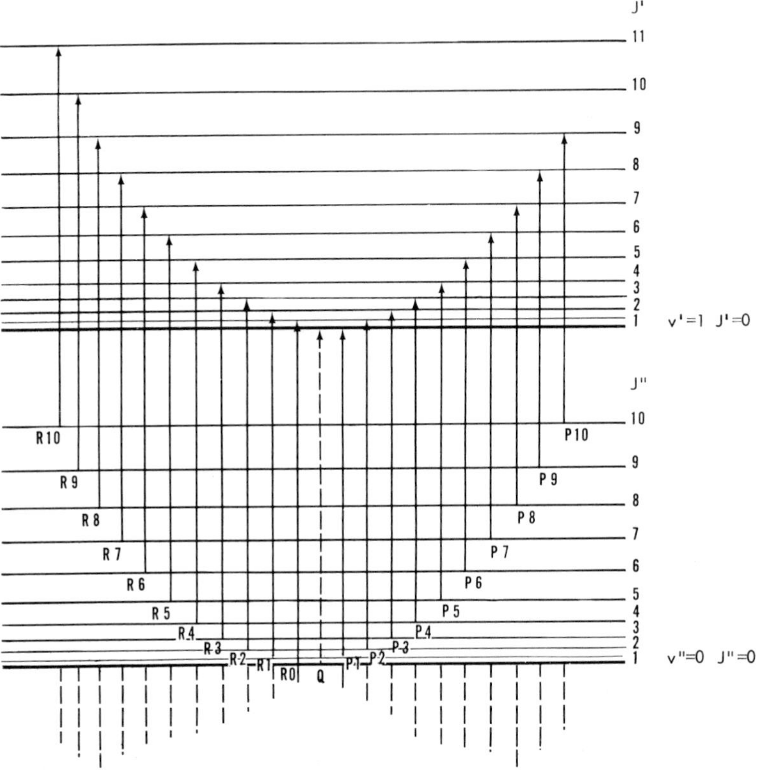

Figure 4-6. Vibrational–rotational energy transitions of the HCl fundamental.

the rotational quantum number which occurs when a molecule absorbs infrared radiation.

The HCl band presented in Figure 4-1 can be related to the rotational and vibrational energy transitions in the manner shown in Figure 4-6. The illustration represents the energy level diagram for HCl and indicates some of the energy transitions which may occur. The band which results from these transitions is shown at the bottom of the figure. The transition marked $R(0)$ represents the changes $\Delta v = +1$ and $\Delta J = +1$, where the energy transition is from the rotational level $J'' = 0$ to the level $J' = 1$, the term (0) indicating that the transition is from the $J'' = 0$ level.

It should be noted that the separation between the rotational levels increases with J; that is, the energy difference between the levels $J = 0$ and $J = 1$ is less than the difference between the levels $J = 1$ and $J = 2$. The reason for this will be developed later. However, the net result of this increasing separation between rotational levels is to make the energy transition $R(1)$ greater than $R(0)$, and each successive energy transition $R(J)$ greater than the preceding transition. This means that although all the $R(J)$ lines result from transitions for which $\Delta J = +1$, these lines do not all represent the same change in energy, but rather a series of energy changes, which gives some width to the observed absorption band. The entire series of transitions $R(J)$ makes up the R-branch of the band. Further, the intensity of each $R(J)$ line is determined by the number of molecules undergoing that particular transition (or, in other words, the number of molecules in the J level before the energy transition occurs). In Figure 4-6, we have drawn the absorption band as it will appear when the greatest number of molecules is undergoing transitions $R(6)$, $R(7)$, and $R(8)$. Beyond these, as fewer and fewer molecules undergo each transition, the intensity of the lines begins to diminish, and the band, as shown in the figure, gradually decreases in intensity.

Similar considerations apply to the P-branch of an absorption band. The energy transitions $P(1)$, $P(2)$, etc., form a series of lines which give the P-branch its width, and the intensity, which is a function of the number of molecules undergoing each transition, is presented as gradually increasing as we approach $P(8)$.

For HCl, transitions with $\Delta J = 0$ are forbidden, and this is true for most diatomic molecules; consequently, no Q-branch is seen in the band of HCl.

Finally, we can now relate the energy transitions which make up the P-, Q-, and R-branches of a band to the fine-line structure shown in Figure 4-1B for the HCl band. This fine-line structure represents the individual $P(J)$ and $R(J)$ lines which, when unresolved, make up the band envelope. The individual lines are seen only when the resolution is sufficiently high.

4.2C. Mathematical Relations for Rotational and Vibrational Energy

Rotations and Moments of Inertia. A molecule may rotate about a number of different axes. For example, a simple molecule such as HCl can rotate about an axis perpendicular to a line drawn through the two nuclei and passing through the center of mass (i.e., the axis will be nearer the heavier of the two atoms, in this case the chlorine atom).

When a molecule rotating at a certain frequency absorbs infrared radiation and begins to rotate at a higher frequency, the frequency of the absorbed radiation gives the position of a rotational absorption band of the molecule. For most molecules these bands occur at long wavelengths (50 μ or higher). The change in rotational energy is also present at lower wavelengths, but since it there occurs in conjunction with changes in the vibrational energy, the spectrum produced does not represent pure rotation but rather the combination of changes in vibrational and rotational energies. It is this rotational–vibrational spectrum which appears in the 2–50 μ region.

We may relate the rotational energy of any molecule to its moment of inertia I_B, which for a rigid body is defined as

$$I_B = \Sigma_i\, m_i r_i^2 \tag{4-5}$$

where r_i is the perpendicular distance of the mass m_i from the axis of rotation. For a diatomic molecule such as HCl, the moment of inertia is $I = m_1 r_1^2 + m_2 r_2^2$, where m_1 and m_2 are the masses of the hydrogen and chlorine atoms, and r_1 and r_2 are their respective perpendicular distances from the axis of rotation. If we let $r_e = r_1 + r_2$, the expression for the inertia becomes

$$I_B = \frac{m_1 m_2}{m_1 + m_2}\, r_e^2 = \mu r_e^2 \tag{4-6}$$

where μ is the *reduced mass* (defined by the fraction) and r_e is the internuclear separation at equilibrium.

If we assume that the rotation of the HCl molecule is similar to that of two balls connected by a rigid bar, we can obtain an equation for the rotational energy. For such a rigid system, called a *rigid rotor*, the energy of rotation is given by

$$E = 2\pi^2 v^2 I_B \tag{4-7}$$

where v is the frequency of rotation.

Equation (4-7) describes a rigid rotor with a continuous energy spectrum, and since it is known that the rotational energy of a molecule is quantized, in order to solve for the discrete energies we must resort to quantum mechanics. The allowed energies can be given in terms of the rotational quantum number J (which can have values $J = 0, 1, 2, \ldots$), and to calculate the rotational energy for a diatomic or linear polyatomic molecule, which can be thought of as a rigid rotor, we solve the Schrödinger wave equation for permitted discrete values (eigenvalues) of E:

$$E = \frac{h^2 J(J + 1)}{8\pi^2 I_B} \tag{4-8}$$

Equation (4-8) is valid whenever the model of a rigid rotor is a sufficiently good approximation for the motion. It is useful to put this equation into a form where the *term value* E/hc is used instead of the energy. This converts the energy in ergs to the term value in cm^{-1}. The equation for the rotational term value F then becomes

$$F = \frac{E}{hc} = BJ(J + 1) \tag{4-9}$$

where

$$B = \frac{h}{8\pi^2 c I_B} \tag{4-10}$$

and c is the velocity of light.

If the term values of a rigid rotor are plotted from Equation (4-9), the levels will form a series, as shown in Figure 4-7. It should be noted that the spacing between the levels has the value $2BJ$ and increases with J.

A diatomic molecule such as HCl has three principal moments of inertia. The one about the internuclear axis is practically zero, and the other two, about axes perpendicular to this axis, are equal. A molecule such as CH_4 has three principal moments of inertia about the three mutually perpendicular principal axes which are equal, i.e., $I_A = I_B = I_C$. Such a molecule with three equal principal moments of inertia is called a spherical rotor (or spherical top); a molecule with two of the three moments equal is called a symmetric top, and one with three unequal principal moments is called an asymmetric rotor (or asymmetric top). For each of these models, a rotational term value equation can be developed. These equations

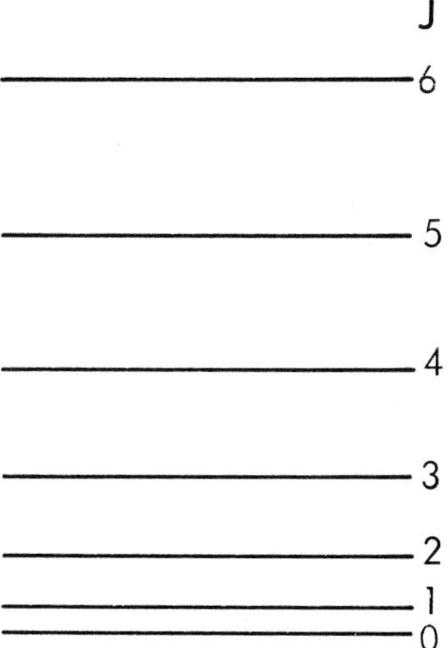

Figure 4-7. Rotational energy levels.

TABLE 4-I. Rotational Energy Equations for Various Rotors

Moment of inertia	Designation	Rotational term value	Examples
$I_A = 0;\ I_B = I_C \neq 0$	Linear or diatomic	$F = BJ(J+1)$ $B = \dfrac{h}{8\pi^2 c I_B}$	CO_2, HCl
$I_A = I_B = I_C \neq 0$	Spherical rotor	$F = BJ(J+1)$	CH_4
$I_B = I_C \neq I_A$ $I_A, I_B, I_C \neq 0$	Symmetric rotor	$F = BJ(J+1) + (A - B)K^2$ $B = \dfrac{h}{8\pi^2 c I_B} \quad A = \dfrac{h}{8\pi^2 c I_A}$	NH_3, CH_3Cl
$I_A \neq I_B \neq I_C$ $I_A, I_B, I_C \neq 0$	Asymmetric rotor	Several approximate equations exist	H_2O, H_2CO

are listed in Table 4-I, together with some examples of molecules which have been analyzed using these models. It is necessary in one of these equations to introduce a second quantum number, K, which is an integer that cannot be greater than J. This number therefore can take on the values $K = 0, 1, 2, \ldots, J$. This quantum number appears in the rotational term value equations of the symmetric rotor.

Vibrational Energy Equations for a Harmonic Oscillator. A diatomic molecule such as HCl can be visualized as vibrating in a manner similar to a harmonic oscillator. The stretching of the chemical bond between the hydrogen and chlorine atoms is a periodic motion and the equations derived for a harmonic oscillator appear to be quite appropriate.

Application of the model of the harmonic oscillator to an HCl molecule leads to the following picture. The masses of the hydrogen and chlorine atoms may be considered as connected by a spring with a restoring force constant f. The frequency of vibration ν_{osc} (or the frequency of the radiation emitted or absorbed) is given in terms of f and the reduced mass μ of the hydrogen and chlorine atoms by the relation

$$\nu_{osc} = \frac{1}{2\pi}\sqrt{\frac{f}{\mu}}\,[\text{cps}] \qquad (4\text{-}11)$$

By making use of the relationship $\omega_e = \nu_{osc}/c$ the wavenumber can be expressed in cm^{-1}. Further, the constant $1/2\pi$ and other constants can be combined to yield the simple form

$$\omega_e = 1307\sqrt{\frac{f'}{\mu'}}\,[\mathrm{cm}^{-1}] \tag{4-12}$$

where f' is the force constant in millidynes per angstrom, ω_e is the wavenumber of the radiation emitted or absorbed by the oscillator, and μ' is the reduced mass in atomic mass units (amu). Equations (4-11) and (4-12) indicate that if the vibration of HCl is assumed to be similar to that of a harmonic oscillator, it is possible to calculate the wavenumbers of the fundamental absorption band if a value for the force constant f' is available. Badger [1] and Gordy [2] have suggested empirical formulas for f'. The formula suggested by Gordy is

$$f' = aN\left[\frac{X_a \cdot X_b}{d^2}\right]^{3/4} + b \tag{4-13}$$

where a and b are constants determined by the positions in the periodic table of the two atoms making up the harmonic oscillator, N is the number of covalent bonds between the two atoms of the harmonic oscillator, d is the internuclear distance, X_a and X_b are the electronegativities of the two atoms making up the harmonic oscillator, and f' is the force constant in millidynes per angstrom (md/Å).

For HCl, equation (4-13) becomes

$$f' = 1.67\left[\frac{2.0 \times 3.0}{1.27^2}\right]^{3/4} + 0.3 = 5.1\ \mathrm{md/\AA}$$

This value of f' can now be substituted into equation (4-12). The wavenumber for the fundamental band of HCl is thus found to be

$$\omega_e = 1307\ \frac{\mathrm{sec}}{\mathrm{cm}}\sqrt{\frac{5.1\ \mathrm{md/\AA}}{\dfrac{35.5 \times 1.0}{35.5 + 1.0}\ \mathrm{amu}}} = 2993\ \mathrm{cm}^{-1}$$

The observed band position is 2885.9 cm^{-1}, so that, while the agreement between the observed and calculated wavenumber is not too good, at least the correct order of magnitude has been obtained.

The relationship between the wavenumber and the masses involved in a harmonic vibration can be used to calculate the wavenumber shift which will occur when an isotope is substituted for one of the atoms of the harmonic oscillator. For a calculation of this type, we start with equation (4-11) and proceed as follows. It is assumed that the substitution of the isotope does not change the force constant f. Let us suppose that the two vibrations of interest are for the molecules HCl and DCl. The ratio of the wavenumbers of these two molecules is then given by

$$\frac{\omega_{\mathrm{DCl}}}{\omega_{\mathrm{HCl}}} = \sqrt{\frac{\mu_{\mathrm{HCl}}}{\mu_{\mathrm{DCl}}}} \tag{4-14}$$

Since the μ's are known, the wavenumber shift can be calculated. It is of interest that if the atoms H and D are the two isotopes, then the factor relating the position of the two bands is approximately $\sqrt{2}$.

Let us next consider the energy levels of a harmonic oscillator. Since vibrational energy is quantized, discrete energy levels exist. Quantum-mechanical treatment of the harmonic oscillator leads to the equation

$$E = h\nu_{\mathrm{osc}}(v + \tfrac{1}{2})[\mathrm{ergs}] \tag{4-15}$$

where E is the energy of the harmonic oscillator, h is Planck's constant, ν_{osc} is the frequency in cycles per second, and v is the vibrational quantum number.

Equation (4-15) indicates that the energy levels of a harmonic oscillator are equally spaced. For the lowest level, where $v = 0$,

$$E = \tfrac{1}{2}h\nu_{\mathrm{osc}}$$

The fundamental band was defined earlier as the band for which the vibrational quantum number changes from $v'' = 0$ to $v' = 1$. We can now describe it using equation (4-15). For example, if we substitute the quantum numbers for the fundamental band into equation (4-15), we obtain

$$E' - E'' = h\nu = h\nu_{\mathrm{osc}}[(1 + \tfrac{1}{2}) - (0 + \tfrac{1}{2})] \tag{4-16}$$

$$\nu = \nu_{\mathrm{osc}}$$

Thus, the calculated wavenumber corresponding to the transition is equal to the wavenumber of the harmonic oscillator. This is true regardless of the value of v for the initial energy state, as long as $\Delta v = +1$ for the transition. For a more thorough analysis of this model, the reader should examine the theoretical paper which recently appeared on the harmonic oscillator model in atomic and molecular physics [25].

The Harmonic Oscillator—Rigid Rotor Equations. We have seen that an absorption band represents molecular transitions that are related to both vibrational and rotational energy changes. The vibrational–rotational energy levels for a molecule which vibrates like a harmonic oscillator and rotates as a rigid rotor are then given by combining equations (4-15) and (4-8). Thus,

$$E = \underbrace{h\nu_{\text{osc}}(v + \tfrac{1}{2})}_{\text{vibration}} + \underbrace{\frac{h^2 J(J+1)}{8\pi^2 I_B}}_{\text{rotation}} \tag{4-17}$$

Equation (4-17) can be used to describe the lines of a vibrational–rotational band such as the one illustrated in Figure 4-1B. We discussed this figure qualitatively earlier, while describing the fine-line structure in terms of the energy transitions $R(J)$, $P(J)$, and $Q(J)$. We can now use equation (4-17) to illustrate how the rigid rotor model predicts an equal spacing of the R and P lines in an absorption band.

Equation (4-17) represents the energy levels that exist for a molecule. If upon absorbing infrared radiation a molecule undergoes a transition from an energy level represented by quantum numbers v'' and J'' to a new level v' and J', then the change in energy associated with the transition is given by

$$E' - E'' = h\nu_{\text{osc}}[(v' + \tfrac{1}{2}) - (v'' + \tfrac{1}{2})] + \frac{h^2}{8\pi^2 I_B}[J'(J'+1) - J''(J''+1)] \tag{4-18}$$

To determine the wavenumber that corresponds to this energy change, we must divide by hc (since $E' - E'' = hc\omega_e$). This gives

$$\frac{E' - E''}{hc} = \tilde{\nu} = \omega_e[(v' + \tfrac{1}{2}) - (v'' + \tfrac{1}{2})] + B[J'(J'+1) - J''(J''+1)] \tag{4-19}$$

where $\tilde{\nu}$ is the wavenumber in cm^{-1}, ω_e is the wavenumber (cm^{-1}) of the molecule vibrating as a harmonic oscillator, and $B = h/8\pi^2 c I_B$.

Equation (4-19) can be used to derive a relation for the wavenumber positions of the lines of the P-, Q-, and R-branches. For example, for the R-branch, where the rotational quantum number change is given by $J' - J'' = \Delta J = +1$, the positions of the lines are given by

$$\tilde{\nu}_R = \omega_e + 2B + 2BJ'' \tag{4-20}$$

where $J'' = 0, 1, 2, 3, \ldots$, while the positions of the lines of the P-branch, where $J' - J'' = \Delta J = -1$, are given by

$$\tilde{\nu}_P = \omega_e - 2BJ'' \tag{4-21}$$

where $J'' = 1, 2, 3, \ldots$. In these equations $\tilde{\nu}_R$ is the wavenumber of a line of the R-branch in cm^{-1}, ω_e is now the wavenumber of the Q-branch in cm^{-1}, and $\tilde{\nu}_P$ is the wavenumber of a line of the P-branch in cm^{-1}.

Let us use equation (4-20) to calculate the position of the line $R(0)$. For this line the J'' value is 0. Hence, from equation (4-20)

$$\tilde{\nu}_{R(0)} = \omega_e + 2B$$

From this result we see that the $R(0)$ line is spaced $2B$ from the Q-branch line ω_e. Further, it can be shown that the $R(1)$ line is $4B$ from ω_e, and the $R(2)$ line $6B$ from ω_e. Thus, the spacing between rotational lines of the R-branch is $2B$. Equation (4-21) can be used similarly to show that the interlinear spacing in the P-branch is also $2B$.

Thus, the harmonic oscillator—rigid rotor model leads to equations which predict that an absorption band will be made up of lines spaced $2B$ apart, with a single line of wavenumber ω_e at the band center. This ω_e line is not observed for diatomic molecules because it represents a forbidden transition. Instead, the band center is observed simply as an intensity minimum in the contour of the band.

It is possible to calculate the internuclear distance in a diatomic molecule by measuring the interlinear spacing in either the P- or the R-branch. If we assume that the equations for a harmonic oscillator—rigid rotor are valid, then this spacing is $2B$. The internuclear distance is calculated from the definition of the rotational constant B. If we combine the factors in the definition of the rotational constant B as given by equation (4-10) with the definition of the moment of inertia for diatomic molecules given in equation (4-5), the internuclear distance r_e of a diatomic molecule is given by

$$r_e = \sqrt{\frac{h}{8\pi^2 c\mu B}} \qquad (4\text{-}22)$$

All the factors in this equation have already been defined, and it can therefore be used to calculate the equilibrium value r_e of the internuclear separation of a diatomic molecule from the observed value of $2B$ in the molecular spectrum. For more exact calculations, the difference in the moments of inertia in the lower and higher energy levels must be taken into account. Bond distance and the position of the stretching vibration also have been correlated [26].

The Nonrigid Rotor Model. Contrary to our earlier approximation, the fine-line structure of the infrared absorption band of a diatomic molecule does not show exactly equal spacing of the P- and R-lines. Instead, as the J'' value increases, we observe a slight crowding of the lines of the R-branch and a slight spreading of the lines of the P-branch. This unequal spacing can be accounted for by a slight modification of the equations describing the rigid rotor.

The rigid rotor model of a diatomic molecule implies that there is no change in the internuclear separation when the molecule undergoes a transition from one vibrational–rotational energy level to another. Let us picture a diatomic molecule (such as HCl) vibrating while it rotates in space. When infrared radiation is absorbed, the molecule vibrates with a larger amplitude and has a slightly larger average internuclear separation than before the absorption. Thus, the molecule has a slightly larger moment of inertia in the new vibrational state. We can describe this effect in terms of the rotational constant $B = h/8\pi^2 cI_B$ by noting that if the moment of inertia I_B is larger in the higher vibrational state, then the rotational constant B' must be smaller for that state, or $B' < B''$.

We can now write equations describing the lines of the P- and R-branches which take into account the change in the moment of inertia which occurs when a diatomic molecule undergoes a transition from a lower to a higher vibrational–rotational energy level. The wavenumbers of the lines in the R-branch can be given by a more general form of equation (4-19):

$$\tilde{\nu}_R = \omega_e + 2B' + (3B' - B'')J'' + (B' - B'')(J'')^2 \qquad (4\text{-}23)$$

where $J'' = 0, 1, 2, 3, \ldots$; similarly, for the P-branch,

$$\tilde{\nu}_P = \omega_e - (B' + B'')J'' + (B' - B'')(J'')^2 \qquad (4\text{-}24)$$

where $J'' = 1, 2, 3, \ldots$.

With these equations it is easy to show that as the value of J'' increases, the spacing between the lines of the R-branch decreases, while the interlinear spacing of the P-branch increases.

The Anharmonic Oscillator Model. The harmonic oscillator model for diatomic molecules predicts that the vibrational energy levels of a molecule will be equally spaced. If this were true, an overtone band would appear at a frequency (or wavenumber) exactly twice the fundamental. What actually occurs is the appearance of an overtone band at a frequency slightly lower than twice the fundamental and we must therefore modify the simple equations for a harmonic oscillator to take this observation into account.

The anharmonic oscillator model has been suggested for this purpose; its energy levels are described by the equation

$$E = hc\omega_e(v + \tfrac{1}{2}) - hc\omega_e\chi_e(v + \tfrac{1}{2})^2 \qquad (4\text{-}25)$$

This equation is similar to equation (4-15), which described the energy levels of the simple harmonic oscillator, and differs only in the addition of the term $[hc\omega_e\chi_e(v + \tfrac{1}{2})^2]$. It is possible to add still other modifying terms, but we will restrict this discussion to the equation for an anharmonic oscillator as it appears in (4-25), i.e., we will limit ourselves to a single modification. In this equation $\omega_e\chi_e$ is called the anharmonicity constant since it is a measure of how much the harmonic oscillator equation must be modified to account for the anharmonic character of a vibration. The equation also predicts that the spacing between energy levels of the anharmonic oscillator will decrease as v increases.

Dividing equation (4-25) by hc, we obtain for the term value F in cm^{-1}

$$F = \frac{E}{hc} = \omega_e(v + \tfrac{1}{2}) - \omega_e\chi_e(v + \tfrac{1}{2})^2 \qquad (4\text{-}26)$$

where ω_e is the wavenumber a diatomic molecule would have if it had vibrations of very small amplitude and behaved like a harmonic oscillator. However, since in reality we are dealing with an anharmonic vibration, ω_e will no longer correspond to any *observed* wavenumber. For example, consider the band which appears when the quantum number changes from $v'' = 0$ to $v' = 1$. The position of this band for an anharmonic oscillator which has term values (E/hc) given by equation (4-26) would be

$$\tilde{\nu}_0 = \frac{E' - E}{hc} = \omega_e[(1 + \tfrac{1}{2}) - (0 + \tfrac{1}{2})] - \omega_e\chi_e[(1 + \tfrac{1}{2})^2 - (0 + \tfrac{1}{2})^2]$$

$$= \omega_e - 2\omega_e\chi_e \qquad (4\text{-}27)$$

Thus, we see that the actual wavenumber of the absorption band center is $\tilde{\nu}_0$, which is not the same as the wavenumber of the harmonic oscillator, ω_e, but is lower by an amount $2\omega_e\chi_e$.

The position of the overtone of an anharmonic oscillator can be calculated from equation (4-26) by substituting the quantum number change from $v'' = 0$ to $v' = 2$. The position obtained for the overtone is

$$\tilde{\nu}_0 = 2\omega_e - 6\omega_e\chi_e \qquad (4\text{-}28)$$

We can see from this result that the position of the overtone is not at exactly $2\omega_e$, but rather displaced to a lower wavenumber by an amount $6\omega_e\chi_e$. By substituting the observed values of the fundamental and overtone bands into equations (4-27) and (4-28), respectively, we can calculate both the anharmonicity constant $\omega_e\chi_e$ and the wavenumber ω_e.

For example, if we substitute observed values of 2885.9 cm^{-1} for the wavenumber of the fundamental of HCl and 5668.0 cm^{-1} for that of the first overtone, we find ω_e to be equal to 2989.7 cm^{-1} and obtain a value of 51.9 cm^{-1} for the anharmonicity constant $\omega_e\chi_e$.

4.2D. The Intensity Distribution of a Band

The distinct shape of an absorption band for a diatomic molecule is related to the numbers of molecules that undergo the various energy transitions. Thus, the most intense line in a P-, Q-, or R-branch is that which represents the largest number of molecules undergoing a particular transition. Since the intensity of each line depends on the number of molecules undergoing a certain energy transition, the band envelope as a whole is representative of the total number of molecules involved. As the temperature of the molecules is increased, the contour of the absorption band will change, but the area under the envelope will remain constant, provided that the increase in temperature does not change the number of molecules initially in the vibrational level $v = 0$. This has been confirmed by observation; the area of a band remains constant for fairly large changes in temperature although the band widens and flattens as the temperature is increased. This widening and flattening is the result of increases in the population of higher rotational levels and consequent decreases in the lower levels. Since the transitions from higher rotational levels are associated with lines further away from the band center, the higher temperature results in an increase in the envelope's intensity away from the band center with a corresponding decrease in the intensity of lines near the band center.

4.3. POLYATOMIC MOLECULES

4.3A. The Relationship Between Observed Absorption Bands and Motions of the Atoms in a Molecule

Infrared absorption bands can be related to the motions of the individual atoms that comprise a molecule. It seems reasonable that distinct motions should result in distinct absorption bands. In the simplest approach it is assumed that isolated parts of a molecule can vibrate independently of the remainder of the molecule. For example, an OH group that forms part of a larger group of atoms exhibits a stretching motion, i.e., the bond between the oxygen and hydrogen is periodically stretched and released. This motion, called a *stretching vibration*, produces a distinct infrared absorption band that is observed for all molecules that contain this grouping. Such vibrations in isolated parts of the molecule provide the basis for the concept of *group frequencies*, which will be discussed in detail in Chapter 5.

A more general picture involves complex motion of all the atoms of a molecule. This over-all complex motion can be resolved into a small number of basic motions, which can be designated as the *fundamental vibrations* of the molecule. It can be shown that in order to resolve the complex motion of a nonlinear molecule of N

atoms it is necessary to describe $3N - 6$ fundamentals. This formula can be simply obtained as follows: $3N$ coordinates are required to describe the degrees of freedom of a molecule. If three coordinates are required to describe rotational degrees of freedom, and three to describe translational degrees of freedom, then $3N - 3 - 3$ or $3N - 6$ are required to describe vibrational degrees of freedom of a nonlinear molecule. For a linear molecule $3N - 5$ are required as there is one less degree of rotational freedom.

This description of atomic motion in terms of fundamental vibrations can be carried further. For many molecules it is possible to diagram the motions atoms undergo for each fundamental vibration. The observed infrared (and Raman) absorption bands can then be correlated with these fundamental vibrations. It is also possible to assign *overtone bands*, which are multiples of the fundamentals. Finally, bands can be described that are combinations of fundamentals and—quite logically—are called *combination bands*. Fundamentals are given the notation ν_1, ν_2, ..., etc., the symbol ν_1 indicating fundamental number one. Overtones are designated as $2\nu_n$ to indicate that they appear at approximately twice the fundamental frequency ν_n. Combination bands can be written as $\nu_n + \nu_m$ (a *sum band*) or $\nu_n - \nu_m$ (a *difference band*). Bands may also appear at frequencies $\nu_n + \nu_m - \nu_l$ or $\nu_n - \nu_m + \nu_l$, i.e., at combinations of sums and differences of fundamental frequencies. The combination $\nu_n + \nu_m - \nu_m$ also represents a band distinct from ν_n.

It is difficult to describe the atomic motion in a molecule during a vibration and interpret the meaning of a fundamental vibration until a model or structure is chosen for the molecule. Many molecular structures have been confirmed by infrared and Raman spectroscopy, and we shall describe the methods used for such confirmations in this chapter. While the methods of Raman spectroscopy do not fall within the scope of this work, we shall refer to Raman spectra from time to time. It is sufficient for the reader to know that Raman spectra are similar to infrared spectra in some respects but differ in that the absorption of radiation occurs as a result of a different process. Further comparisons between Raman and infrared spectroscopy will be developed in subsequent sections.

Before discussing how a model is chosen for a molecule and how the motions of atoms are related to fundamental vibrations (as well as to combinations and overtones), let us first introduce the nomenclature used to describe atomic motion and the concept of change in the molecular dipole moment.

When two atoms are held together by a chemical bond, we encounter a vibratory motion, i.e., alternate stretching and compression along the bond, which is described as a stretching vibration.

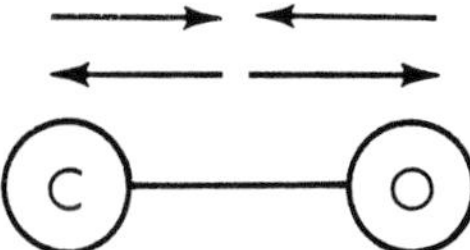

Figure 4-8. A stretching vibration.

To diagram such a motion, arrows indicating the direction of motion can be placed above symbols of the atoms and the bond between them, as shown in Figure 4-8. If three atoms are linked together linearly and a stretching occurs, it can be a symmetric motion, where the two end atoms move in and out in phase, or it can be an asymmetric motion, where one atom moves in and the other moves out in alternation.

A system of three atoms can also undergo a bending motion. For example, for a linear triatomic molecule, the end atoms can alternately move off and back to the axis of the molecule. A similar motion, for a nonlinear molecule, is diagrammed in Figure 4-9. As in the linear case, the three atoms form a plane, with the central atom being bound to the other two. A motion of this nature is called a *bending vibration.*

All motions other than stretching are classified as deformation vibrations, and bending vibrations are one type of deformation. If the total number of vibrations for a nonlinear molecule is $3N - 6$, it will be found that $N - 1$ of these are stretching motions and $2N - 5$ are deformations. For linear molecules $N - 1$ are stretching motions and $2N - 4$ are deformations.

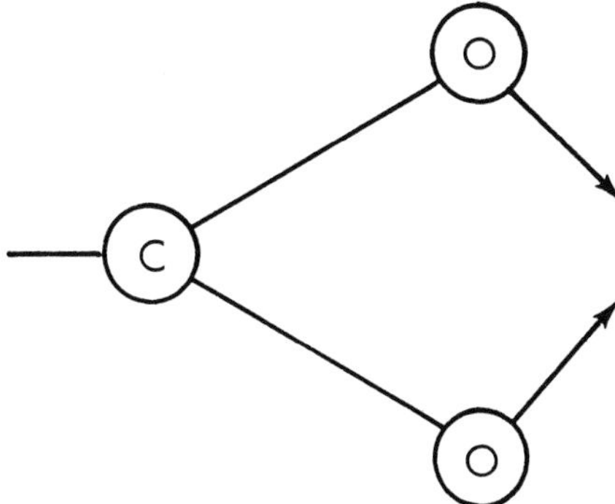

Figure 4-9. A bending or scissoring vibration.

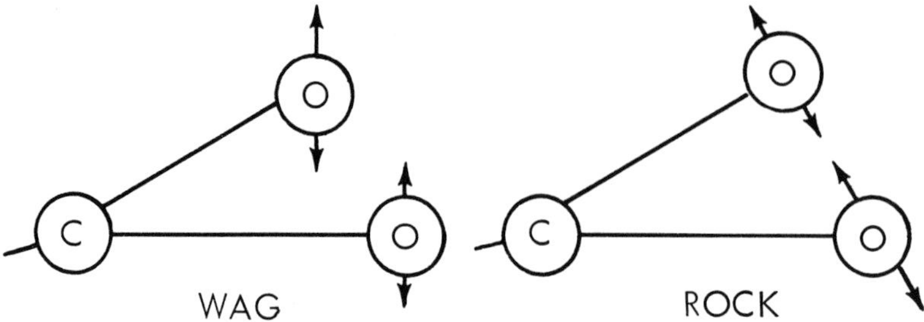

Figure 4-10. Wagging and rocking vibrations.

Besides bending, we encounter wagging, twisting, and rocking deformation vibrations. These can be diagrammed as follows. Consider a configuration of three atoms in a plane, one of which is bound to a fourth atom. It is possible for the two atoms not bound to the fourth atom to move in phase above and below the plane; this is called *wagging*. It is also possible for the two O atoms to move in the plane in such a way that the C—O angle is changed while the two O atoms move in phase; this is called *rocking*. Wagging and rocking motions are illustrated in Figure 4-10. *Twisting* is the motion of the plane of three atoms as a unit about the bond to the fourth atom, as shown in Figure 4-11.

Qualitatively, for the same atoms we can generally expect the stretching vibrations to give absorption bands at higher frequencies (or wavenumbers) than the deformation vibrations. For example, the stretching vibrations for the CH_3 group are found near 2800 cm^{-1}

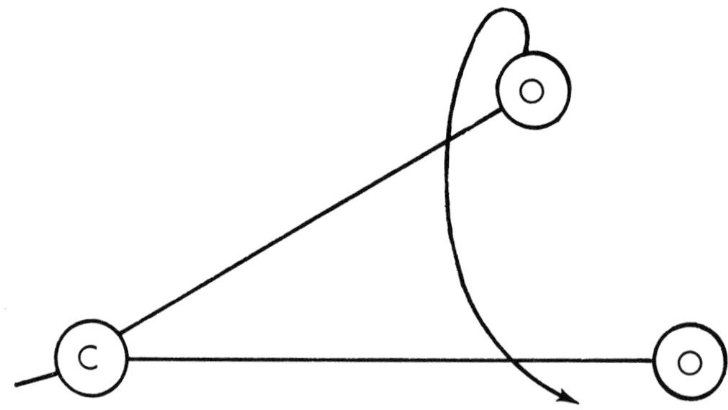

Figure 4-11. A twisting vibration.

and the deformation vibrations below 1500 cm^{-1}. While asymmetric stretching and bending vibrations are usually located at higher frequencies (or wavenumbers) than the corresponding symmetric vibrations, some exceptions are known.

4.3B. The Relationship Between Dipole Moment Changes and Infrared Absorption Bands

If a molecule is to absorb infrared radiation, the radiation must be of the correct frequency to cause a quantum jump in its vibrational (and, in most instances, rotational) energy. A vibrating molecule will interact with electromagnetic radiation if an oscillating dipole moment accompanies the vibration. A change (oscillation) in dipole moment occurs for a molecule whenever a change in position of the centers of positive and negative charge resulting from atomic motion occurs.

A simple explanation (see Figure 4-12) can be used to illustrate how, in some instances, electromagnetic radiation can excite vibratory motion of a molecule. Consider the molecule N_2, which, owing to the existence of charge symmetry, has no permanent dipole moment. It can be assumed that equal positive and negative charges are present on each N atom. Electromagnetic radiation will cause positive charges to move in one direction and negative charges to move in another. For such a symmetric molecule, this type of interaction produces no vibratory motion; however, for an asymmetric molecule, such as NO, where the negative charge on the oxygen atom is greater than that on the nitrogen atom, electromagnetic

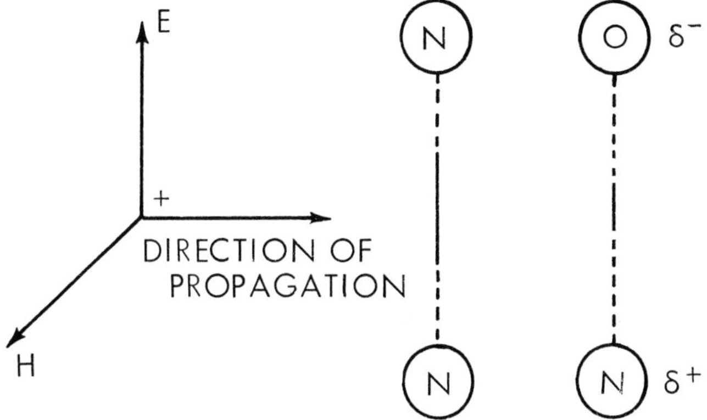

Figure 4-12. The interaction between electromagnetic radiation and the molecules N_2 and NO.

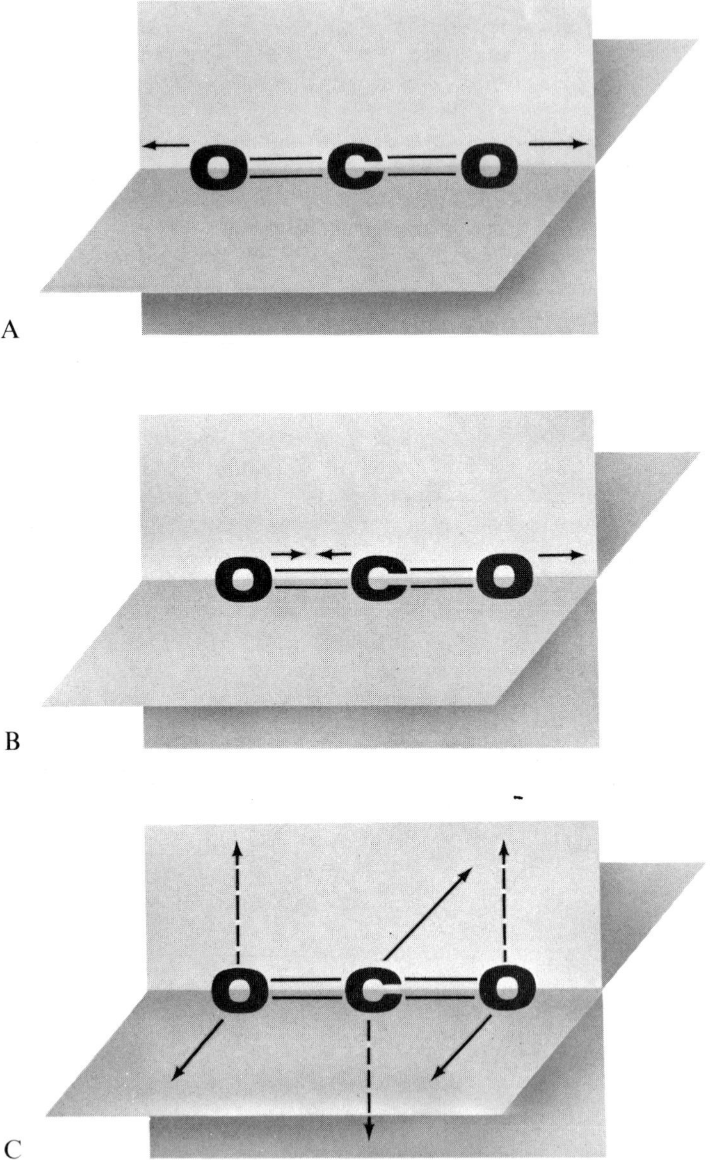

Figure 4-13. Vibrational motions of CO_2. (A) ν_1 symmetric stretch, parallel band; (B) ν_3, asymmetric stretch, parallel band; (C) ν_2, bend, perpendicular band.

radiation of the correct frequency can produce a stretching of the bond between these atoms as the negative charge on the oxygen atom moves in a direction opposite to that of the positive charge on the nitrogen atom. This type of interaction results in a stretching of the chemical bond between the nitrogen and the oxygen, with consequent absorption of infrared radiation, giving rise to an absorption band.

When a variation in dipole moment accompanies an atomic motion in a manner similar to that just described for NO, then that motion is capable of being excited by infrared radiation and is called "infrared active." Some motions that are not infrared active may be Raman active, since Raman activity is present when the electron cloud symmetry is changed. A Raman spectrum is obtained using ultraviolet or visible radiation and observing the change in frequency produced when the molecule is irradiated. Radiation will be scattered either as Rayleigh or Raman scattering. In Rayleigh scattering the scattered radiation is of the same frequency as the incident radiation; in Raman scattering the frequencies differ. We can illustrate these concepts by using carbon dioxide as an example. Three kinds of vibratory motion are present in the carbon dioxide molecule. These are shown in Figure 4-13. The first motion (Figure 4-13A) is a symmetric stretching vibration since the two oxygen atoms move on a line toward and away from the carbon atom in a symmetrical manner. This motion is designated as the fundamental ν_1. Such a motion does not result in a change of dipole moment since, if we consider that negative charges reside on the oxygen and positive charges on the carbon, the motion produces no change in the location of the centers of charge. Thus, the ν_1 vibration is not infrared active, but it is a Raman-active vibration, since the symmetry of the electron cloud is changed. The asymmetric stretch (Figure 4-13B) is infrared active since it does cause a change in dipole moment. Clearly in this case, the relative motion of the atoms brings the carbon alternately closer to one and then to the other oxygen. This necessarily causes a corresponding relative motion of the charge centers. The bending vibration shown in Figure 4-13C also is infrared active. It is said to be a doubly degenerate vibration, i.e., two superimposed identical fundamentals. Thus, to resolve the complex motions of some molecules, it is often necessary to describe degenerate motions.

One further concept can be introduced at this juncture. The asymmetric and symmetric stretching vibrations of carbon dioxide are described as parallel ($\parallel$) vibrations, that is, the change in dipole moment which occurs during these vibrations is parallel to the axis of symmetry of the molecule. The bending vibration is described as a perpendicular ($\perp$) motion since the change of dipole moment

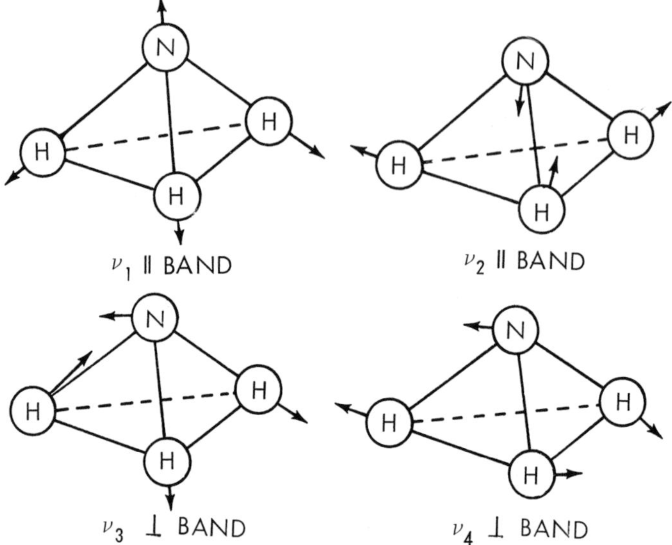

Figure 4-14. Vibrations of NH_3.

during the vibration is perpendicular to the major axis of symmetry. For many molecules, the observed band contours can be related to these concepts of parallel and perpendicular vibrations. For nonlinear molecules three axes of symmetry can be described, the major, intermediate, and minor axes. We can therefore describe a parallel, perpendicular, or intermediate vibration, depending on whether the dipole moment change is parallel to the major axis, perpendicular to it, or intermediate to it.

4.3C. Vibrations for Molecules Containing Four or More Atoms

While it is generally easy to picture the vibrations of a simple molecule, it is often more difficult to visualize the complex motions of large and/or asymmetric molecules. In this section, we present a simple method for visualizing molecular motion involving four or more atoms.

Let us begin by diagramming the movements for the nonlinear molecule NH_3. Since there are four atoms, there should be $3N - 6$ or six fundamentals. The structure of NH_3 is pyramidal, with the N atom at the apex and the three H atoms forming the base. The vibrations for NH_3 are shown in Figure 4-14. Only four of the six motions are shown since two of them (ν_3 and ν_4) are two-dimensional oscillations and are degenerate, which results in two degenerate frequencies. We search therefore for only four fundamentals in the infrared and Raman spectra of this molecule. If NH_2D were substituted for NH_3, the two degenerate frequencies would appear separately and all of the six vibrations would give absorption bands in the infrared region.

In a figure such as 4-14, the magnitudes and positions of the arrows may be used to indicate the relative amplitudes and directions of the vibrational motions. (The directions of vibrational motion must be such that the movements result in no net translation or rotation of the molecule as a whole.) In this text, in cases where the clarity of illustrations comes into conflict with precise scaling, we shall not always draw the arrows exactly proportional to the relative amplitudes, but instead shall sacrifice some exactness in detail in order to present the nature of the atomic motions more clearly. In Figure 4-14, however, the arrows are drawn to indicate approximate amplitude.

For NH_3, the first type of vibration, ν_1, consists of a motion in which the nitrogen moves up and down the axis of symmetry of the molecule and the hydrogens move back and forth along the axes of the chemical bonds between the nitrogen and the hydrogens. This is a symmetric motion which produces a parallel band and is classified as an a_1 vibration. (The significance of this and other classifications or types will be discussed later in this chapter.)

In the second vibration, ν_2, the base of three hydrogens moves up and down in a symmetric motion while expanding and contracting in rhythm with the up-and-down motion. As the base moves toward the nitrogen, the nitrogen atom moves toward the base, and as the base moves away, the nitrogen also recedes from the base in a symmetric fashion. ν_2 is also a parallel, a_1 vibration.

In the third vibration, ν_3, the nitrogen oscillates in a plane parallel to the base formed by the three hydrogens. The three hydrogens oscillate in their own plane, and this motion gives rise to a perpendicular vibration. ν_3 is doubly degenerate and is classified as an e vibration.

In ν_4 the nitrogen moves in the same manner as for the ν_3 vibration; however, the base of three hydrogens tips in an oscillatory motion, with one point of the base plane moving up and the other two ends moving down, and *vice versa*. This motion gives rise to a perpendicular band. The vibration is doubly degenerate and is again an e vibration.

Having described these motions for NH_3, we now find it possible to discuss movement in certain molecular structures involving five atoms, such as CH_3X, where X is a halogen. The symmetry of this molecule can be considered to be similar to that of NH_3, except that another atom is positioned above the apex atom of the pyramid. The number of vibrations for this five-atom molecule will be nine; six of these will be similar to the motions described for NH_3; of these six, there will be two degenerate frequencies, so that only four motions can be drawn, as for NH_3. Because of the similarity to the NH_3 case, we shall not draw these six vibrations for CH_3X. Of the three remaining vibrations for CH_3X, one will be a parallel vibration in which the X atom moves symmetrically toward and away from the pyramid CH_3, while a second will result in a degenerate vibration, thus accounting for the two remaining vibrations of CH_3X. This latter motion can be visualized if we consider the C—X bond to form a rigid bar about which the pyramid CH_3 oscillates as if it were held to the bar by a flexible joint.

A molecule such as CH_4, which takes the form of a tetrahedron with H's at the corners and the C at the center, may now be thought of as a special case of the CH_3X model, and due to symmetry, the nine vibrations described for CH_3X reduce to four distinct motions of the tetrahedron for CH_4. Using the four vibrations of NH_3 as reference, we can describe the four vibrations for CH_4 as follows: (1) the symmetric vibration ν_1 becomes the totally symmetric a_1 vibration; (2) ν_2 and the degenerate ν_3 combine to give the second vibration, which is now triply degenerate; (3) the doubly degenerate ν_4 remains a doubly degenerate vibration; and finally, (4) the three vibrations newly described for CH_3X combine to give one triply degenerate vibration for CH_4.

We can diagram this transition of vibration types in the following manner:

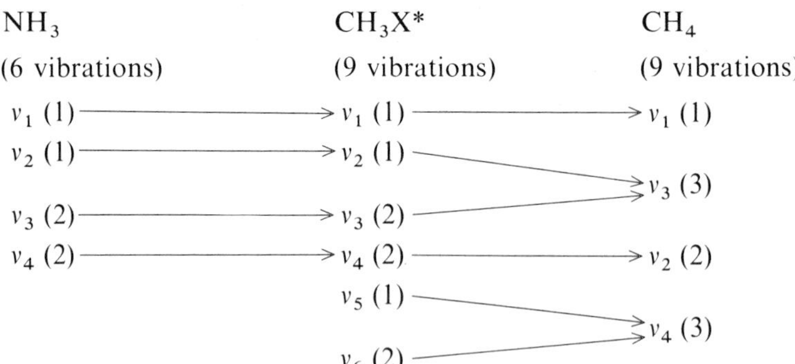

Thus, CH_4 has what appear to be four fundamental absorption bands; CDH_3, which has the same symmetry as CH_3X, would have

* Some authors number these vibrations in a different order.

six, and CD_2H_2 would have nine fundamental frequencies since for this molecule none of the frequencies are degenerate.

The concept of double and triple degeneracy of vibrations is well illustrated by the vibrations of CH_4.

To diagram the motions of a six-atom molecule, such as C_2H_4, it would be necessary to show at least 12 motions. This will not be done here, but the reader can find the motions diagrammed in several standard texts. In diagramming these motions it is possible to indicate the groups involved in the vibrations as well as planes in which the vibrations are symmetric or antisymmetric. This requires both superscripts and subscripts on the vibration symbol. For example, a vibration designated as $\nu_{\beta\gamma}^{CH}$ indicates that the motion involves the C—H bond and is symmetric with respect to the two planes β and γ.

In numbering fundamentals, the totally symmetric vibration of highest frequency is called ν_1, the second highest ν_2, and so on. One exception to this rule is found with linear molecules of the form XY_2 or XYZ, for which ν_2 is the perpendicular vibration. For types of vibrations (species) of a given symmetry, the vibrations are numbered in order of decreasing wavenumber. Some authors use the symbols ν, δ, ω, and τ to denote stretching, bending, wagging, and twisting vibrations, respectively.

The vibrations for ethane (C_2H_6) are not difficult to visualize if we compare C_2H_6 with NH_3 since each of the CH_3 groups of ethane can be thought of as having vibrations similar to those of the NH_3 molecule. We can thus assign the following vibrations to ethane: Each CH_3 group can have the four vibrations diagrammed for NH_3; however, since there are two CH_3 groups these four vibrations can be either symmetric, when the two CH_3 groups are in phase with each other, or antisymmetric, when the two groups are out of phase. This gives a total of eight types of vibrations involving the CH_3 groups of ethane. To these eight we must add a C—C stretch, two bending motions of the structure R—C—C—R, and a torsional oscillation or twist of the two CH_3 groups around the C—C bond.

4.4. THE RELATIONSHIP BETWEEN SYMMETRY OF MOLECULES AND OBSERVED ABSORPTION BANDS

4.4A. Introduction

To relate the observed infrared bands to the vibratory motions of the atoms in a molecule it is necessary to assume a structural model for the molecule. Vibrations can then be discussed as fundamentals, overtones, or combination bands in terms of this model. Since the description of vibrations is in terms of a molecular model, it will be possible to predict whether a vibration will be infrared active, that is, if a change in dipole moment occurs during the vibration. By assuming various models for a molecule and examining the associated infrared spectra, it is possible to choose that model which results in the spectrum that agrees most closely with the observed infrared spectrum. Thus, infrared spectroscopy (combined in most instances with Raman spectroscopy) finds one of its more important applications in structural assignment. In order to understand the use of infrared spectroscopy for structural assignment we must introduce group theory and the concept of symmetry properties of molecules. The following sections are not intended to be a comprehensive treatment of these concepts but rather are designed to provide the beginning spectroscopist with the background he requires for a qualitative approach to these methods.

4.4B. Point Symmetry and Point Groups

Symmetry is best described in terms of the operations by which the atoms of a molecule can be interchanged without changing the configuration of the molecule in space. We must consider four such *symmetry elements*:

1. Center of symmetry (or inversion center)
2. Rotational axes
3. Mirror plane
4. Alternating or inversion axes

These are described in some detail below.

Center of Symmetry (C_i). A center of symmetry exists in a molecule if one half of the molecule is obtained from the other by inversion through the center of symmetry; there may or may not be an atom at the center of symmetry. There are a few molecules with a center of symmetry as the sole symmetry element. One such molecule would be a substituted ethane HXYC–CYXH, where each pair of similar atoms is in the *trans* position. If a line is drawn from one atom to the center of symmetry and continued on, it would intercept a similar atom at the same distance from the center of symmetry as the original atom. The molecule HXYC–CYXH would be described as belonging to the point group C_i, since it has only the symmetry element C_i.

Rotational Axes (C_n). Axes of rotation are symmetry elements if, when the molecule is rotated about these axes, similar atoms are interchanged so that the resulting configuration is indistinguishable from the original one. The angles of rotation may be designated by $2\pi/n$ or $360°/n$, where $n = 1, 2, \ldots$. In practice, only values of n from 1 to 6 and the value ∞ are encountered. We call n the *order* of the rotational axis, or else, we may speak of *n-fold axes* of rotation. Thus C_1 implies a onefold rotational axis (a rotation of 360° required for coincidence), while C_2 denotes a twofold axis (or a rotation of 180° required). A molecule such as CHFClBr would have C_1 symmetry since a 360° rotation is the only symmetry element that exists for it. Nonplanar H_2O_2 has C_2 symmetry. The rotations C_1 and C_2 are also the point groups for these molecules. (Actually the point group C_2 includes the rotation of 180° and the identity I described below.)

For reasons that will become apparent later, the primary rotation axis chosen for a molecule is the one that represents the highest order of symmetry in the molecule; this axis is designated the *vertical axis* of the molecule. Other rotational axes, perpendicular to the first, are often found for a molecule; they are given a special designation, which will be described later.

Mirror Plane (σ). A molecule can have mirror planes of symmetry, i.e., the geometric arrangement on one side of the plane is a mirror image of that on the other. The sole symmetry element of the molecule CH_2BrCl is a mirror plane, which includes the carbon, bromine, and chlorine atoms, while the hydrogens form mirror images of each other.

If a molecule has a vertical rotational axis C and a mirror plane can be drawn containing this axis, the mirror plane is given the designation σ_v, indicating a vertical mirror plane. If the mirror plane is perpendicular to the vertical rotational axis, it is designated as σ_h. A third designation, σ_d, represents a plane containing the rotational axis and diagonal to other symmetry elements. (σ_v and σ_d planes often differ only in that an atom is included in one but not in the other.)

Alternating or Inversion Axes (S). If a rotation of $360°/n$ is followed by a reflection in a mirror plane perpendicular to the axis of rotation and the combination of these operations transforms the molecule into itself, then we have a symmetry element designated as S_n.

Identity Symmetry Element (I). There is an additional symmetry element which all molecules have, called the identity, given the symbol I. It is the operation in which the molecule remains in its original position. While such an operation appears trivial, it must be considered in applying the principles of point groups.

4.4C. Point Groups for Various Molecules

The molecules most commonly encountered generally possess a number of symmetry elements. Every molecule is associated with a point group, where the point group is defined as the collection of symmetry operations which when applied about a point leave the molecular configuration unchanged in space, while the point itself remains at rest, or more precisely, transposes into itself. There is a

TABLE 4-II. Species of Vibration for Various Symmetry Groups

Group	C_i	C_2	C_3	C_4	C_5	C_6	S_2	S_4	S_6	S_8	σ_v	σ_h	σ_d	Species of vibration	Example
C_1	1													a	CHFClBr
C_2		1												a, b	Nonplanar H_2O_2
C_3			1											a, e	
C_4				1											
C_5					1										
C_6						1								a, b, e_1, e_2	
S_2 (C_i)														a_g, a_u	
S_4		1						1							
S_6	1		1						1						
C_{1v} (C_s)											1			a', a''	NOCl
C_{2v}		1									2			a_1, a_2, b_1, b_2	H_2O
C_{3v}			1								3			a_1, a_2, e	NH_3
C_{4v}				1							4			a_1, a_2, b_1, b_2, e	IF_5
C_{6v}						1					6			$a_1, a_2, b_1, b_2, e_1, e_2$	

TABLE 4-II (continued)

Group	C_i	C_2	C_3	C_4	C_5	C_6	S_2	S_4	S_6	S_8	σ_v	σ_h	σ_d	Species of vibration	Example
$C_{\infty v}$											∞			π, σ^+	HCN
C_{2h}	1	1					1					1		a_g, a_u, b_g, b_u	$(HF)_2$
C_{3h}			1									1		a', a'', e', e''	H_3BO_3
C_{4h}	1			1								1		$a_g, a_u, b_g, b_u, e_g, e_u$	
C_{6h}	1					1						1		$a_g, a_u, b_g, b_u, e_{1g}, e_{1u}, e_{2g}, e_{2u}$	
D_2		3												a, b_1, b_2, b_3	
D_3		3	1											a_1, a_2, e	
D_4		4		1											
D_6		6				1									
D_{2d} (V_d)		3						1					2	a_1, a_2, b_1, b_2, e	B_2Cl_4
D_{3d}	1	3	1						1				3	$a_{1g}, a_{1u}, a_{2g}, a_{2u}, e_g, e_u$	C_2H_6
D_{4d}		4		1						1			4	$a_1, a_2, b_1, b_2, e_1, e_2, e_3$	S_8
D_{2h}	1	3									3 planes			$a_g, a_u, b_{1g}, b_{1u}, b_{2g}, b_{2u}, b_{3g}, b_{3u}$	C_2H_4
D_{3h}		3	1								3	1		$a_1', a_1'', a_2', a_2'', e', e''$	BCl_3

TABLE 4-II (continued)

Group	C_i	C_2	C_3	C_4	C_5	C_6	S_2	S_4	S_6	S_8	σ_v	σ_h	σ_d	Species of vibration	Example
D_{4h}	1	4		2			1				4	1		$a_{1g}, a_{1u}, a_{2g}, a_{2u}, b_{1g}, b_{1u}, b_{2g}, b_{2u}, e_g, e_u$	C_4H_8
D_{5h}		5			1						5	1		$a_1', a_1'', a_2', a_2'', e_1', e_1'', e_2', e_2''$	
D_{6h}		6				1			1		6	1		$a_{1g}, a_{1u}, a_{2g}, a_{2u}, b_{1g}, b_{1u}, b_{2g}, b_{2u}, e_{1g}, e_{1u}, e_{2g}, e_{2u}$	C_6H_6
T_d		3	4					3			6 planes			a_1, a_2, e, f_1, f_2	CH_4
T_h	1	3	4								3 planes			a, e, f	
O_h	1	6	4	3				3			9 planes			$a_{1g}, a_{1u}, a_{2g}, a_{2u}, e_g, e_u, f_{1g}, f_{1u}, f_{2g}, f_{2u}$	SF_6
$D_{\infty h}$	1	∞									∞	1		$\sigma_g^+, \sigma_u^+, \pi_g, \pi_u$	CO_2

limited number of point groups, and every molecule must belong to one of them.

We have seen the simple point groups $C_1, C_2, \ldots, C_n$, which involve only the rotational axis of symmetry, and the point group C_i, with only a center of symmetry. The point group S_n has also been introduced, but the symbol S_n will be used only where n is even, since for odd n the S_n group is equivalent to the C_n group plus a mirror plane of symmetry perpendicular to the rotational axis. This combination of C_n and a mirror plane is given another symbol (e.g., C_2 and σ_v are combined to give C_{2v}). In addition, S_2 is equivalent to the center of symmetry group C_i.

Point groups which combine several symmetry elements are described below. Common point groups and their symmetry elements are summarized in Table 4-II.

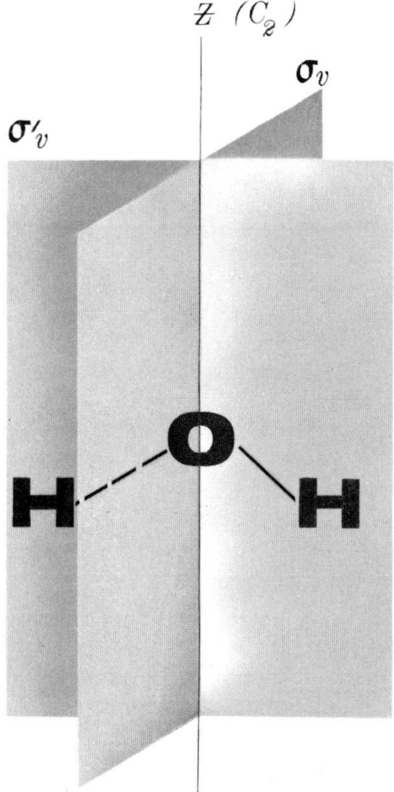

Figure 4-15. The symmetry of H_2O and the C_{2v} point group.

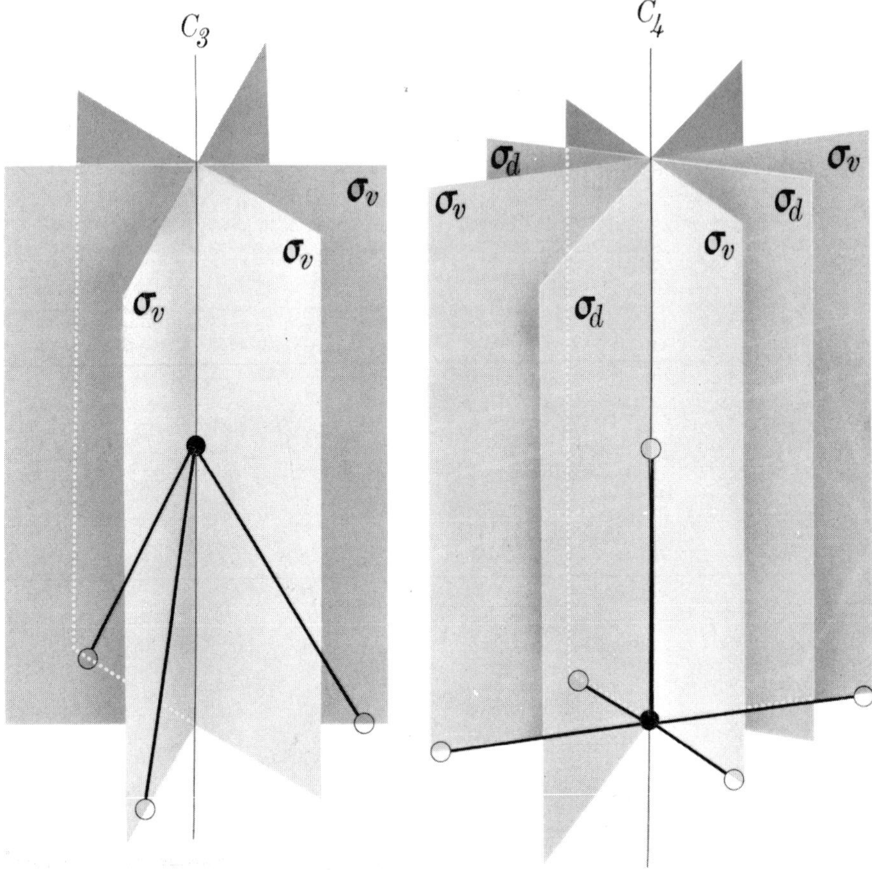

Figure 4-16. The C_{3v} point group.

Figure 4-17. The C_{4v} point group.

Point Groups C_{nv} and C_{nh}. Molecules having a rotational axis of symmetry and mirror planes containing the rotational axis are designated as C_{nv}. The number of σ_v planes is determined by the value of n, since these planes are parallel to the rotational axis and are symmetrically arranged about it at angles $360°/n$.

The point group C_{1v} has one plane of symmetry and the identity I as its symmetry elements. It is often given the symbol C_s. An example of a C_{1v} symmetry is the nonlinear molecule NOCl.

Many molecules belong to the point group C_{2v}; H_2O is an example. To illustrate this symmetry for H_2O, let us examine

Figure 4-18. The $C_{\infty v}$ point group.

Figure 4-15. The rotational axis C_2 is shown as the Z axis. Planes parallel to this axis include the one containing all three atoms and the one passing through the oxygen atom and midway between the two hydrogens. The two planes are 90° from each other and are

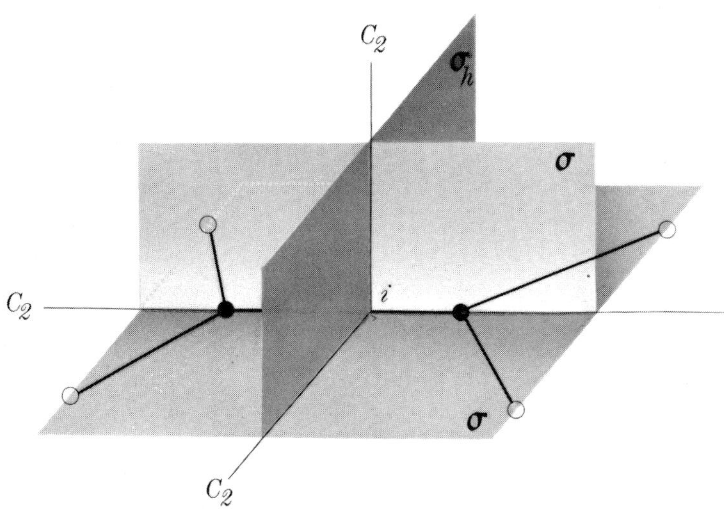

Figure 4-19. The D_{2h} point group.

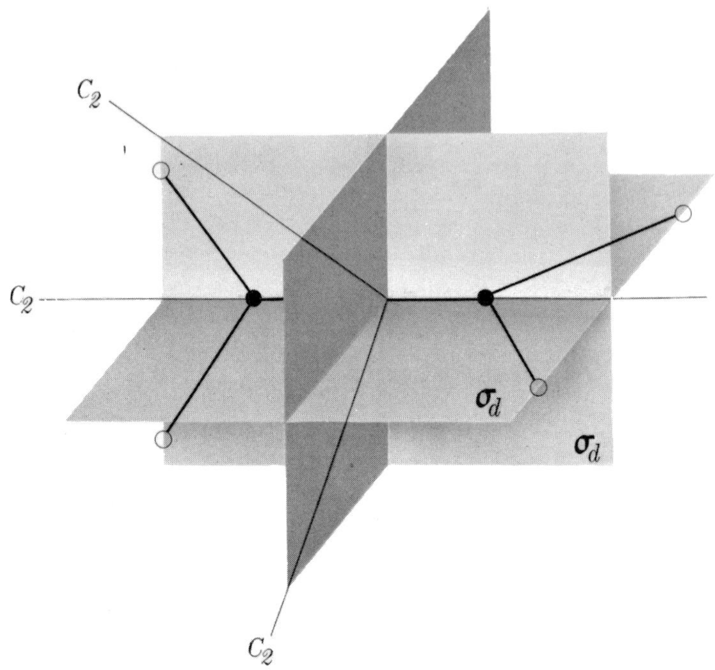

Figure 4-20. The D_{2d} point group.

designated as σ_v and σ_v'. Thus the symmetry elements of the C_{2v} group are a rotation of 180°, two planes σ_v and σ_v', and the identity I. Other point groups are presented in Figures 4-16 to 4-24 and will be described below.

The C_{3v} group can be assigned to molecules such as NH_3 and CH_3Cl. The rotational axis C_3 passes through the nitrogen atom (for NH_3) and intersects the plane of the three hydrogens. Rotations of 120° and 240° are possible. Three reflection planes σ_v can be drawn for the C_{3v} group. For NH_3 each plane includes the nitrogen and one hydrogen and passes midway between the remaining two hydrogens. The planes are parallel to the C_3 axis.

Many molecules also belong to the point groups C_{4v}, C_{5v}, and C_{6v}.

One other C_{nv} group of importance is $C_{\infty v}$. This group has an infinite number of planes through the vertical rotational axis, as well as an infinite number of rotations. A linear molecule such as HCN belongs to this group. The rotational axis passes through all

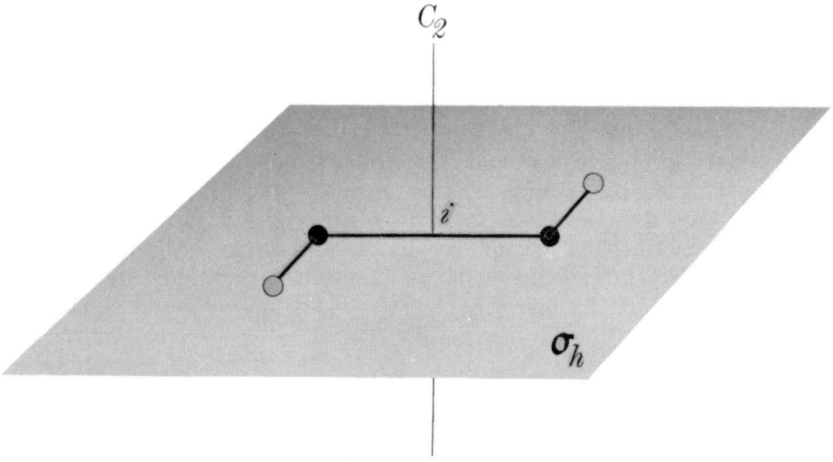

Figure 4-21. The C_{2h} point group.

three atoms. Any heteronuclear diatomic or asymmetric linear molecule would belong to this group.

C_{nh} point groups combine a (vertical) rotational axis C_n with horizontal mirror planes σ_h. C_{1h} is equivalent to C_{1v} or C_s since there is only one plane of symmetry. An example of a C_{2h} group is *trans*-$C_2H_2Cl_2$. The twofold axis C_2 is perpendicular to the plane of the molecule, and the σ_h plane is the plane of the molecule. C_{2h} molecules contain a center of symmetry.

Point Groups D_n, D_{nd}, and D_{nh}. A molecule can have rotational axes perpendicular to each other. A D_n point group has an n-fold axis C_n and n twofold axes C_2 perpendicular to C_n at equal angles to one another. The D_1 group is equivalent to the C_2 group, and the D_2 group has two twofold axes mutually perpendicular to one another as its sole symmetry elements. A molecule of the form CX_2CX_2 with the two CX_2 groups not aligned but rotated from each other by an angle other than 90° would have D_2 symmetry. No such molecule is known to exist.

If we combine the concept of a D_n group with that of diagonal mirror planes σ_d, we can conceive of a D_{nd} group. This will involve n diagonal planes which bisect the angles between two adjacent twofold axes, and which contain the n-fold axis.

A molecule which has an n-fold axis of symmetry C_n, n twofold perpendicular axes, and n vertical planes of symmetry σ_v, and which also has a horizontal plane of symmetry σ_h perpendicular to C_n belongs to the D_{nh} point group.

By utilizing three models of ethane (C_2H_6) it is possible to illustrate the three point groups D_3, D_{3d}, and D_{3h}. Figures 4-22 to 4-24 illustrate these three groups, but the reader may wish to use space models to help him understand the following discussion.

Ethane consists of two CH_3 groups joined together. Three configurations are theoretically possible: (1) If the three hydrogens of one CH_3 group are opposite the three of the second one (eclipsed form), then the point group is D_{3h}. (2) If one CH_3 group is rotated with respect to the second so that the hydrogens are no longer opposite each other but the angle of rotation is neither 60° nor 120°,

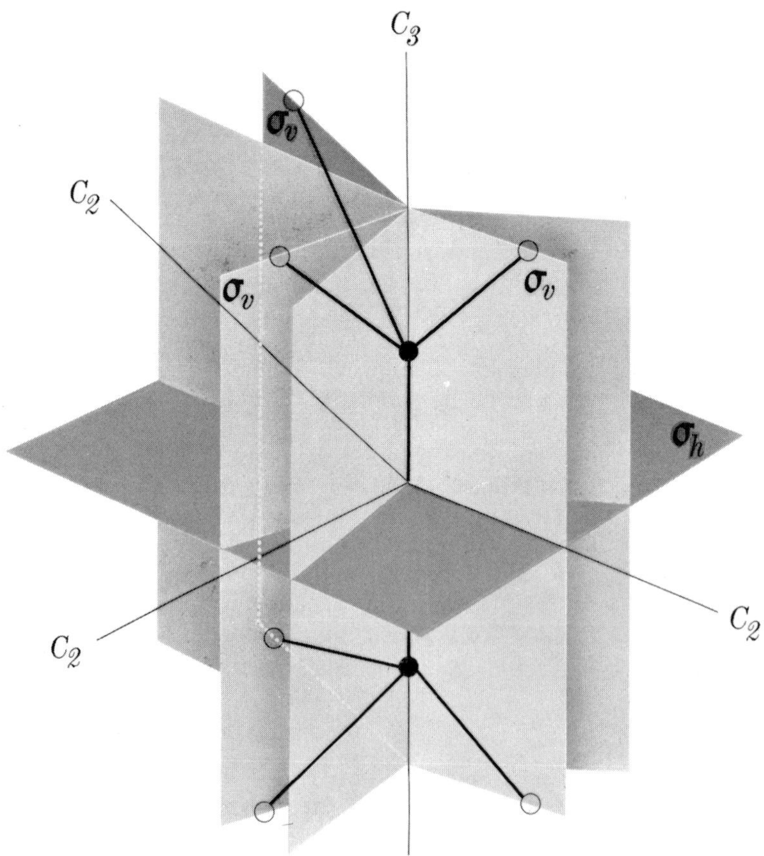

Figure 4-22. The D_{3h} point group.

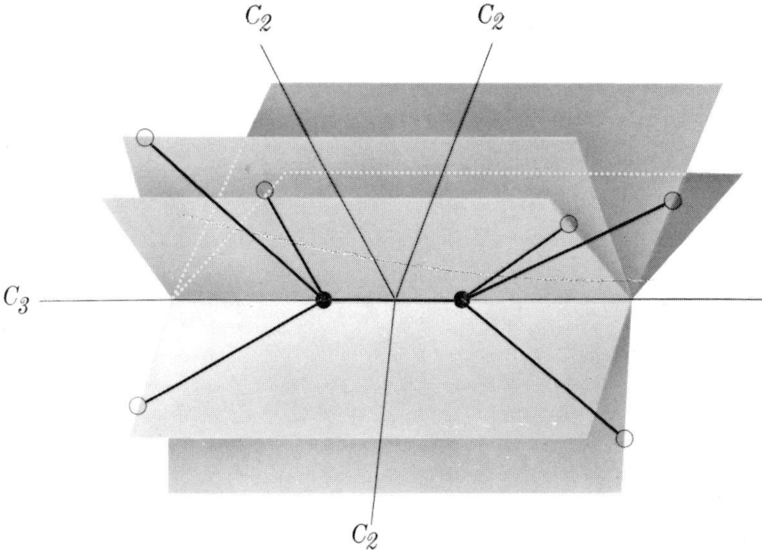

Figure 4-23. The D_3 point group.

then the point group is D_3. (3) If the three hydrogens are rotated so that the angle is 60°, then the structure is in the so-called staggered form and the point group is D_{3d}. The point group for ethane is known to be D_{3d}.

Reference to space models or Figures 4-22 to 4-24 reveals the symmetry elements for these three point groups. For the D_{3h} group the C_3 axis passes through the C—C bond. Three rotations of 120° are possible about this axis. There are three C_2 axes at right angles to the C_3 axis and three σ_v planes, each passing through one of the C_2 axes and the C_3 axis. Finally there is the σ_h mirror plane perpendicular to the C_3 axis. All of these symmetry elements are indicated in Figure 4-22. The D_3 group has the C_3 axis and the three C_2 axes but no σ planes since the hydrogens are not opposite each other (see Figure 4-23). In the point group D_{3d} the hydrogens are not opposite each other but three diagonal planes can be drawn (see Figure 4-24). A center of symmetry is also present for ethane in this configuration.

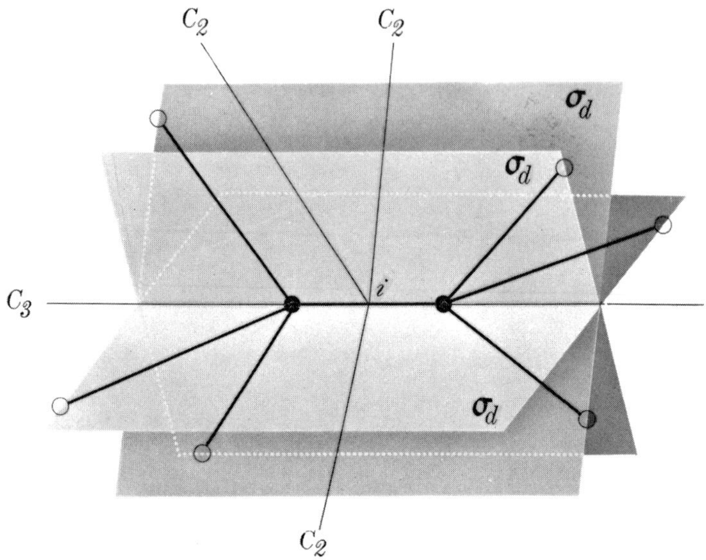

Figure 4-24. The D_{3d} point group.

Point Group T_d. The symmetry of methane (and all molecules of tetrahedral symmetry) is that of point group T_d. There are three mutually perpendicular twofold axes and four threefold axes. Each pair of threefold axes is associated with a σ_d mirror plane. It is of interest that the T_d group has no center of symmetry.

Point Group O_h. Octahedral molecules belong to the O_h point group. This group has three mutually perpendicular fourfold axes C_4, four threefold axes C_3, and a center of symmetry. SF_6 is an example of a molecule having O_h symmetry.

4.5. CHARACTER TABLES AND TYPES (SPECIES) OF VIBRATIONS

The character table of a point group can be derived in a rigorous manner using the concepts of matrix algebra and the geometric model of a molecule. Since this is not the purpose of this text, the following explanation of character tables must be somewhat artificial. The definitions, however, will serve a useful purpose in further discussion of infrared spectroscopy.

Let us consider the character table for the C_{2v} point group presented in Table 4-IV. Listed as column headings are the operations I, C_2, σ_v, and σ_v'. These groupings which are used for column headings are referred to as classes, since all the operations in each grouping behave in a related manner. The row headings for the C_{2v} group are labeled a_1, a_2, b_1, and b_2. We shall refer to these as species or types of vibrations. In group theory calculations they are referred to as "irreducible representations." It is the species or types that are used to describe the vibrations of a molecule.

Examination of Table 4-IV shows that the entries for the species a_1 for a C_{2v} group are $+1$ under each class, while for the a_2 vibration the entries are 1, 1, -1, and -1, respectively. What is the significance of these numbers? A vibration can be either symmetric, antisymmetric, or degenerate with respect to the symmetry elements in a molecule. The entries in the character table indicate which is the case for each class. For example, a C_{2v} molecule such as H_2O has a symmetric stretching vibration in which the hydrogens move symmetrically toward and then away from the oxygen. If we examine the symmetry elements of H_2O, which include the C_2 axis, two mirror planes, and of course the identity I, we see that the symmetric stretching vibration of H_2O is a motion which is symmetric with respect to all of these symmetry elements. The a_1 vibration is therefore described as totally symmetric and the character table entries are $+1$ for all classes. If the entry for a certain type of vibration is -1 for a given class, this indicates that the motion is antisymmetric with respect to the symmetry elements in that class. Degeneracy is indicated by entries such as 2 or 0, depending on the motion. The only other type of entry is one listing the cosine of an angle or some related function. These latter types of entries are utilized in group theory calculations and will not be explained here.

The number of species is not great and it will be of value to discuss the more important ones. In Table 4-II, the species found for each point group are listed together with the symmetry elements. Most of the character tables are presented in the Appendix, and it is from these character tables that the qualitative explanation presented below was obtained.

The accepted notation for species of vibration employs lower-case symbols if the vibration is a fundamental and capital letters if it is an overtone or combination band. However, in general discussions of species, we shall use the lower-case symbols consistently. In Table 4-II the species are listed in this manner.

4.5A. Species for Point Groups C_1, C_2, C_s, and C_i

For the point group C_1 only one species of vibration a for the normal vibrations is listed since no symmetry element exists for C_1 molecules except the identity I.

Each of the groups C_2, C_s, and C_i has one symmetry element: C_2 has one twofold axis, C_s has one mirror plane, and C_i has an inversion center. For all three of these groups there are two species, one of which is symmetric with respect to the element of symmetry and one of which is antisymmetric. The symmetric species for C_2, C_s, and C_i are designated as a, a', and a_g, respectively, and the antisymmetric species as b, a'', and a_u. The subscripts g and u will be found on all species symbols of groups with a center of symmetry element, the g always indicating symmetric and the u antisymmetric vibrations relative to the inversion center.

Molecular vibrations can be described from the species of the point group to which the molecule has been assigned. For example, the planar, nonlinear molecule N_3H belongs to the point group C_s. It will have a' vibrations, where the atoms move in the plane of the molecule, and a'' vibrations, where they move along lines perpendicular to the plane. A triatomic molecule belonging to the C_s group may not exhibit a'' vibrations, since no such vibrations can be diagrammed that would not result in rotation of the molecule.

4.5B. Species for Point Groups C_{2v}, C_{2h}, and D_2

The groups C_{2v}, C_{2h}, and D_2 have three symmetry elements each; however, only two need be considered in obtaining the species of vibration since the third symmetry element is identical with the combination of the other two. Since only two elements of symmetry need be considered, it follows that only four species of vibration (a symmetric and an antisymmetric one for each symmetry element) need be listed. For the C_{2v} group these species are designated as a_1, a_2, b_1, and b_2. For C_{2h} they are designated as a_g, a_u, b_g, and b_u since the C_{2h} molecules have a center of symmetry. For D_2 they are designated as a, b_1, b_2, and b_3. For each of these groups the a designation indicates the symmetric and the b the antisymmetric species. For a clearer picture of the actual motions of the atoms of a molecule during a vibration, it will be of value to discuss each species in detail.

Vibrations of Species a_1, a_g, and a. Symmetry species a_1 is found not only for the C_{2v} point group, but also for C_{3v}, C_{4v}, C_{6v}, D_3, D_{2d}, D_{4d}, T_d. For these groups, a vibration of a_1 species is totally symmetric, that is, the motion of the atoms is symmetric with respect to all symmetry elements. If we consider a molecule such as $CF_2{=}CH_2$, which belongs to the group C_{2v}, then the CF_2, CH_2, and C═C symmetric stretching and bending frequencies are of species a_1. Another C_{2v} molecule, H_2CO, has three fundamentals, ν_1, ν_2, and ν_3, which are of species a_1. These are described as parallel vibrations and are diagrammed in Figure 4-30. We will deal with them in greater detail in Section 4.16. Two vibrations of species a_1 are present for NH_3, and these are shown in Figure 4-14.

The vibrations of species a_1 can have distinct band contours. The Q-branch is generally fairly strong in these bands. The type A band for asymmetric rotors that have moments of inertia such that they are nearly symmetric rotors can be considered to be an a_1 vibration.

The a_g species for the C_{2h} group and the a species of the D_2 group can be considered to be similar to the a_1 species. The a_g is symmetric with respect to all symmetry elements of the C_{2h} group, and will occur for C_{4h}, C_{6h}, D_{2h}, and C_i groups as well. Ethylene belongs to the D_{2h} group, and if we follow the species designation for ethylene employed in Herzberg's book, then there are three a_g fundamentals for this molecule, all of which are Raman active. The C═C stretching frequency in the Raman spectrum is an a_g vibration.

Vibrations of Species a_2 and a_u. For the C_{2v} group the a_2 vibration is symmetric with respect to the axis of rotation but antisymmetric with respect to planes of reflection. Vibrations of this species often are neither Raman nor infrared active. The torsional mode of CH_2═CF_2 is a vibration of species a_2 and is Raman active. The a_u species of the C_{2h} group is antisymmetric with respect to the σ_h planes as well as to the center of symmetry. Often, this vibration is not active for the C_{2h} group; for example, although one a_u vibration is predicted for ethylene, it is neither Raman nor infrared active.

Vibrations of Species b_g and b_u. The b_g and b_u vibrations of the C_{2h} group are similar in that they are both antisymmetric with respect to the C_2 axis; however, the b_g species is also antisymmetric with respect to the σ_h plane, while the b_u species is antisymmetric with respect to the center of inversion.

4.5C. Species of the D_{2h} Point Group

The D_{2h} point group has the symmetry elements I, σ_1, σ_2, σ_3, C_i, and three C_2's. However, only three of these are considered "necessary," since the others can all be obtained by the performance of two of the three "necessary" symmetry operations in succession. The number of necessary symmetry elements can be used to calculate the number of species a D_{2h} group will have. If we consider that for the three necessary elements the motion of the atoms can either be symmetric (+) or antisymmetric (−), then there are only eight ways groups of three + or − signs can be arranged. These are

$+++$, $++-$, $+-+$, $-++$, $+--$, $-+-$, $--+$, and $---$. There are therefore only eight species of vibration for the D_{2h} group: a_g, a_u, b_{1g}, b_{1u}, b_{2g}, b_{2u}, b_{3g}, b_{3u}. The molecule C_2H_4 belongs to the D_{2h} group, and has twelve ($3N - 6$) fundamentals, distributed as follows: three a_g (Raman active), one a_u (inactive), two b_{1g} (Raman active), one b_u (infrared active), one b_{2g} (Raman active), two b_{2u} (infrared active), and two b_{3u} (infrared active). In order to be brief, we shall not discuss each vibrational species in detail. However by analyzing the character table for the D_{2h} group given in the Appendix, the reader can readily ascertain the species which are symmetric or antisymmetric with respect to each symmetry element.

4.5D. Species of Vibration for C_{3v} and D_3 Groups

In considering the species of vibration of C_{3v}, D_3, and other higher groups, a new factor must be introduced. These point groups have some species of vibration which are degenerate, that is, more than one vibration is associated with the same frequency. To picture the motions involved in a degenerate vibration, it is necessary to consider the molecule in three dimensions. Consider for the moment the NH_3 molecule, which belongs to the point group C_{3v}. We shall discuss the vibrations of this molecule in greater detail later; let us at this point, however, consider the vibrations of this molecule as diagrammed in Figure 4-14. The motions for the fundamentals ν_3 and ν_4 are both doubly degenerate vibrations of species *e*. Each of these vibrations is considered to be doubly degenerate because the same motions diagrammed for the fundamentals could also be diagrammed for the molecule in a direction perpendicular to that shown, i.e., by rotating the figure 90° about a vertical axis the second degenerate vibration would be obtained. If a third direction needed to be considered, then the vibration would be triply degenerate and would carry the species notation *f*. The bending vibration ν_2 of CO_2, discussed earlier, is another example of a doubly degenerate vibration.

The entries for species *e* and *f* in character tables will not be $+1$ or -1 as it was for *a* and *b* species. The symmetry of degenerate vibrations is discussed in terms of the symmetry each component of the degeneracy has with respect to the symmetry elements. The entries can be generally explained as follows: First, for doubly degenerate vibrations, if for the symmetry element one of the components of the vibration is symmetric ($+1$) and the second antisymmetric (-1), the entry is 0; if both components are symmetric, the entry is $+2$; and if both are antisymmetric, the entry is -2.

If the vibration is triply degenerate the entries can be $+3$ if all three components are symmetric, -3 if all are antisymmetric, $+1$ if

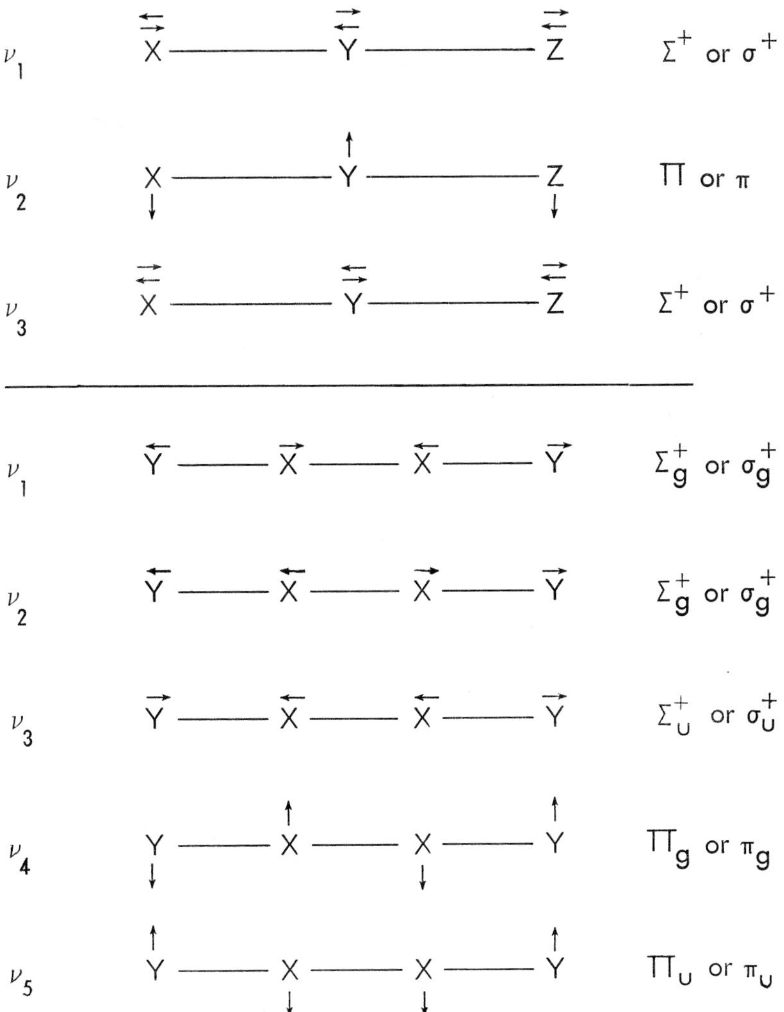

Figure 4-25. Vibrations of linear molecules.

two are symmetric and one antisymmetric, and -1 if two are antisymmetric and one symmetric. There are other possible entries for degenerate vibrations, which are best explained in terms of group theory, but we will not deal with these here.

The C_{3v} and D_3 point groups have one doubly degenerate vibration, which is given the species symbol e. The two fundamentals ν_3 and ν_4 of NH_3 just discussed are of this type, and if the motions during this vibration are examined, it will be found that they are antisymmetric with respect to the C_3 axis of the molecule.

The nondegenerate vibrations of the C_{3v} and D_3 groups are a_1 and a_2 species. The relationship between these species and the symmetry elements of the molecule can be obtained from the character tables for these groups given in the Appendix.

4.5E. Species of Vibration for Other C_{nv} Groups

For purposes of brevity we shall not discuss species for other specific C_{nv} groups in any further detail. Instead we shall turn to a brief discussion of some general concepts. More specific details can be obtained from the character tables given in the Appendix.

All C_{nv} groups with n equal to or greater than three will have doubly degenerate e vibrations, in addition to nondegenerate vibrations. If more than one doubly degenerate vibration occurs, subscripts are used to differentiate between them.

A special notation is assigned the species for the point group $C_{\infty v}$, to which heteronuclear diatomic and asymmetric linear molecules belong. The types of vibration for some examples of this point group are presented in Figure 4-25. The species of vibration are listed with each vibration and will not be discussed further, except to note that Greek rather than roman letters are used to denote species of all types of linear molecules. For example, the bending degenerate vibration of the $C_{\infty v}$ group of linear molecules of the asymmetric form XYZ are designated as π species.

4.5F. Species of Vibration of Other Groups

The species symbols utilized for the point groups not yet discussed are similar to those presented thus far. The relationship between the symmetry elements and the species symbol can be obtained from the character table of the group. For molecules having an inversion center, both degenerate and nondegenerate vibrations that are symmetric with respect to the center of symmetry are given the subscript g and those that are antisymmetric are given the subscript u. For molecules without an inversion center, a single prime is used with the symbol to indicate a symmetric species and a double prime to indicate an antisymmetric vibration.

The remaining sections of this chapter will deal with specific examples of molecules belonging to various point groups and the types of vibrations found for them.

4.6. MOLECULES WITH CENTERS OF SYMMETRY

If we examine Table 4-II, we can list those point groups which have a center of symmetry. These include the groups C_{2h}, C_{4h}, C_{6h}, D_{3d}, D_{2h}, $D_{\infty h}$, T_h, and O_h. The existence of a center of symmetry for all molecules belonging to these groups carries an important implication for their infrared spectra, since it has been shown that for molecules with an inversion center transitions allowed in the infrared are forbidden in Raman spectra and *vice versa.* This is the so-called "rule of mutual exclusion," which states that fundamentals appearing in the infrared spectra of these molecules will not appear in the Raman spectra, and conversely, those appearing in the Raman spectra will not appear in the infrared. It is also possible for certain transitions to be forbidden in both.

It is also generally true that for molecules without centers of symmetry most transitions occur in both the Raman and the infrared spectra. Groups such as D_{5h}, O, and D_{7h} form the only exceptions to this rule.

The rule of mutual exclusion has been used extensively to establish the structure of many molecules. For example, planar C_2H_4, which belongs to group D_{2h}, has a center of symmetry. The number and species of the vibrations and their activities are as follows:

$$3a_g(R),\ a_u(\text{inactive}),\ 2b_{1g}(R),\ b_{1u}(IR),\ b_{2g}(R),\ 2b_{2u}(IR),\ 2b_{3u}(IR)$$

It will be noted that no one fundamental is active in both the Raman spectrum and the infrared spectrum, while one is neither infrared nor Raman active.

The activity of overtone and combination bands can also be related to the point group of the molecule. However, it is not possible simply to state which will be active without first considering in some detail the implications of *group theory*.

4.7. CORRELATION OF SPECIES OF RELATED MOLECULES

The types of vibrations (species) of molecules with related symmetry can be correlated and these correlations used to assign vibrations. For example, ethane and its various derivatives have somewhat similar symmetry elements. Ethane belongs to point group D_{3d}, while CH_3CCl_3 has C_{3v} symmetry. The D_{3d} groups have

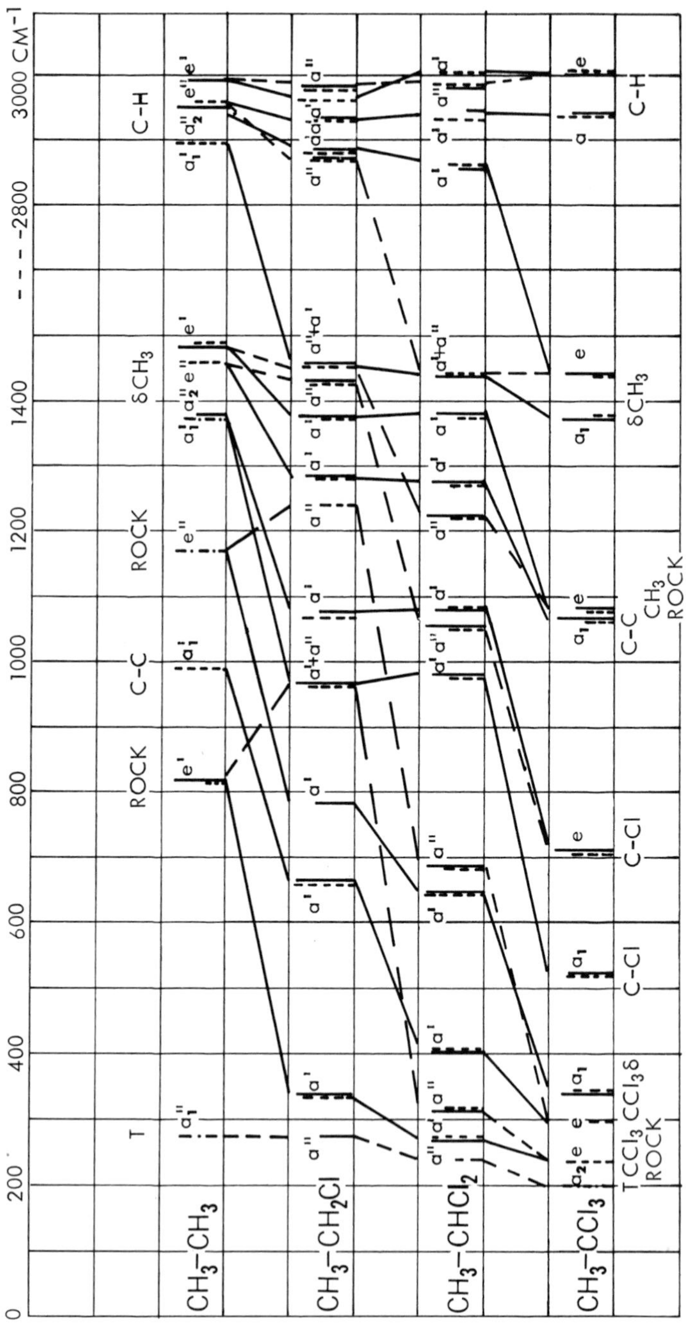

Figure 4-26. Correlation of the fundamentals of CH_3—CH_3, CH_3—CH_2Cl, CH_3—$CHCl_2$, and CH_3—CCl_3. [Reproduced with permission from *J. Chem. Phys.* **22**(8): 1293 (1954).]

centers of symmetry, and by the mutual exclusion rule infrared active fundamentals are not Raman active and *vice versa*. In Figure 4-26, the species of vibration of ethane are traced through the series CH_3CH_2Cl, CH_3CHCl_2, and CH_3CCl_3. In the correlation chart shown in Figure 4-26, solid vertical lines denote infrared active bands, dashed vertical lines denote Raman bands, and dash-dotted vertical lines denote fundamental vibrations whose position is calculated from overtone or combination bands. The heights of the vertical lines are a measure of the relative intensities of the bands. It can be seen from this figure that for ethane the Raman active bands are not infrared active and *vice versa*. Related species for the various molecules are connected either by solid lines for symmetric or broken lines for asymmetric vibrations. Solid lines connecting species do not cross each other. This noncrossing rule is useful in relating similar vibrations of molecules and in assigning vibration species to observed absorption bands. For example, the assignment of the vibrations of CH_3CHCl_2 is facilitated by noting those of CH_3CH_2Cl and CH_3CCl_3. Connecting species lines from these molecules to those of CH_3CHCl_2 and applying the noncrossing rule, we find it possible to make fairly unambiguous assignments.

It can also be seen that the degenerate *e* vibrations of ethane separate into two distinct nondegenerate vibrations in the less symmetrical ethanes.

In utilizing infrared spectroscopy for qualitative analysis, use is made of so-called *group frequencies*. These are discussed in further detail in Chapter 5. However, the relationship between group frequencies and correlation tables such as those shown in Figures 4-26 and 4-27 can best be presented here. In these figures the atoms involved are listed above the respective vibrations. The CH stretching vibrations are found in the region near 3000 cm^{-1}, while the CH_3 deformations are found near 1400 cm^{-1}. If the species for these CH vibrations are traced through a series of compounds, it will be found that the location of these vibrations is fairly constant and the atoms involved in the vibration remain the same. Such vibrations are considered "good" group frequencies, since they retain their identity in related compounds. This is not true for all vibrations. For example, consider the CH_3 rocking mode of CH_3CF_3. If the species of vibration of this mode is followed through the series of molecules given in Figure 4-27, then for CH_3CCl_3 this vibration is more properly classified as a C—Cl stretch. The vibration has changed its character from a CH_3 rock to a C—Cl stretch, with intermediate characteristics for members of the series between CH_3CF_3 and CH_3CCl_3. The CH_3 rock does not make a "good group frequency" because of this change in character.

TABLE 4-III. Fundamentals of $CH_2{=}C{=}CH_2$

Species	Vibration number	Description	Infrared activity
a_1	1	C—H stretch	none
a_1	2	CH_2 deformation	none
a_1	3	C=C=C stretch	none
b_1	4	torsional	none
b_2	5	C—H stretch	parallel band
b_2	6	C=C=C stretch	parallel band
b_2	7	CH_2 deformation	parallel band
e	8	CH stretch	perpendicular band
e	9	CH_2 rock and wag	perpendicular band
e	10	CH_2 rock and wag	perpendicular band
e	11	C=C=C bend	perpendicular band

As one final example of the tracing of species of vibrations, consider the molecule $CH_2{=}C{=}CH_2$, which has the symmetry D_{2d}. It has fifteen fundamentals, four of which are doubly degenerate. The fundamentals are listed in Table 4-III. If two of the hydrogens of this molecule are replaced by deuterium to give the molecule $CD_2{=}C{=}CH_2$, the symmetry becomes C_{2v}. There are no longer degenerate vibrations and, except for the torsional mode, all the fundamentals are allowed in the infrared. The change in species of vibration could be described as follows:

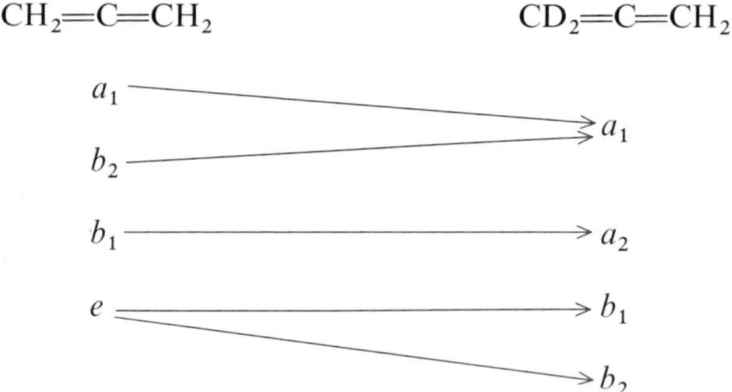

4.8. CALCULATION OF ALLOWED BANDS USING GROUP THEORY

A molecule having axes of symmetry can be classified into one of the several dozen symmetry groups. The simplest of the symmetry groups are those with one major axis of symmetry, such as the C_{2v} and C_{3v} groups. Water is an example of a molecule belonging to the C_{2v} group, while NH_3 is an example of the C_{3v} group. Using

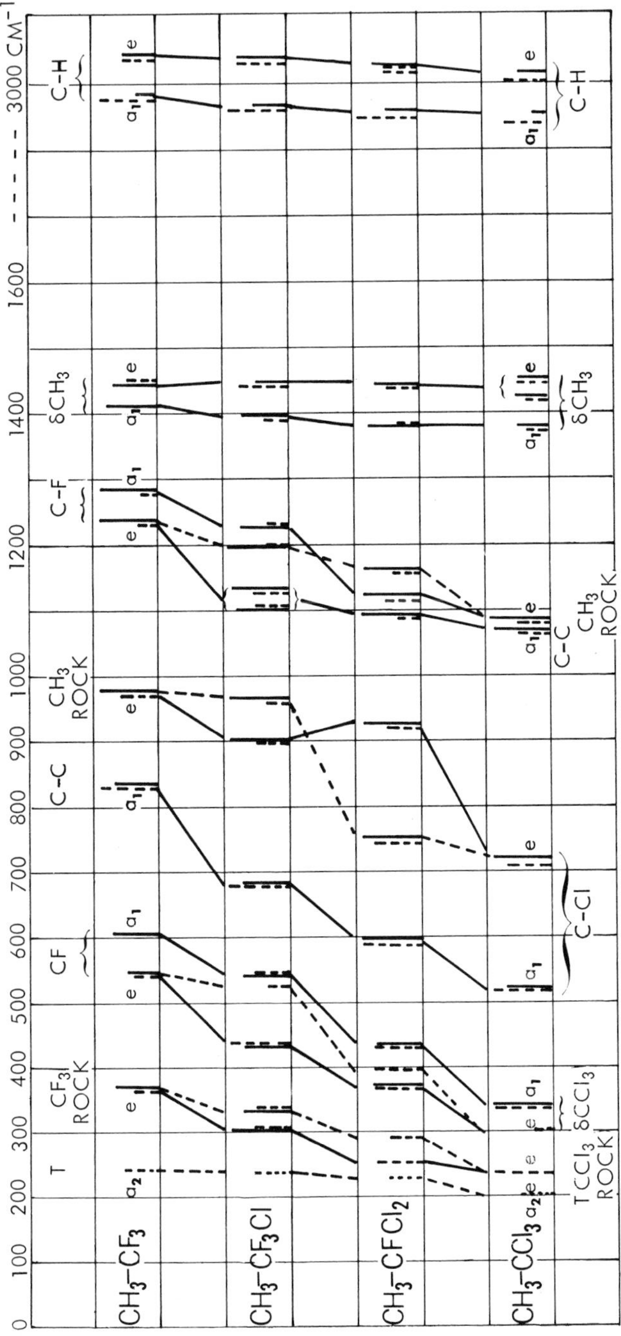

Figure 4-27. Correlation of the fundamentals of CH_3—CF_3, CH_3—CF_2Cl, CH_3—$CFCl_2$, and CH_3—CCl_3. [Reproduced with permission from *J. Chem. Phys.* **20**(3): 473 (1952).]

group theory methods, we can show that the vibrations allowed in the infrared for a molecule belonging to the C_{2v} group are those of species a_1, b_1, and b_2, while for a molecule of C_{3v} symmetry the species allowed in the infrared are the a_1 and e vibrations.

To find the number of each type of vibration for a specific molecule (such as H_2O or NH_3), further calculations are required. The selection rules for overtones and combination bands can also be determined for any molecule belonging to the symmetry group.

We shall use the method suggested by Cleveland and his coworkers [3] to illustrate how these deductions are made. The most recent paper on normal coordinate analysis as a tool is that listed in Reference 27.

4.8A. Calculation of Allowed Fundamentals. The Character Table

As we know, each symmetry group has associated with it a character table. For example, the tables for the C_{2v} and C_{3v} groups are given in Table 4-IV and Table 4-V, respectively. We shall now use the character table in a group theory calculation. At the left in the table are listed the species of vibration associated with the symmetry group. For example, for the C_{2v} group, vibrations of a_1, b_1, a_2, and b_2 species can be visualized, though not all of these are necessarily allowed in the infrared and Raman spectra of all molecules of symmetry C_{2v}. We must perform calculations to determine which of the vibrations are allowed. For the C_{3v} group, vibrations of species a_1, a_2, and e are possible (Table 4-V).

In each character table the geometric (or symmetry) operations that can be performed on the molecule are listed at the top, collected into classes. These involve the now familiar symmetry elements discussed earlier, such as the identity I, rotations C_n, and reflection planes σ_v. The number of symmetry operations in each class is listed. For example, for a C_{3v} molecule the following operations and numbers are found:

$$I(1) \qquad C_3(2) \qquad \sigma_v(3)$$

Let us now consider a specific molecule of C_{3v} symmetry, such as the methylacetylene molecule shown in Figure 4-28. The principal axis of symmetry of this molecule is the H—C≡C—C axis, which also passes through the center of the triangle formed by the three H atoms of the —CH_3 group. The only possible symmetry operations are a rotation of 0°, two rotations of ±120°, and reflection in three planes. The rotation of 0°, which of course is the identity I, is obviously unique, and there is thus only one element in this class. Two elements are listed for the ±120° rotations, and three for the reflections. Each of the reflection planes passes through the major axis (z axis) of the molecule, through one of the three hydrogens of

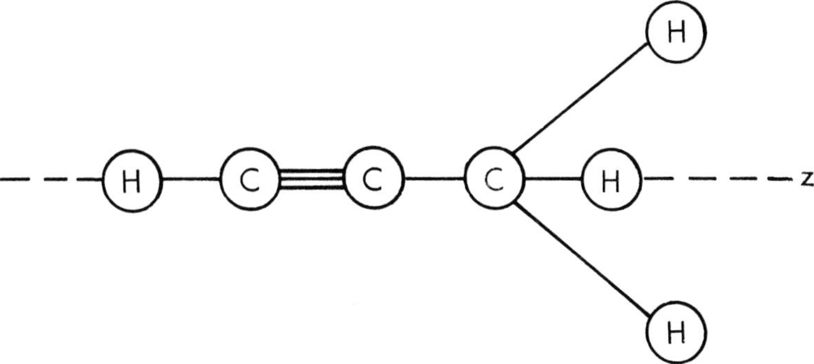

Figure 4-28. The symmetry of methylacetylene.

the —CH_3 group, and through the midpoint of the line joining the other two hydrogens.

It should be noted that for the C_{2v} molecules the character table has four columns, since there are four classes of symmetry operations, while for C_{3v} molecules there are only three. If one considers, for example, an a_1 vibration for each of these groups, then for the C_{2v} groups the a_1 vibration has the four characters listed as 1, 1, 1, 1 under the four classes, while for the C_{3v} there are three. These are used to calculate the allowed infrared and Raman vibrations.

To determine which vibration types are allowed or forbidden in the infrared, it is necessary to calculate the character $\chi_m(R)$ of the dipole moment for the operation R. This is given by

$$\chi_m = \pm 1 + 2 \cos \phi \qquad (4\text{-}29)$$

where ϕ is the angle associated with the symmetry operation. If we

TABLE 4-IV. Character Table for C_{2v} Group

	$I(1)$	$C_2(1)$	$\sigma_v(1)$	$\sigma_v'(1)$
a_1	1	1	1	1
b_2	1	−1	−1	1
a_2	1	1	−1	−1
b_1	1	−1	1	−1
Variables and Characters for Calculating Allowed Infrared Bands				
ϕ	0°	180°	0°	0°
χ_m	3	−1	1	1
U_R	3	1	1	3
θ	3	1	1	3

TABLE 4-V. Character Table for C_{3v} Group

	$I(1)$	$C_3(2)$	$\sigma_v(3)$
a_1	1	1	1
a_2	1	1	-1
e	2	-1	0

Variables and Characters for Calculating Allowed Infrared Bands

ϕ	0°	120°	0°
χ_m	3	0	1
U_R	7	4	5
θ	15	0	5

limit ourselves to molecules of symmetry C_{3v}, ϕ will be 0° for the identity operation, $\pm 120°$ for the two rotations, and 0° for each of the three reflections. The following definitions will be useful: *proper rotation*—a rotation of $\pm\phi$ about some axis of symmetry; *improper rotation*—a rotation followed by a reflection in a plane perpendicular to the axis of rotation. Thus, for C_{3v} molecules the identity I and the rotations C_3 are proper rotations, while the reflections σ_v are improper rotations of 0°. In equation (4-29) the plus sign is used for proper rotations and the minus sign for improper rotations. Using this formula, the reader can show that the character of χ_m is 3, 0, 1 for the identity, rotation, and reflection, respectively. We have listed the values for ϕ and χ_m in the proper column in the character table of the C_{3v} molecules (Table 4-V).

We shall now illustrate that the character χ_m has components of the types of vibrations which are allowed in the infrared spectra of C_{3v} molecules. The character χ_m is always a linear combination of the characters of the vibration type of that group. A reduction formula is used to calculate how many times each of the vibrational types a_1, a_2, and e occurs in χ_m. In the following sections we shall frequently resort to reduction formulas to determine the number of times a vibrational type occurs.

The reduction formula for χ_m is as follows:

$$N_i = \frac{1}{N_g} \sum N_n \chi_m(R) \chi_i(R) \tag{4-30}$$

where N_i is the number of times the character χ_i appears in χ_m, N_g is the total number of elements in the group, $\chi_i(R)$ is the character of the vibration type for the operation R, R is one of the symmetry operations, and N_n is the number of elements in each class.

Each of these terms is obtained from the character table of the C_{3v} group. The value of N_g is the sum of the number of operations in each class $(1 + 2 + 3 = 6)$ and the χ_i are the characters of the vibration types and are obtained from the character table. $\chi_m(R)$ has previously been found to be 3, 0, 1. Thus,

$$N_{a_1} = (1/6)[1 \cdot 3 \cdot 1 + 2 \cdot 0 \cdot 1 + 3 \cdot 1 \cdot 1] = 1$$
$$N_{a_2} = (1/6)[1 \cdot 3 \cdot 1 + 2 \cdot 0 \cdot 1 + 3 \cdot 1 \cdot (-1)] = 0$$
$$N_e = (1/6)[1 \cdot 3 \cdot 2 + 2 \cdot 0 \cdot (-1) + 3 \cdot 1 \cdot 0] = 1$$

From the above calculation it can be seen that the characters of the a_1 and e vibrations appear once in χ_m and those of a_2 do not appear at all.

We can prove that this reduction is correct by adding the characters of the a_1 and e vibrations once and showing that they give the character χ_m:

	I	C_3	σ_v	
χ_{a_1}	1	1	1	
χ_e	2	-1	0	(add)
χ_m	3	0	1	

The preceding calculation also indicates that only a_1 and e vibrations will be allowed in the infrared for C_{3v} molecules since theirs are the only characters that appear in χ_m.

To determine allowed Raman bands a character χ_α is calculated in a manner similar to that used for χ_m. It can be shown that the a_1 and e vibrations also are the only ones allowed in the Raman spectra for C_{3v} molecules. We shall not illustrate the calculation of Raman allowed fundamentals.

4.8B. Calculation of the Allowed Combination Bands in the Infrared

To determine the combination bands that are allowed in the infrared, it is necessary to form the direct product of the characters of the vibrations making up the combination. For example, if a combination band is due to the sum of fundamentals ν_1 and ν_2, the combination band will have a character made up of the direct product of the characters of ν_1 and ν_2. Let us consider a combination band $\nu_{a_1} + \nu_e$, which is due to an a_1 and an e fundamental. This combination band for a C_{3v} molecule has a character calculated as follows:

	I	C_3	σ_v	
χ_{a_1}	1	1	1	
χ_e	2	−1	0	(multiply)
$\chi_{a_1 \times e}$	2	−1	0	

Since the combination band has the same character as an e fundamental, it will be allowed in the infrared.

Often the character obtained for a combination band is not simply that of a fundamental. In this case, we can use a reduction formula to determine which species of vibrations have characters present in the combination band. For example, let us again consider the combination band $\nu_{a_1} + \nu_e$; its character is 2, −1, 0. These values can be reduced by the following formula:

$$N_i = \frac{1}{N_g} \sum N_e \chi_{a_1 \times e} \chi_i \tag{4-31}$$

All the terms of this equation were used and defined in connection with the previous reduction formula (4-30), except $\chi_{a_1 \times e}$, which is the character of the combination band. If we substitute into (4-31), the following results are obtained:

$$N_{a_1} = (1/6)[1 \cdot 2 \cdot 1 + 2 \cdot (-1) \cdot 1 + 3 \cdot 0 \cdot 1] = 0$$
$$N_e = (1/6)[1 \cdot 2 \cdot 2 + 2 \cdot (-1) \cdot (-1) + 3 \cdot 0 \cdot 0] = 1$$
$$N_{a_2} = (1/6)[1 \cdot 2 \cdot 1 + 2 \cdot (-1) \cdot 1 + 3 \cdot 0 \cdot (-1)] = 0$$

We see that only e vibrations make up the character of the combination $\nu_{a_1} + \nu_e$. Thus, since e vibrations are allowed in the infrared, the combination band is also allowed.

If this calculation is repeated for all possible binary combination bands, the results presented in Table 4-VI are obtained. In this table, if a certain type of vibration, for instance a_1, has characters in the combination band, a 1 appears in the proper column, for instance, under a_1. If the species does not make a contribution, a 0 is placed in the a_1 column. The data in the table indicate that the only binary combination band forbidden in the infrared is $\nu_{a_1} + \nu_{a_2}$.

4.8C. Calculation of the Allowed Overtone Bands in the Infrared

The calculation of the allowed overtones is performed in a manner similar to that used for combination bands, except that a degenerate type of vibration requires special consideration. An overtone of a vibration may be considered as a special case of a combination

TABLE 4-VI. Selection Rules for Binary Combination Frequencies of a Molecule with C_{3v} Symmetry

Combination	Number of times the vibration type appears in the character of the combination			Activity*	
	a_1	a_2	e	Raman	Infrared
$a_1 \times a_1$	1	0	0	+	+
$a_1 \times a_2$	0	1	0	–	–
$a_1 \times e$	0	0	1	+	+
$a_2 \times a_2$	1	0	0	+	+
$a_2 \times e$	0	0	1	+	+
$e \times e$	1	1	1	+	+

*+ indicates active; – indicates inactive.

band since the overtone can be thought of as the summation $\nu_1 + \nu_1$. On this basis, we can proceed to calculate the infrared activity of the overtone in the same manner as was used for combination bands. For example, if we again consider a C_{3v} molecule, then the overtone of an a_2 vibration has the following character:

	I	C_3	σ_v	
χ_{a_2}	1	1	-1	
χ_{a_2}	1	1	-1	(multiply)
$\chi_{a_2}^2$	1	1	1	

i.e., the overtone of the a_2 fundamental has the character 1, 1, 1, which is the same as the character of an a_1 vibration. Since the a_1 vibration is allowed in the infrared, this overtone is also allowed, even though the a_2 fundamental of which it is made up is forbidden.

If the character of the overtone is comprised of the characters of several types of vibrations, then a reduction formula similar to (4-30) and (4-31) is used to determine the characters present. We can illustrate this for the a_2 overtone calculated above, even though we have already shown that the character of the first a_2 overtone is 1, 1, 1 (i.e., the same as that of an a_1 fundamental).

To show that only the a_1 characters are present in the $\chi_{a_2}^2$ overtone, we perform the following calculation:

$$N_{a_1} = (1/6)[1 \cdot 1 \cdot 1 + 2 \cdot 1 \cdot 1 + 3 \cdot 1 \cdot 1] = 1$$

$$N_{a_2} = (1/6)[1 \cdot 1 \cdot 1 + 2 \cdot 1 \cdot 1 + 3 \cdot 1 \cdot (-1)] = 0$$

$$N_e = (1/6)[1 \cdot 1 \cdot 2 + 2 \cdot 1 \cdot (-1) + 3 \cdot 1 \cdot 0] = 0$$

TABLE 4-VII. Selection Rules for the Overtones of a Molecule with C_{3v} Symmetry

Overtone	Number of times the vibration appears in the character of the overtone			Activity*	
	a_1	a_2	e	Raman	Infrared
a_1^n	1	0	0	+	+
a_2^n (n even)	1	0	0	+	+
a_2^n (n odd)	0	1	0	−	−
e^2	1	0	1	+	+
e^3	1	1	1	+	+
e^4	1	0	2	+	+
e^5	1	1	2	+	+

*+ indicates active; − indicates inactive.

If this type of calculation is repeated for other overtones the allowed infrared overtone bands will be found to be those listed in Table 4-VII. In this table, the presence or absence of the character of each type of vibration in the overtone is indicated by 1 or 0, respectively.

To calculate the second overtone of a fundamental, the product of the first overtone and the fundamental is formed. For example, the character of the second overtone of the a_2 vibration is given as follows:

	I	C_3	σ_v	
$\chi_{a_2}^2$	1	1	1	
χ_{a_2}	1	1	-1	(multiply)
$\chi_{a_2}^3$	1	1	-1	

Since the characters of the second overtone of an a_2 fundamental are the same as those of the forbidden fundamental itself, this overtone is not allowed in the infrared. A simple repetition of the procedure used to obtain the characters of the second overtone will lead to the characters of overtones of higher order.

The character of the e overtones is not obtained by multiplying the characters of the e fundamentals. The method used differs only slightly from that used for calculating the overtones of the nondegenerate a_1 and a_2 vibrations, and will not be presented here. The formula for calculating the overtones of the doubly degenerate e vibration is of a slightly different form than that for nondegenerate vibrations, namely,

$$\chi_e^n(R) = \tfrac{1}{2}\chi_e^{n-1}(R)\chi_e(R) + \chi_e(R^n)$$

where $\chi_e^n(R)$ is the character of the nth overtone for the operation R, $\chi_e^{n-1}(R)$ is that of the $(n - 1)$th overtone, $\chi_e(R)$ is the character of the fundamental for the operation R, and $\chi_e(R^n)$ is the character corresponding to the operation R performed n times in succession.

4.8D. Calculation of the Number of Allowed Infrared Fundamentals

In the previous sections the allowed fundamentals, overtones, and combination bands for a C_{3v} molecule were calculated. The number of fundamentals is determined by the number of atoms in the C_{3v} molecule, and for some symmetries the number of atoms also determines whether the allowed vibration exists for the particular molecule in question. For example, a b_1 vibration, although an allowed infrared band for larger C_{2v} molecules, does not exist for a triatomic C_{2v} configuration.

We shall show here the calculations used to determine the number of a_1 and e fundamentals of the C_{3v} molecule methylacetylene.

We must first calculate the quantity θ, which is given as follows:

For proper rotations:

$$\theta = (U_R - 2)(1 + 2\cos\phi) \tag{4-32}$$

For improper rotations:

$$\theta = U_R(-1 + 2\cos\phi) \tag{4-33}$$

All the above quantities except U_R have been defined earlier. For C_{3v} molecules the identity I and rotations C_3 are proper rotations, and the reflections σ_v are improper rotations. We define U_R as the number of nuclei unchanged by the symmetry operation. For methylacetylene the identity operation leaves all seven atoms unchanged; for the rotation C_3, four atoms remain unchanged; and for the reflection σ_v, five atoms remain unchanged. Thus, U_R is 7, 4, 5 for the three classes of operation. θ can now be calculated for each operation from the above equations and is found to be 15, 0, 5.

To determine the number of times each type of vibration appears, a reduction formula is used:

$$N_i = \frac{1}{N_g}\sum N_e\theta\chi_i \tag{4-34}$$

All terms have been defined and the calculation proceeds as follows:

$$N_{a_1} = (1/6)[1\cdot 15\cdot 1 + 2\cdot 0\cdot 1 + 3\cdot 5\cdot 1] = 5$$

$$N_{a_2} = (1/6)[1\cdot 15\cdot 1 + 2\cdot 0\cdot 1 + 3\cdot 5\cdot(-1)] = 0$$

$$N_e = (1/6)[1\cdot 15\cdot 2 + 2\cdot 0\cdot(-1) + 3\cdot 5\cdot 0] = 5$$

There are therefore five a_1 and five e vibrations for methylacetylene. Since e vibrations are doubly degenerate, the total number of fundamentals is 15. The same number would be predicted by the simple formula $3N - 6$ mentioned previously.

4.9. FACTORS INFLUENCING BAND POSITIONS AND BAND CONTOURS IN POLYATOMIC MOLECULES

In the following sections, we shall first describe the band contours and band positions observed for polyatomic molecules and the various factors that influence them, and then proceed to the vibrational analyses for some specific molecules. Except for small molecules it is not possible to observe the individual rotational lines; therefore the overall band contour is examined. One fairly large molecule where the fine-line structure of a band can be seen is butadiene [28].

4.9A. Fermi and Coriolis Perturbations

Perturbations between close-lying energy levels can occur for all types of molecules. These perturbations may be caused by either *Fermi resonance* or *Coriolis interaction.* Both phenomena can produce either vibrational or rotational perturbations. The rotational perturbations will affect only levels of the same over-all species and the same J value. The restriction to the same species also holds true for vibrational perturbations in the case of Fermi resonance, but in the case of Coriolis interaction perturbation can occur between vibrational levels of different species. We shall discuss Fermi resonance first.

The rule limiting perturbation effects of the Fermi resonance type to vibrational levels of the same species restricts the occurrence of this phenomenon in many molecules. Fermi resonance has been observed between fundamental, combination, and overtone bands in some molecules. If two sublevels exist which have the same energy but different species, only the sublevel with the same species as the perturbing vibration will be excited. This will be illustrated below for CO_2.

Fermi resonance can produce a vibrational or a rotational perturbation. The former involves a shift of a vibrational level from its normal position. In addition, the vibrational change will alter the rotational constant B for the two interacting energy levels. Although the change in the rotational constant B could be thought of as a rotational perturbation, we shall restrict the use of this term to cases for which a rotational perturbation occurs without a vibrational disturbance and consider the vibrational level shift as a vibrational perturbation.

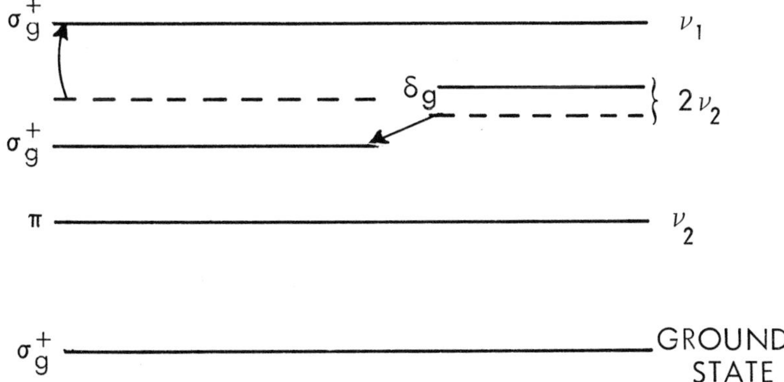

Figure 4-29. Energy level diagram for CO_2. The dotted lines represent the unperturbed levels which go over to the levels indicated by the arrows as a result of Fermi resonance.

***Fermi Resonance* (*Vibrational Perturbation*).** Two vibrational levels in a molecule may have nearly the same energy. If the molecule can be excited by infrared radiation to both these levels, the resulting vibrations will have nearly the same energy. If these two vibrations are of the same species, a perturbation of the vibrations can occur, and the result of this perturbation is the shifting of one of the levels to a higher energy while the other falls to a lower energy. We shall illustrate this further in the following example.

The energy levels of the fundamental ν_1 and the first overtone $2\nu_2$ of CO_2 are nearly equal and are of the same species, so that they can perturb each other. This perturbation can be visualized if we consider the simplified energy level diagram for carbon dioxide shown in Figure 4-29. In this figure, the energy level for the vibration $2\nu_2$ should be nearly twice that of the fundamental ν_2. This would place it near 1334 cm^{-1}; but the observed position for this overtone is at 1285.5 cm^{-1}, and we can see that it has been shifted from the expected position. Further, the ν_1 fundamental appears at a higher frequency than expected, i.e., the energy levels of ν_1 and $2\nu_2$ appear to repel each other. In Figure 4-29, this "repulsion" is indicated by arrows that show the direction in which each level has been moved. There are actually two energy sublevels for the vibration $2\nu_2$, but only the one which has the same species as the ν_1 vibration is perturbed. The other sublevel remains at a frequency position near the unperturbed value and energy transitions to this level could occur under certain conditions. An absorption band for carbon dioxide could therefore be found at the frequency position representing an energy transition to this unshifted level. This has been observed in the Raman spectrum of CO_2. When Fermi resonance occurs for a molecule, it is not always possible to find fundamental, overtone, or combination bands at exactly the positions normally expected for them. Moreover, the intensities of overtone or combination bands that are in Fermi resonance with other bands may be much greater than expected. In the example cited above, the overtone $2\nu_2$ has an intensity near that of the fundamental ν_1 as a consequence of this resonance. It has been observed that levels near each other and of the same species often do not interact. No explanation is available.

Fermi Resonance (Rotational Perturbation). A rotational perturbation of the Fermi resonance type can occur even if the interaction between two vibrational levels of the same species is slight. If the two vibrational energy states are near each other, then some lines of the rotational fine-line structure may be perturbed. The rotational lines which have nearly the same energy will be the ones which perturb each other. We shall not give an example of such a perturbation here; however, examples of this type of perturbation will be found in the spectra interpreted in the later sections of this chapter.

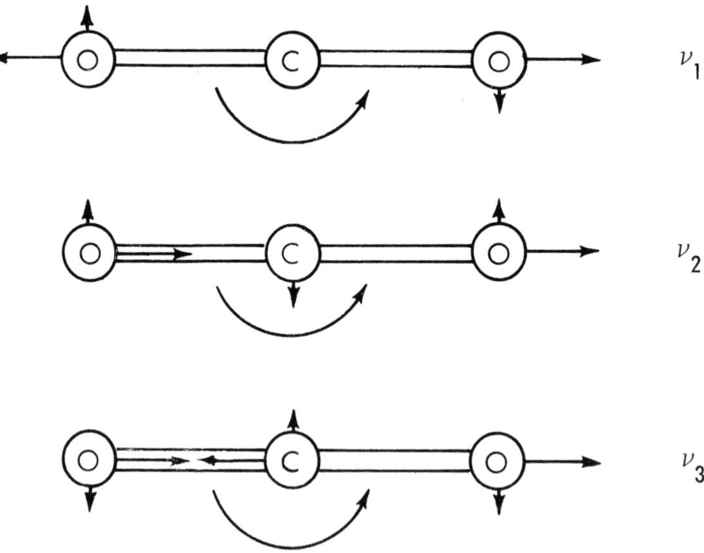

Figure 4-30. Coriolis forces in CO_2.

Coriolis Interaction (Vibrational and Rotational Perturbation). When a molecule rotates and vibrates simultaneously, two forces can appear. One is the apparent *centrifugal force* present whenever a mass is rotated; the second is the *Coriolis force*. The coupling which has been observed between a rotation and a vibration has been related to this Coriolis force.

We can visualize the forces involved between a rotation and a vibration in the following way. The forces which come into play for the various vibrations of carbon dioxide are diagrammed in Figure 4-30. In the figure, the small arrows represent the direction of the Coriolis forces, while the larger arrows indicate the direction of the vibration. During a ν_3 vibration, the Coriolis force excites the ν_2 vibration with the frequency ν_3. If ν_2 and ν_3 are near each other, the excitation is a strong one.

Coriolis interactions can be classified as rotational perturbations, when the interaction between two vibrational levels of different species results in a change in the rotational constant B, or they may be vibrational perturbations, when the energy levels of the band are shifted from their normal positions.

The Coriolis interaction is found to be larger for degenerate vibrational states than for nondegenerate levels. Thus, for symmetric rotor molecules, the doubly degenerate vibrations may be split so that they differ slightly in energy. The fine-line structure of the absorption band will then show lines due to transitions to both of these levels. For spherical rotor molecules, the Coriolis interaction will cause the triply degenerate frequencies to split.

When allowed and forbidden vibrations are in close-lying levels, Coriolis interactions can excite infrared vibrations not normally allowed. However, these forbidden bands are usually weak.

For symmetric rotor molecules of C_{3v} symmetry, the Coriolis perturbations occur between the following pairs of vibrations: (a_1e), (a_2e), (a_1a_2), (ee). For the first two pairs, the perturbation increases with increasing J, while for the latter two it increases with increasing K. For molecules of C_{2v} symmetry, the following pairs of vibrational levels may perturb each other: (a_1a_2), (a_1b_1), (a_1b_2), (a_2b_1), (a_2b_2), (b_1b_2).

An example of these pairs of perturbing vibrations occurs for the molecule H_2CO. The fundamentals $\nu_5(b_1)$ and $\nu_6(b_2)$, which are close together, perturb each other very strongly. (This interaction for H_2CO is discussed in Section 4.16 and the effect on the fine-line structure of these bands is shown in Figure 4-38).

Another example of Coriolis interaction is presented in Section 4.13. J. Watson recently published an excellent paper on the determination of the centrifugal distortion coefficient of asymmetric top molecules. [[29]].

4.9B. Inversion Doubling

Inversion doubling is a phenomenon detectable in the fine-line structure of bands: the rotational lines of a band appear to be split. The intensity of the lines of the doublet can be equal or unequal, depending on the position of the line in the band. The classical example of a molecule which shows such inversion doubling is NH_3, whose infrared spectrum is shown in Figure 4-31A. If we consider NH_3 to be a low pyramidal structure, two configurations are possible, i.e., the nitrogen atom can be thought of as moving through the plane of the hydrogens from one position to another, each position of the nitrogen giving rise to a different configuration. In each configuration the structure is pyramidal. The analogy of an umbrella being turned inside out will help to visualize this motion.

The two nearly equivalent equilibrium positions for the nitrogen nucleus with respect to the three hydrogen nuclei as the configuration changes back and forth results in a doubling of all energy levels of the molecule. Each vibrational level must therefore be considered doubled. This further implies that a molecule will have rotational constants that differ because of this separation. For example, for the symmetric top NH_3, two B and two A rotational constants can be calculated. The vibrational level separation is small in many molecules, but for NH_3 the ground state has a separation of levels of 0.66 cm^{-1}, while that for the vibrational level ν_1 is about 0.9 cm^{-1}; in other vibrational levels it is still higher. We can visualize energy transitions from either one or the other of these double levels. The lower level of any doublet is usually designated by the symbol s and the upper level by the symbol a. Only transitions from s to a or from a to s are allowed. Thus the lines of a fundamental, such as ν_2, will be doubled and could be considered to be made up of energy transitions of both $s \rightarrow a$ and $a \rightarrow s$. The high-resolution spectrum of the ν_2 band is shown in Figure 4-31B (located in pocket on inside back cover). It can be seen that the $s \rightarrow a$ transitions result in a band having a Q-branch near 968 cm^{-1}, while the $a \rightarrow s$ transitions have a Q-branch near 932 cm^{-1}. Thus, the fundamental ν_2 has a double Q-branch. The J values for the $s \rightarrow a$ transitions are listed below the $a \rightarrow s$ values. It is now possible to assign each line of the ν_2 band to a distinct energy transition. The numbers below some of the lines are the corresponding K values. Each line of the $s \rightarrow a$ transitions obeys an equation of the form below, and those of $a \rightarrow s$ obey a similar equation:

$$\tilde{\nu}\left[s_{(J-1),K}a_{J,K}\right] = \frac{(F_a - F_s)}{hc} + (A'_a + A''_s)J + (A'_a - A''_s)J^2$$

$$+ [(B'_a - B''_s) - (A'_a + A''_s)]K^2 + \text{centrifugal distortion terms}$$

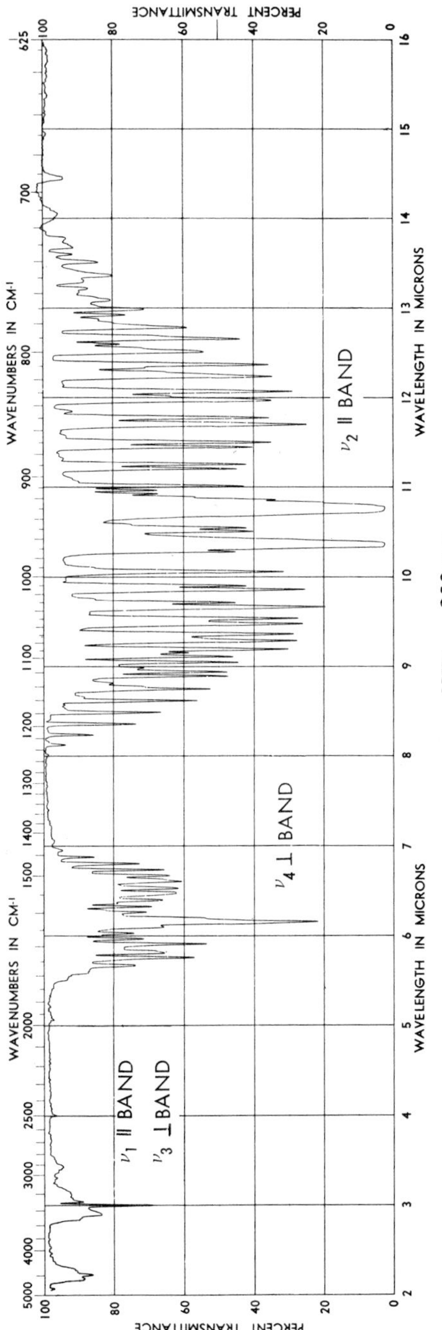

Figure 4-31A. The infrared spectrum of NH_3 under medium resolution.

where $\tilde{\nu}$ is the wavenumber of each line of the $s \rightarrow a$ transitions,

$$A_a = \frac{h}{8\pi^2 c I_A^a}$$

where a indicates that the moment of inertia corresponds to the higher-energy configuration of NH_3,

$$A_s = \frac{h}{8\pi^2 c I_A^s}$$

where s indicates the moment of inertia for the lower-energy configuration of NH_3,

$$B_a = \frac{h}{8\pi^2 c I_B^a}$$

and

$$B_s = \frac{h}{8\pi^2 c I_B^s}$$

The ν_1 fundamental has a series of rotational lines with an intensity distribution similar to that shown in Figure 4-32. It can be seen that we are confronted with alternating intensities; that is, for the first pair the line on the high-frequency side is more intense, while for the second pair the low-frequency line has the greater intensity, and so on. Moreover, the first lines of the P- and R-branches do not split. It is possible to show that both the missing

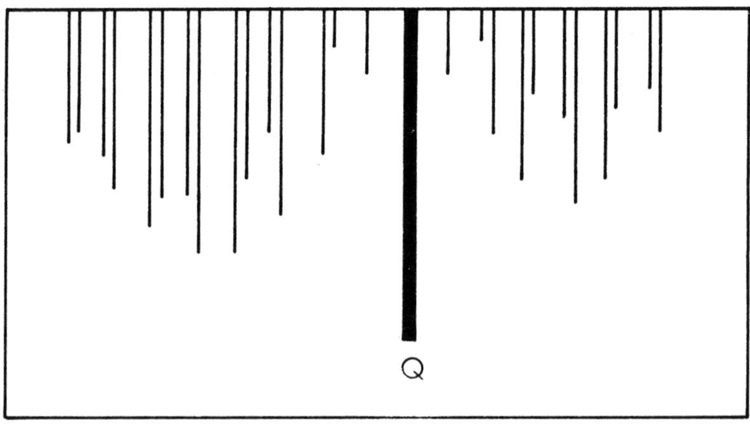

Figure 4-32. Inversion doubling in the ν_1 vibration of NH_3.

lines and the alternation of intensity can be explained in terms of the possible energy transitions, but the explanation will not be presented here. The fundamental ν_1 undergoes far less splitting than ν_2; and ν_3 and ν_4, being perpendicular vibrations, split even less, although their splitting is still in the observable range. The molecule ND_3 would show less splitting of lines than does NH_3.

Such molecules as PH_3, BCl_3, and PF_3 are capable of showing inversion doubling. Theoretically, all molecules except the planar ones could show a form of inversion doubling, but it has not been observed for many molecules. This would seem to indicate that the energy barrier for inversion is quite high.

4.9C. Intensity Alternation of the Line Structure of Bands

The number of molecules which will be present in the rotational level of a given J value depends on a number of factors. Since the population of the rotational levels will determine the possible transitions, upon it will depend the intensity of the fine-line structure and the contours of the band envelope. The levels which will contribute to the band contours are determined by the spins of the nuclei comprising the molecule. For a molecule such as CO_2, where the oxygen nuclei have zero spins, alternate lines are absent. For example, in the fundamental ν_2, the spacing in the R-branch is twice that which would be expected if all rotational energy levels were present. All linear molecules of $D_{\infty h}$ symmetry will show alternating intensity (or alternate absent lines). For example, some of the bands of C_2H_2 show alternating intensity, with lines of odd J having an intensity three times as high as those with even J.

Linear molecules of $C_{\infty v}$ symmetry will not show intensity alternation. Thus, it is possible to distinguish between linear molecules of $C_{\infty v}$ and $D_{\infty h}$ symmetry. The classical example of an application of this concept is the proof that N_2O has the structure N—N—O rather than N—O—N, which is based on the lack of intensity alternation in the bands of this molecule.

Symmetric rotor molecules can also show intensity alternation. For example, for CH_3Br the fundamental ν_6 shows this effect, the lines being strong, weak, strong, weak, and so on. The variation of the line intensity can become quite complex if the number of axes of symmetry increases. Intensity alternation can also occur for asymmetric rotors, but the complexity in such cases precludes any discussion here.

4.9D. *l*-Type Doubling

When a degenerate vibrational–rotational energy level is present in a molecule, a phenomenon termed "*l*-type doubling" is sometimes observed. It makes its appearance in the following manner. In some perpendicular bands the rotational constant B appears to take on two different values, depending on whether it is determined from the Q-branch or from P- and R-branches, and in some parallel bands we are confronted with a doubling of the fine lines of the band. This l-type doubling has been observed for a number of linear molecules including N_2O [4] and C_2H_2 [5].

The source of l-type doubling is the change in the moment of inertia of a molecule that occurs as a result of Coriolis interaction between a vibration and a rotation.

The perpendicular vibrations of linear polyatomic molecules can be doubly degenerate and can therefore show l-type doubling. Some combination bands of linear molecules will not show intensity alternation but will be subject to l-type splitting.

4.10. BAND CONTOURS FOR LINEAR MOLECULES

Linear molecules can belong to either of two point groups. The first, the $D_{\infty h}$ group, consists of molecules which have a horizontal plane of symmetry perpendicular to the vertical internuclear axis, e.g., CO_2, C_2H_2, etc. The second, the $C_{\infty v}$ group, consists of those molecules for which no horizontal plane of symmetry exists, e.g., HCN, CSO, etc. Both of these groups have an infinite number of vertical symmetry planes, all passing through the internuclear axis.

A linear molecule can be assumed to be a rigid rotor, and the rotational term values (energies/hc) can be described by an equation similar to that used for diatomic molecules. The equation will be of the form

$$F = BJ(J + 1) \tag{4-35}$$

However, the moment of inertia I_B associated with the rotational constant B will, of course, be different from that used for diatomic molecules.

If B is defined as

$$B = \frac{h}{8\pi^2 c I_B} \tag{4-36}$$

then the moment of inertia I_B is defined by the more general formula

$$I_B = \sum m_i r_i^2 \tag{4-37}$$

For a symmetric molecule, such as CO_2, this formula reduces to

$$I_B = 2mr_e^2 \tag{4-38}$$

where m is the mass of the oxygen and r_e is the equilibrium value of the internuclear separation of the carbon and the oxygen atoms.

The vibration of a linear molecule can be considered similar to that of a harmonic oscillator, so that the vibrational energy levels can be described by a formula analogous to that used for diatomic molecules. However, the energy equation must be written in a more general form:

$$E = h\nu_i(v_i + \tfrac{1}{2}) \tag{4-39}$$

where ν_i is the frequency of one of the normal vibrations and $v_i = 0$, 1, 2, 3, ... is the vibrational quantum number. Even for a triatomic molecule, the above formula predicts a rather complex energy level diagram, which for that reason will not be presented here.

It is also difficult to introduce simple expressions in the harmonic oscillator equations to account for the anharmonic character of polyatomic molecules. Because of the complexity of this type of calculation, and the relatively small number of simple linear molecules to which the analogy can be reasonably applied, this method is of only limited usefulness and will therefore not be discussed here.

Since stretching and bending vibrations can occur for linear molecules, it is necessary to classify these two types of vibrations. As was indicated earlier, for infrared radiation to be absorbed, there must be a change in the dipole moment. When the change is along the axis drawn through the nuclei, the band is designated as parallel ($\parallel$). All stretching vibrations of linear molecules yield parallel bands. When the change in dipole moment is perpendicular to the internuclear axis, we have a perpendicular ($\perp$) band.

The number of fundamentals for a linear molecule containing N atoms can be calculated from the simple formula $3N - 5$. However, this calculation does not reveal whether or not the fundamental is allowed in the infrared or Raman spectrum. Also, the calculation will not yield the degree of degeneracy. To obtain this information, we must resort to methods developed in group theory. We can, however, predict the infrared selection rules of a simple molecule such as CO_2 by noting the nature of the change in the dipole moment. If the vibration results in an *asymmetric* change in the charge distribution, the frequency is allowed in the infrared. If the vibration results in a *distortion* of the electron cloud *without* an asymmetric change in charge distribution, the frequency will be allowed in the Raman spectrum. This can be illustrated by examining the fundamentals for carbon dioxide.

The vibrational modes for CO_2 were presented earlier in Figure 4-13. If we re-examine this figure, we find we can now describe the vibrations in greater detail.

Figure 4-13A shows the symmetrical stretch ν_1. Since this vibration is along the axis of the molecule, it is a parallel vibration and gives rise to a parallel band, whose shape differs from those of vibrations along other axes. The dipole moment change here is zero and the interaction *merely changes the symmetry* of the electron cloud; this band is therefore allowed only in the Raman spectrum. Another parallel band is produced by the asymmetric stretch ν_3, shown in Figure 4-13B. The figure shows both phases of the vibration. As one oxygen atom approaches the carbon, the other recedes. It is to be noted, of course, that the carbon atom itself must move to the left in order to prevent a change in the center of mass. This band is allowed in the infrared since there is a change in dipole moment during the vibration.

Let us now consider the deformation frequencies for carbon dioxide. Although the formula $3N - 5$ predicts two deformation frequencies for CO_2 the exact nature of the second deformation is not immediately evident from Figure 4-13C. The deformation pictured is a bending of the two oxygen atoms about the carbon. This motion is perpendicular to the axis of the molecule and so gives rise to a perpendicular band. The second deformation frequency is the bending in the plane perpendicular to the plane of the diagram. It is again to be noted that the carbon atom must move to keep the center of mass stationary. Since these two bendings have the same energies, they are degenerate. This vibration is doubly degenerate since two vibrations in planes at right angles to each other are involved.

It is interesting to note that the symmetry of the carbon dioxide molecule is proved by the fact that the symmetrical stretch appears only in the Raman spectrum. If the molecule did not possess this symmetry, similar bands would appear in both the Raman and infrared spectra.

The band contours observed for parallel bands of linear molecules are similar to those found for diatomic molecules; that is, only P- and R-branches are observed. For perpendicular bands, however, the Q-branch is also present and may be quite intense. These aspects are illustrated by the observed spectrum of carbon dioxide shown in Figure 4-33 (located in pocket on inside back cover).

Bending frequencies are usually lower than stretching frequencies, and the perpendicular bands of linear symmetric molecules can be identified by their symmetric P-, Q-, and R-branches, the Q-branch being quite intense. The strong band at 667 cm^{-1} in the spectrum of carbon dioxide must therefore be the degenerate perpendicular band ν_2. Since ν_1 is allowed only in the Raman spectrum, the only other fundamental to be assigned is ν_3, the asymmetric stretch. This parallel band should lie higher than ν_2 and should have no Q-branch. In addition, because it is a fundamental of a linear, symmetric molecule, it should be quite strong and fairly symmetric. We recognize these features in the band at 2349.3 cm^{-1}. The remainder of the spectrum is assigned to combination and overtone bands.

Although most organic molecules have the C═O stretch between 1850 and 1640 cm^{-1}, there is no such peak for CO_2. This is not surprising, however, when one recalls that CO_2 has both a symmetric and an asymmetric stretch, while organic molecules containing but one carbonyl group have only an asymmetric stretch.

In general, there are three types of bands for linear molecules: two types of fundamentals (parallel and perpendicular vibrations) and the combination and overtone bands of these fundamentals. The parallel vibrations have only P- and R-branches. The perpendicular vibrations have P-, Q-, and R-branches, with the Q-branch fairly intense. The combination bands can have P-, Q-, and R-branches, with the Q-branch weak in some instances.

For parallel vibrations, the equations for the wavenumbers of the P- and R-branches are

$$\tilde{\nu}_R = \tilde{\nu}_0 + 2B' + (3B' - B'')J'' + (B' - B'')J''^2 \qquad (4\text{-}40)$$

and

$$\tilde{\nu}_P = \tilde{\nu}_0 - (B' + B'')J'' + (B' - B'')J''^2 \qquad (4\text{-}41)$$

Similar formulas apply for perpendicular vibrations. The wavenumbers of the lines of the Q-branch of perpendicular vibrations are given by

$$\tilde{\nu}_Q = \tilde{\nu}_0 + (B' - B'')J'' + (B' - B'')J''^2 \qquad (4\text{-}42)$$

The lines of the Q-branch are usually very close together, resulting in only a single maximum.

The method of combination differences (described below) can be applied to both parallel and perpendicular vibrations to obtain the rotational constant. If the rotational constant for the Q-branch does not agree with that obtained from the P- and R-branches for a perpendicular vibration, this is possibly due to the presence of *l*-type doubling. For parallel vibrations, we encounter the possibility of intensity alternation, which causes changes in the line spacing.

Asymmetric linear molecules have band contours similar to those observed for symmetric linear configurations. For example, a molecule such as HCN will undergo a symmetric and an asymmetric stretching of the CH group against the nitrogen atom, which give rise to the fundamentals ν_1 and ν_3, respectively. These two vibrations are both parallel and are of the same species. They will therefore show P- and R-branches, with the Q-branch absent. Both are infrared and Raman active, since the molecule is asymmetric, but ν_1 will have very low intensity in the infrared because the associated change in dipole moment is very small.

The fundamental ν_2 is the bending frequency and it is found in both Raman and infrared. It is a perpendicular vibration and the band contours show P-, Q-, and R-branches.

Linear molecules of $D_{\infty h}$ symmetry have fine-line band structures subject to the intensity alternation described earlier.

The existence of distinct band contours in linear molecules makes it possible to recognize these molecules from an examination of their spectra. Although the missing Q-branch in parallel vibrations and the strong Q-branch for perpendicular vibrations are distinctive in linear molecules, the presence of such bands does not conclusively prove a linear structure. On the other hand, the correct fine-line structure of a band with the characteristic intensity alternation can be considered as fairly conclusive evidence of the linearity of a symmetric linear molecule.

4.11. THE METHOD OF COMBINATION DIFFERENCES USED TO OBTAIN ROTATIONAL CONSTANTS OF DIATOMIC AND LINEAR MOLECULES

The rotational constants can be calculated from the fine-line structure of bands. For the R-, P-, and Q-branches, we could use, respectively, equations (4-40), (4-41), and (4-42); and if the observed wavenumbers are substituted in these equations, the values of B' and B'' could be found. But this is not the best technique for obtaining the values of these rotational constants. The method of *combination differences* is usually better. This method is as follows. The difference is taken between lines in the P- and R-branches corresponding to the same J'' value. This difference, obtained from equations (4-40) and (4-41), is

$$R(J'') - P(J'') = (B' + B'')(J'' + 1) + (B' - B'')(J'' + 1)^2 - (B' + B'')J'' - (B' - B'')J''^2 = 4B'(J'' + \tfrac{1}{2}) \qquad (4\text{-}43)$$

For the difference between a line in the P-branch, where the value of the quantum number of the lower vibrational energy state is $J'' + 1$, and a line in the R-branch, where this quantum number is $J'' - 1$, we can write

$$R(J'' - 1) - P(J'' + 1) = (B' + B'')J'' + (B' - B'')J''^2 + (B' + B'') \times (J'' + 1) - (B' - B'')(J'' + 1)^2 = 4B''(J'' + \tfrac{1}{2}) \qquad (4\text{-}44)$$

Thus, one difference equation has the rotational constant B' while the other has B''. Plotting the differences *versus* J'' results in a straight line whose slope is $4B'$ or $4B''$, depending on which equation is used. Thus, the slopes of these lines can be used to calculate the value of the rotational constant for either the lower or the upper vibrational energy state. A plot of this type gives a nearly straight line with a slight deviation from linearity at high values of J''. This deviation is due to the increase in the internuclear separation as J'' increases, which has not been taken into account.

The difference

$$R(J'' + 1) - P(J'') = 2\nu_0 + 2(B' - B'')J''^2 \tag{4-45}$$

can also be used to calculate the rotational constants B' and B''. In addition, if the left-hand side of this equation is plotted against J''^2, the intercept is $2\nu_0$, and thus an accurate value of the band center ν_0 can be obtained.

Variations of this technique of combination differences can be used to obtain internuclear separations in many molecules. It is important, however, to recognize that many factors influence the spacing of the lines in an absorption band and should be accounted for in any calculation.

For a linear molecule having $I_B = I_C$ and $I_A \approx 0$, the spacing of the fine-line structure in both the parallel and the perpendicular bands should be similar (assuming no *l*-type doubling or inversion doubling occurs). For nonlinear molecules having no moment of inertia equal to zero, the spacing of the fine-line structure will depend on the nature of the vibration. For the symmetric linear molecule CO_2, which has bands that show *l*-type doubling (if the combination difference is formed for such a band), the value obtained for the rotational constant B will be different from the value obtained from the Q-branch of the same band. This difference is a measure of the *l*-type doubling.

For a linear, symmetric, triatomic molecule, since $I_B = I_C$ and $I_A \approx 0$, it is possible to calculate an internuclear distance directly from the rotational constant B. For a linear molecule of this type, if the atoms are represented by Y—X—Y, the moment of inertia is given by

$$I_B = I_C = 2m_Y r_{XY}^2 \tag{4-46}$$

where m_Y is the mass of the atom Y and r_{XY} is the distance between the X and Y nuclei.

For the asymmetric linear molecule, where $I_B = I_C$ and $I_A \approx 0$ (I_B is about an axis through the center of mass, but without a nucleus at the origin), it is possible to substitute an isotope for one of the atoms and obtain rotational constants B for both the isotopic and the normal molecule. By assuming that the corresponding internuclear separations are the same in the normal and isotopic molecules, it is possible to relate the two rotational constants to each other and thus determine the internuclear distances in the molecules.

Combination differences of the observed wavenumbers of the lines in the absorption bands of symmetric, spherical, and asymmetric rotors can be obtained. However, the rotational constant B to be calculated will depend on the type of vibration the band represents. In considering a symmetric rotor, the rotational constant for one of the molecule's moments of inertia is obtained from a parallel vibration, while a second moment of inertia is determined from the perpendicular vibration. Thus, the spacing of lines in the bands of a symmetric rotor is dependent upon the type of vibration. Moreover the interlinear spaces are not equal, as they are for linear or symmetric triatomic molecules. The perturbations of the fine-line structure of bands that have been discussed in other sections must also be taken into account before a combination difference is formed and a rotational constant calculated.

The interlinear spacing of spherical rotors can differ for each type of molecular vibration. The moment of inertia of a spherical rotor therefore cannot be calculated from a single band. It is possible to obtain a value for the moment of inertia only if the interactions in the fine-line structure of the bands are known so that allowance can be made for them.

For asymmetric rotors the problem of calculating a rotational constant from the fine-line structure is even more complicated than for the examples listed above, although such a simple asymmetric rotor as H_2O has been analyzed and all of the three moments of inertia have been calculated [16]. A method related to the combination difference method can be used; however, it will not be developed here.

4.12. VIBRATIONAL–ROTATIONAL ENERGY EQUATIONS FOR THE SYMMETRIC ROTOR

The term values of the rigid symmetric rotor are given by

$$F[\mathrm{cm}^{-1}] = BJ(J + 1) + (A - B)K^2 \qquad (4\text{-}47)$$

where

$$A = \frac{h}{8\pi^2 cI_A} \qquad B = C = \frac{h}{8\pi^2 cI_B} = \frac{h}{8\pi^2 cI_C} \qquad (4\text{-}48)$$

By definition, of the three moments of inertia of symmetric rotor molecules, two are equal and differ from the third. For the equations above,

$$I_B = I_C \qquad (4\text{-}49)$$

In addition, the moment I_A is taken about the axis of inertia which coincides with the principal axis of symmetry (the C axis) of the molecule.

In equation (4-47), the quantum numbers J and K are integers for which

$$K \leq J \tag{4-50}$$

The spacing of the energy levels depends on the relative magnitudes of I_A and I_B. We shall discuss two particular cases. The first is that for which $I_A = I_B < I_C$. Molecules with this property are called *oblate* symmetric rotors. Ammonia is such a molecule. The second case is that for which $I_A < I_B = I_C$; the molecule here is termed a *prolate* symmetric rotor. Methyl iodide is an example of a prolate rotor.

It is possible to write selection rules for energy transitions in terms of the quantum numbers J and K. As indicated earlier, the type of band observed will depend on the nature of the change in the dipole moment: when the change is along the principal symmetry axis, the band is parallel; when the change is normal to the axis, the band is perpendicular. The selection rule for the energy transitions is as follows:

Parallel Bands

$$\Delta J = \pm 1 \qquad \Delta K = 0 \qquad (\text{for } K = 0)$$

$$\Delta J = 0, \pm 1 \qquad \Delta K = 0 \qquad (\text{for } K \neq 0)$$

Perpendicular Bands

$$\Delta J = 0, \pm 1 \qquad K = \pm 1$$

Inversion doubling imposes further restrictions on permitted energy transitions, but these cases will not be discussed here. Hybrid bands, i.e., bands of partly parallel and partly perpendicular character, are also possible for symmetric rotor molecules.

The fine-line structure of symmetric rotors may be subdivided: For example, when $\Delta J = 0, +1, -1$, the rotational lines associated with $\Delta K = 0$ are termed Q_Q-, Q_R- and Q_P-branches; for $\Delta K = +1$, the lines are R_Q-, R_R-, and R_P-branches; and for $\Delta K = -1$, they are P_Q-, P_R-, and P_P-branches. These branches are known as *subbands*. Under medium or low resolution the subbands are not resolved from each other and the maxima observed for the total band will represent collections of the subbands. The subbands can be considered to have small interlinear spacings, which are given by equations similar to those developed for linear molecules. For example, the R-branch of a parallel vibration can be described by an equation of the type used to describe the R-branch of a perpendicular vibration of a linear molecule. The wavenumber of the branch of the subband would be given by an equation of the form

$$\tilde{\nu}_R^{\text{sub}} = \tilde{\nu}_0^{\text{sub}} + 2B' + (3B' - B'')J + (B' - B'')J^2 \qquad (4\text{-}51)$$

The spacing of the lines in the subband would be approximately $2B$ if $B' \simeq B''$, just as it was in the linear molecule. However, the value of the origin, $\tilde{\nu}_0^{\text{sub}}$, would be the subband origin, not the total band origin. In addition, the rotational lines seen in each subband would be limited by the condition $J \geq K$; if intensity alternation occurred, it would influence the lines observed. Each subband would be designated by a quantum number K. The total band observed would be the sum of all of these subbands.

Frequently no simple band contours are observed for either the parallel or the perpendicular bands of symmetric rotor molecules. The contours depend on many factors, including (1) the relative difference between the terms A and B; (2) the value of $(A - B)$ in the initial state (before the energy transition) compared to its value in the final state; and (3) the extent of the Coriolis interaction between the rotational and vibrational levels.

For some symmetric rotors, the parallel band shows well-formed P-, Q-, and R-branches, similar to the band contours found for the perpendicular band of the linear molecule CO_2. Let us assume that there is no Coriolis resonance in such a band and that the values of $(A - B)$ in the initial and final energy states are similar. To construct this well-formed band, a series of subbands is drawn. The subbands, designated as $K = 0, 1, 2, 3, \ldots$, are then added together to give the total band. The band is shown in Figure 4-34. Although only the subbands $K = 0, 1, 2, 3, 4$ are shown, others could be drawn. However, their intensity would decrease for larger K values; consequently, they would not greatly influence the over-all contour of the band. The transitions associated with the Q-branch are not designated by any symbol in the subbands, but are shown as a grouping of closely spaced lines beginning in the subband $K = 1$, The sum of all the subbands gives the contour of the total band, shown at the bottom of the figure.

For other parallel bands of symmetric rotors, the subbands may add differently, giving a total band in which the Q-branch transitions are distributed in the P- and R-branches. For such a band, only the P- and R-branches are seen, and the band center is a minimum. The P- and R-branches may have somewhat uneven intensity distributions since the regions where both Q_Q and Q_P, or Q_Q and Q_R, lines occur may be more intense than the regions where there are only Q_P and Q_R lines.

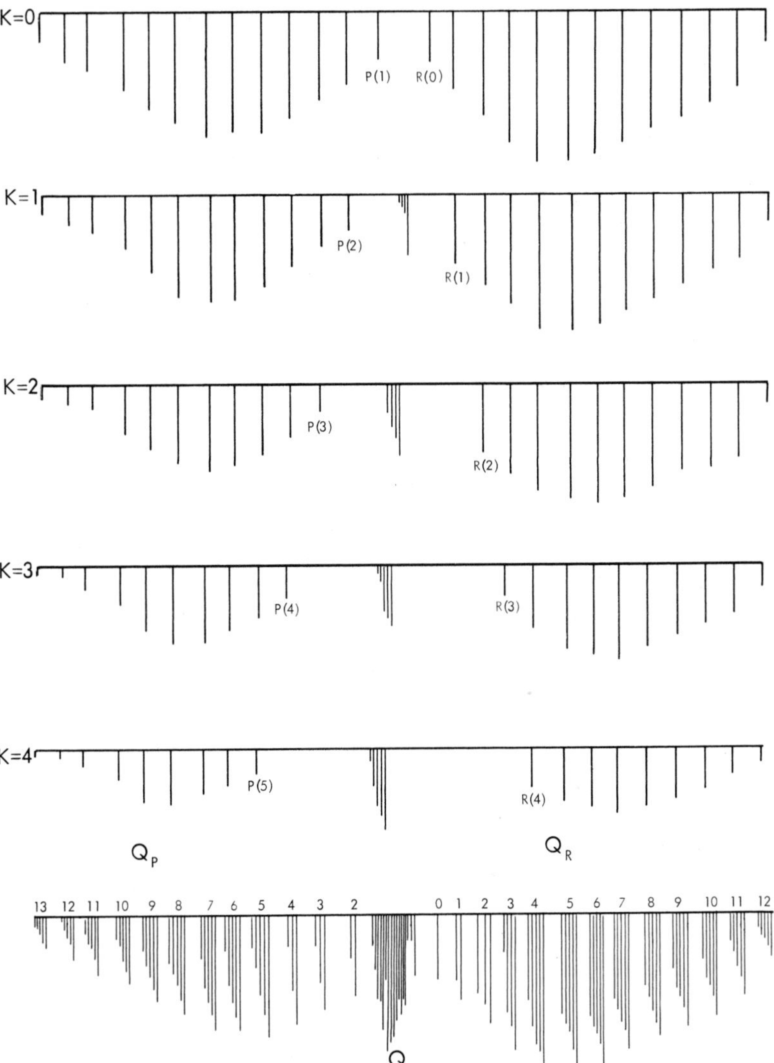

Figure 4-34. Parallel band of a symmetric top.

Band contours for perpendicular vibrations of symmetric rotors are more complex than those for parallel vibrations. Coriolis interactions can give abnormal intensities to subbands. Generally, under medium resolution the band contours show a number of maxima where the Q-branches of the subbands are distributed throughout the P- and R-branches of the total band. There are exceptions to this general conclusion. For example, some prolate symmetric rotor molecules have only *one* strong maximum at the band center, and the P- and R-branches appear with only low or medium intensity.

In summary, we can say that for some symmetric rotors, quite regular and distinct contours can be seen; for others the band contours may be quite irregular. Under high resolution, when the fine structure becomes visible, it is sometimes possible to identify lines with particular energy transitions within a subband.

In the following section the spectra of such symmetric rotors as NH_3, H_2CO, etc., will be presented. An excellent summary of band contours found for *e* type vibrations of these rotors is given in the work of Rao and Polik [20].

4.13. ANALYSIS OF THE AMMONIA SPECTRUM (SYMMETRIC ROTOR)

The spectrum of the molecule NH_3 is of interest because the fine-line structure of some of the absorption bands can be seen even with an NaCl prism. An additional feature of interest is the presence of inversion doubling.

Since NH_3 has four atoms it should have six fundamentals ($3N - 6$ vibrations). The fundamentals and the types of vibration are given in Table 4-VIII (see also Figure 4-14).

Two fundamentals are of the type a_1, while two are of the doubly degenerate type *e*. The doubly degenerate vibrations ν_3 and ν_4 account for the fifth and sixth fundamentals.

The a_1 vibrations are totally symmetric, so that the molecule remains a pyramid during the vibration. The *e* vibrations are asymmetric and distort the pyramid. Although an a_2 vibration might be expected for a molecule having C_{3v} symmetry, no such vibration exists for NH_3. All four fundamentals are allowed in both the infrared and Raman spectra.

TABLE 4-VIII. The Fundamentals of NH_3

Fundamental	Type	Position (cm^{-1})
ν_1	$a_1(\parallel)$	3336.0
ν_2	$a_1(\parallel)$	950.24
ν_3	$e(\perp)$	3414.9
ν_4	$e(\perp)$	1627.77

The inversion doubling of the ν_1 and ν_2 fundamentals was discussed in Section 4.9B.

The spectrum observed for ammonia was presented in Figure 4-31. The distinct fine-line structure is useful in calibrating spectrophotometers [6]. An extensive analysis of the spectrum of this compound has been carried out by H. H. Nielsen and co-workers [7].

The fundamental ν_4 is perturbed by the overtone $2\nu_2$. This overtone has high intensity in the R-branch, which suggests that there is a Coriolis interaction between this overtone and the ν_4 rotational levels.

The fundamental ν_2 shows two distinct Q-branches, with a few lines of moderate intensity between them. These lines do not belong to either Q-branch.

4.14. SPHERICAL ROTOR MOLECULES

The number of spherical rotors, i.e., molecules with three equal moments of inertia, is quite limited. Methane has been the one most extensively studied, and its band contours have been analyzed by a number of workers [9,10]. Coriolis perturbations have been observed to occur between the fundamentals ν_4 at 1306 cm^{-1} and ν_2 at 1526 cm^{-1}.

The fundamental ν_4, measured under high resolution, is shown in Figure 4-35 (located in pocket on inside back cover). The Coriolis interaction interferes with the regular spacing of the fine-line structure of the P-branch, and causes the high-frequency side of the R-branch to become quite irregular. The fundamental ν_2 for methane has distinct P-, Q-, and R-branches and, under medium resolution, resembles the well-formed band of CO_2 at 667 cm^{-1}.

If the Coriolis interaction did not split the lines of a spherical rotor, it would be expected that the line structure would be similar to that of a perpendicular vibration of a linear molecule, for which the spacing of the lines is $2B$. There are no subband designations similar to those discussed for the symmetric rotors.

4.15. THE ASYMMETRIC ROTOR

Since asymmetric rotor molecules have three unequal moments of inertia, unless some symmetry can be found, the band contours for these tops are too complicated to permit any conclusions to be drawn concerning the structure of the molecule.

For asymmetric rotors that have at least the symmetry associated with one of the groups C_{2v}, D_2, or D_{2h}, it is possible to recognize three types of band contours. These are called A, B, or C bands, depending on whether the change in dipole moment is along the direction of the axis of least (I_A), intermediate (I_B), or largest (I_C) principal moment of inertia. As might be expected, band perturbation can occur quite frequently for asymmetric rotors, and both Coriolis and Fermi resonances have been observed for many bands.

Inversion doubling and intensity alternation can occur in the fine-line structure of asymmetric rotor molecules. If the fine lines of the bands were not perturbed or split, we should expect the interlinear spacing to be related to the three rotational constants of the molecule. Thus, for a C band of an asymmetric rotor in which the ratio of the two moments of inertia I_A/I_B is nearly 1, the band would have the fine-line spacing $2A' = 2B' = 4C'$, where the three principal moments of inertia of the rotor are related so that $A > B > C$ ($I_A < I_B < I_C$) are the rotational constants.

An A band for a fairly symmetric molecule resembles (under medium dispersion) a parallel band of a symmetric rotor, that is, it shows somewhat symmetric P- and R-branches, with a medium to strong Q-branch. The ν_3 band for H_2O at 3756 cm^{-1} is an example of this type of band. The band for H_2CO at 1746 cm^{-1} is another example. The spectrum of H_2CO is shown in Figures 4-36 and 4-37. Using the nomenclature introduced in Section 4.4, we can describe both A bands of H_2CO as a_1 vibrations.

The presence of a B band, with its distinctive shape, is often indicative of an asymmetric molecule. The band consists of a double Q-branch, separated by a maximum at the band center, and has P- and R-branches as well. If the Q-branches are weak, a series of four approximately equal maxima can be seen: a P-branch, two Q-branches, and an R-branch. An example would be the band at 1247 cm^{-1} for H_2CO, which is also designated as a b_1 vibration for molecules of symmetry C_{2v}.

Finally, C bands can have a strong Q-branch, and P- and R-branches of moderate intensity. A band of this type occurs only for asymmetric rotors having more than three atoms. A typical C band is seen for H_2CO at 1163.5 cm^{-1}; it is a b_2 vibration for a molecule of C_{2v} symmetry.

No simple equation is presently known to describe the energy levels of the asymmetric rotor; hence no simple expressions for the fine-line structure of absorption bands can be given. Several approximate equations have been suggested, but we shall not present them at this time.

4.16. SPECTRAL ANALYSIS OF FORMALDEHYDE [11] (NEAR-SYMMETRIC ROTOR)

The formaldehyde molecule (H_2CO) can be assumed to have a structure in which the two hydrogen nuclei are located symmetrically

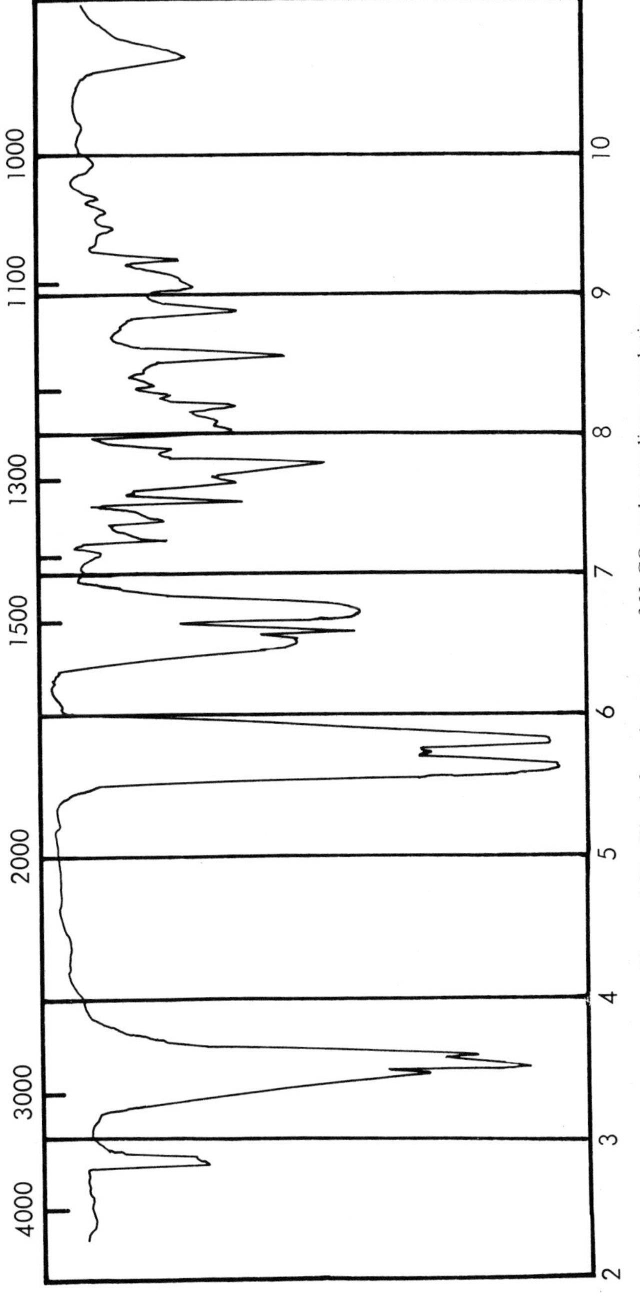

Figure 4-36. The infrared spectrum of H_2CO under medium resolution.

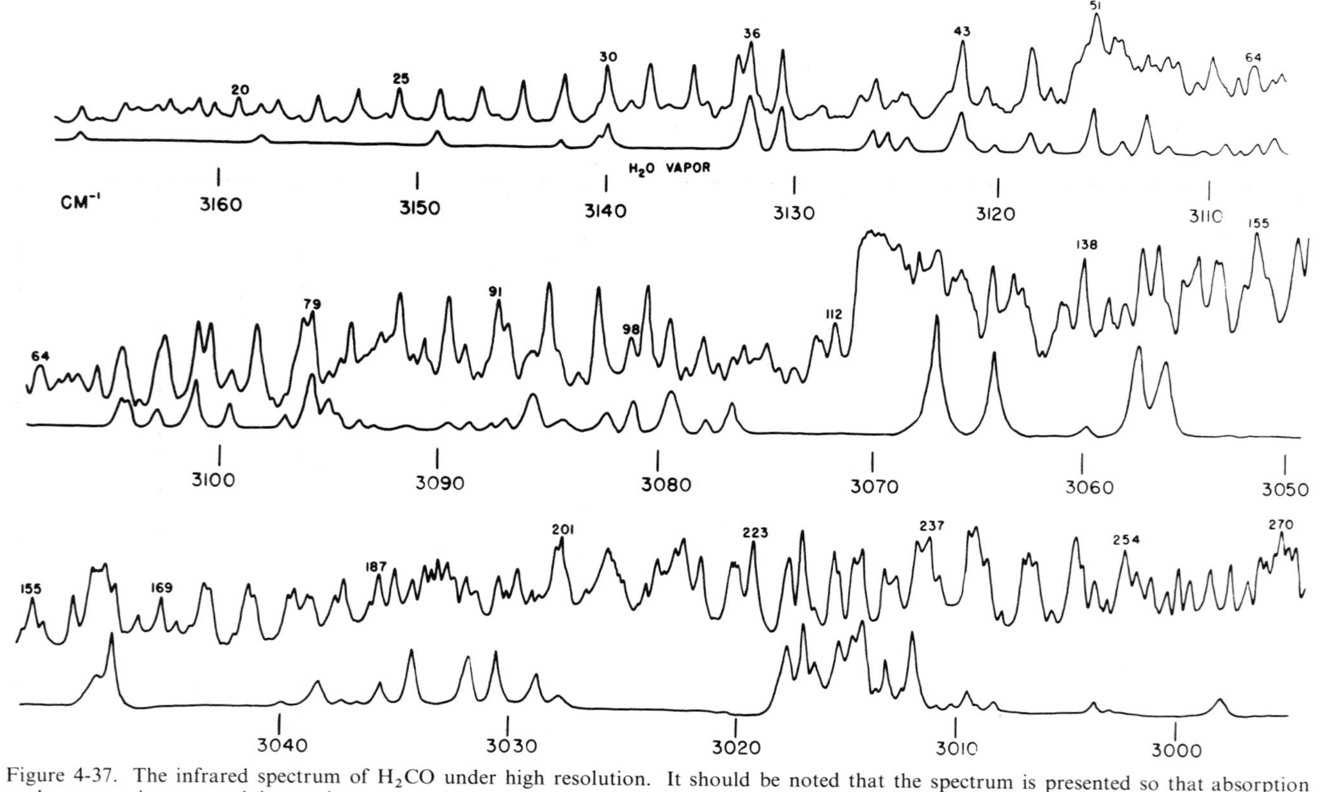

Figure 4-37. The infrared spectrum of H_2CO under high resolution. It should be noted that the spectrum is presented so that absorption peaks are maxima, not minima as in most previous spectra. [Reproduced with permission partly from the Ph.D. thesis of H. H. Blau, Ohio State University, and partly from *J. Mol. Spectroscopy* **1**(2): 124 (1957).]

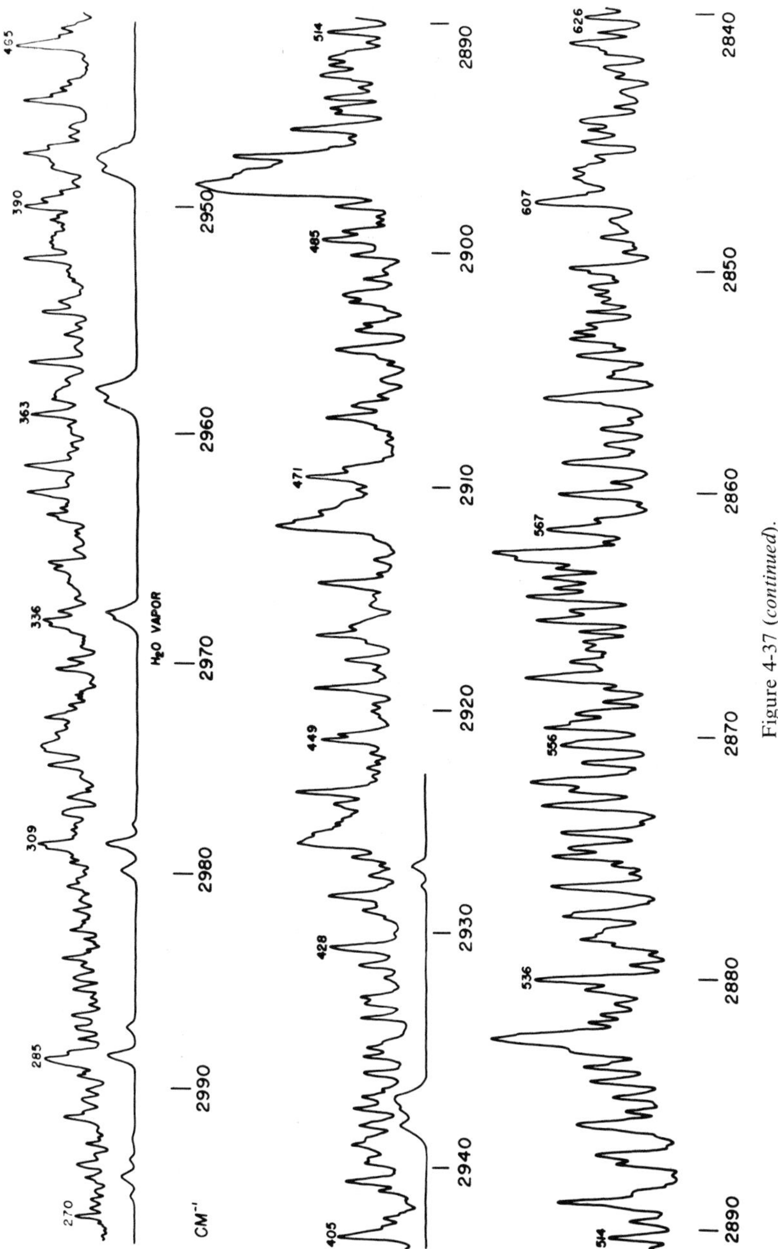

Figure 4-37 (*continued*).

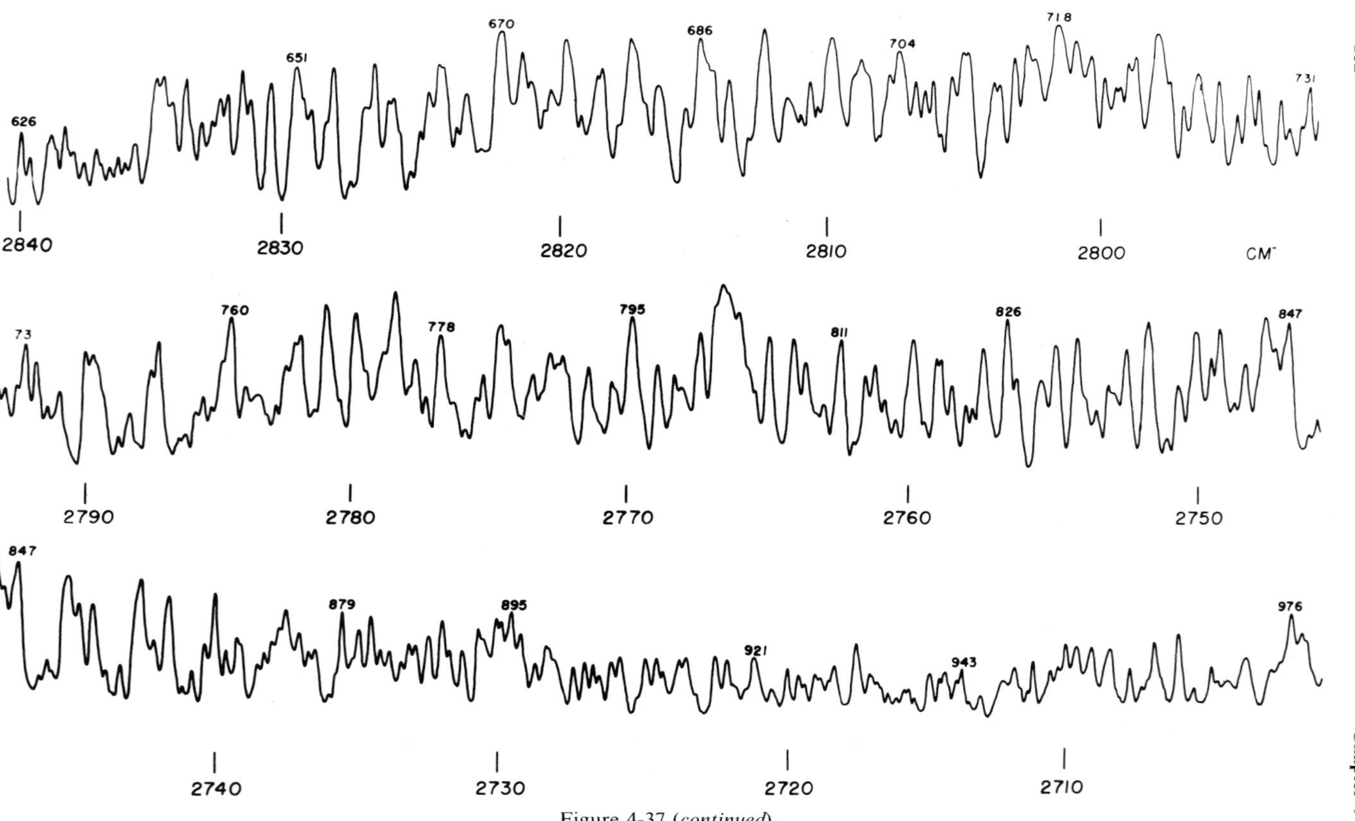

Figure 4-37 *(continued)*.

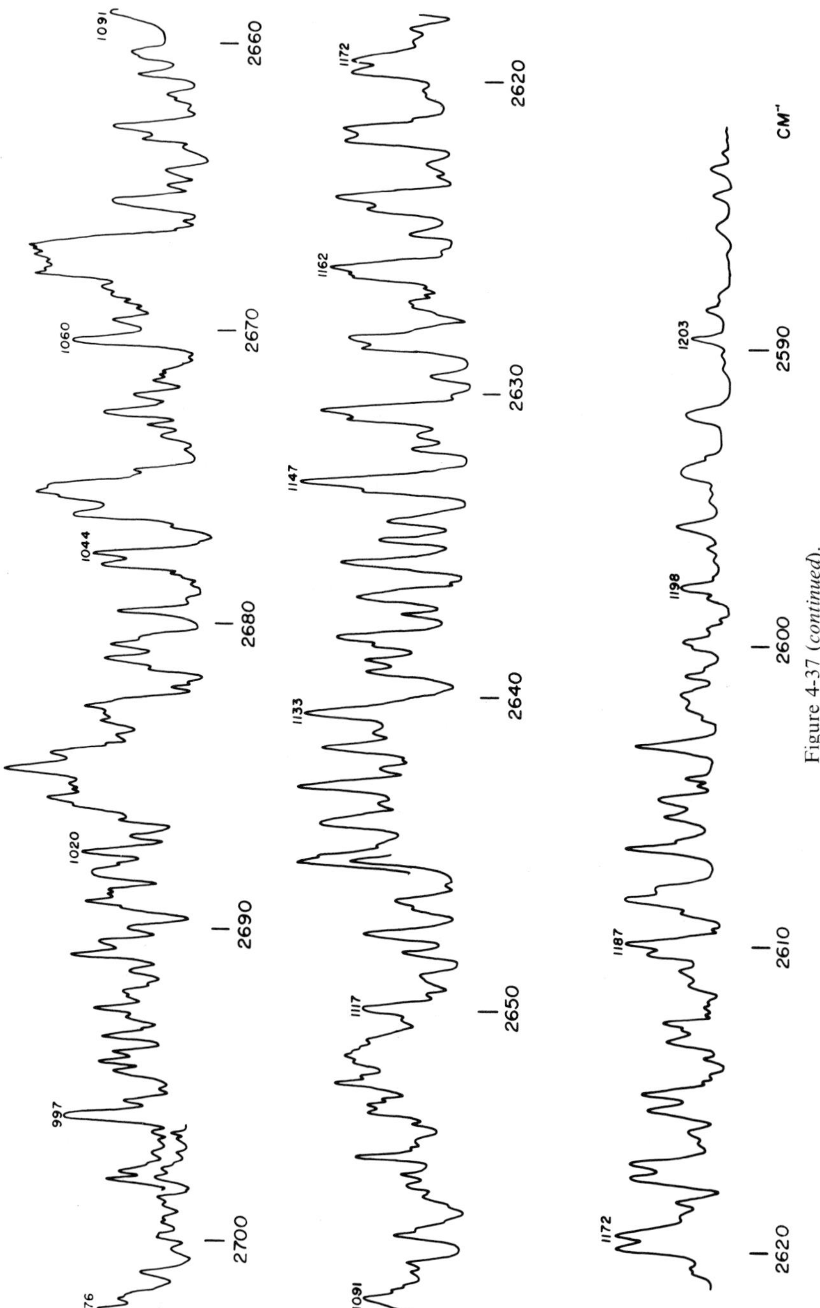

Figure 4-37 *(continued)*.

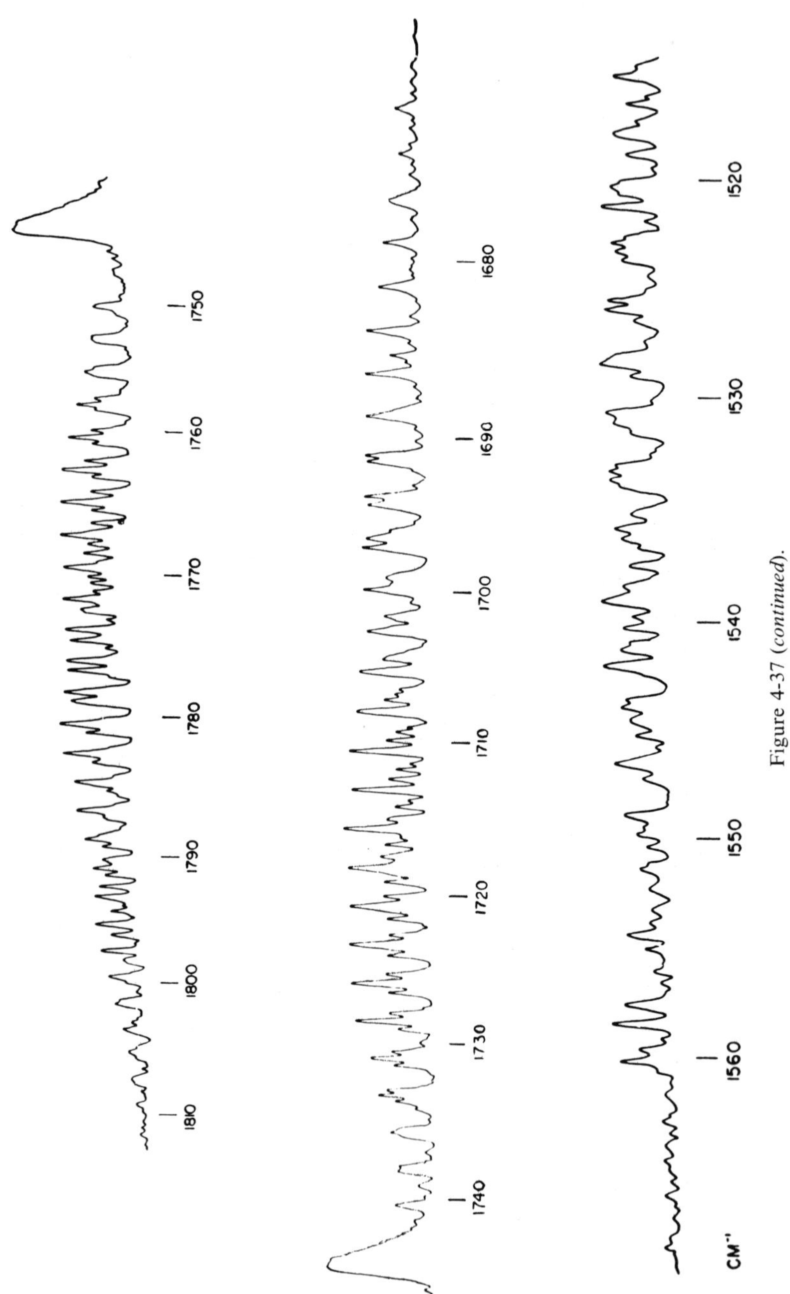

Figure 4-37 (*continued*).

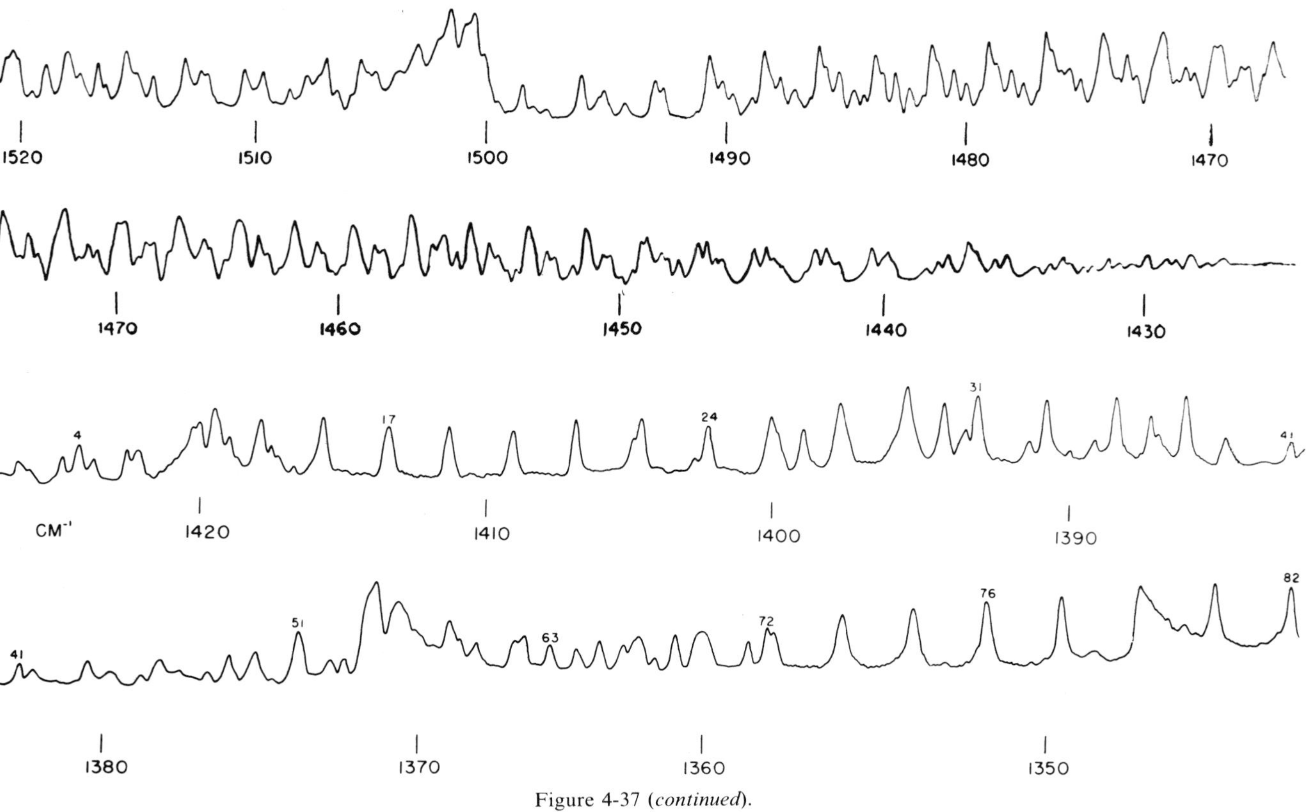

Figure 4-37 (*continued*).

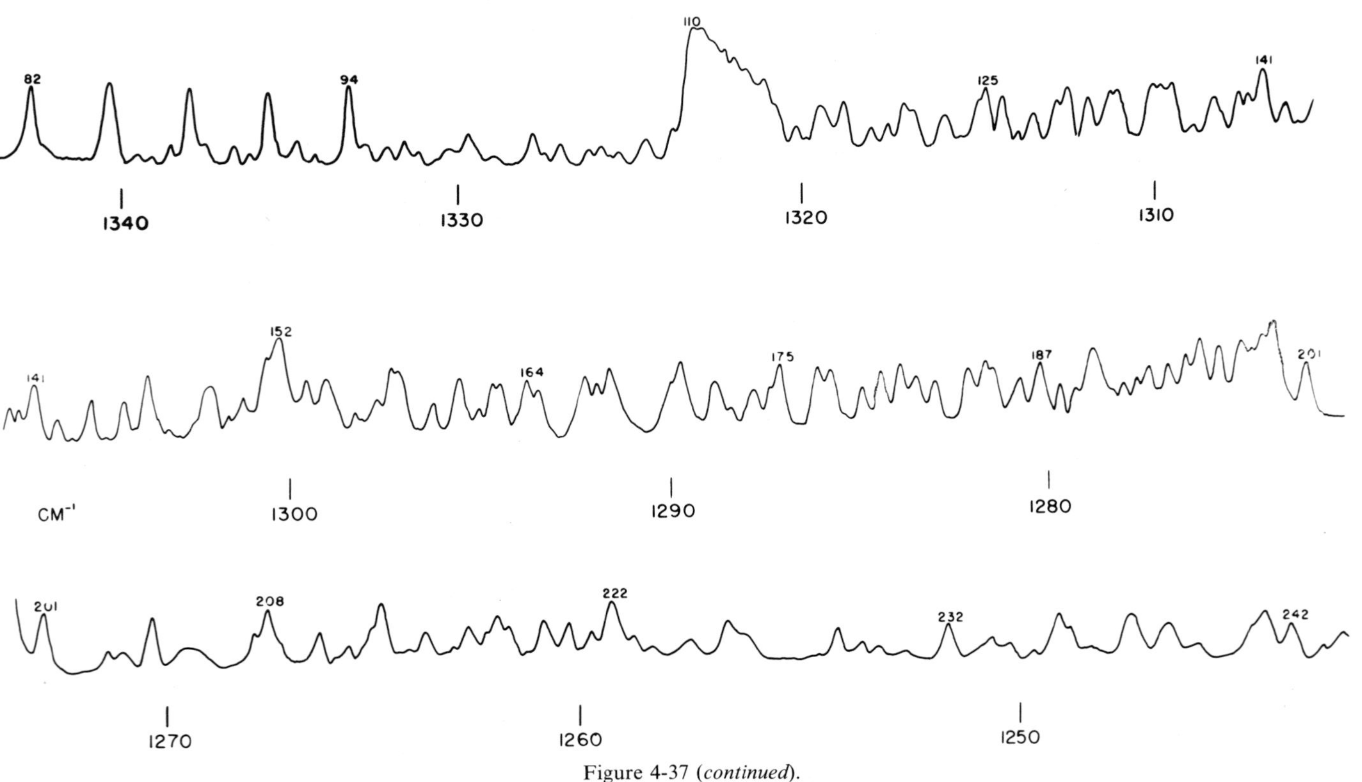

Figure 4-37 (*continued*).

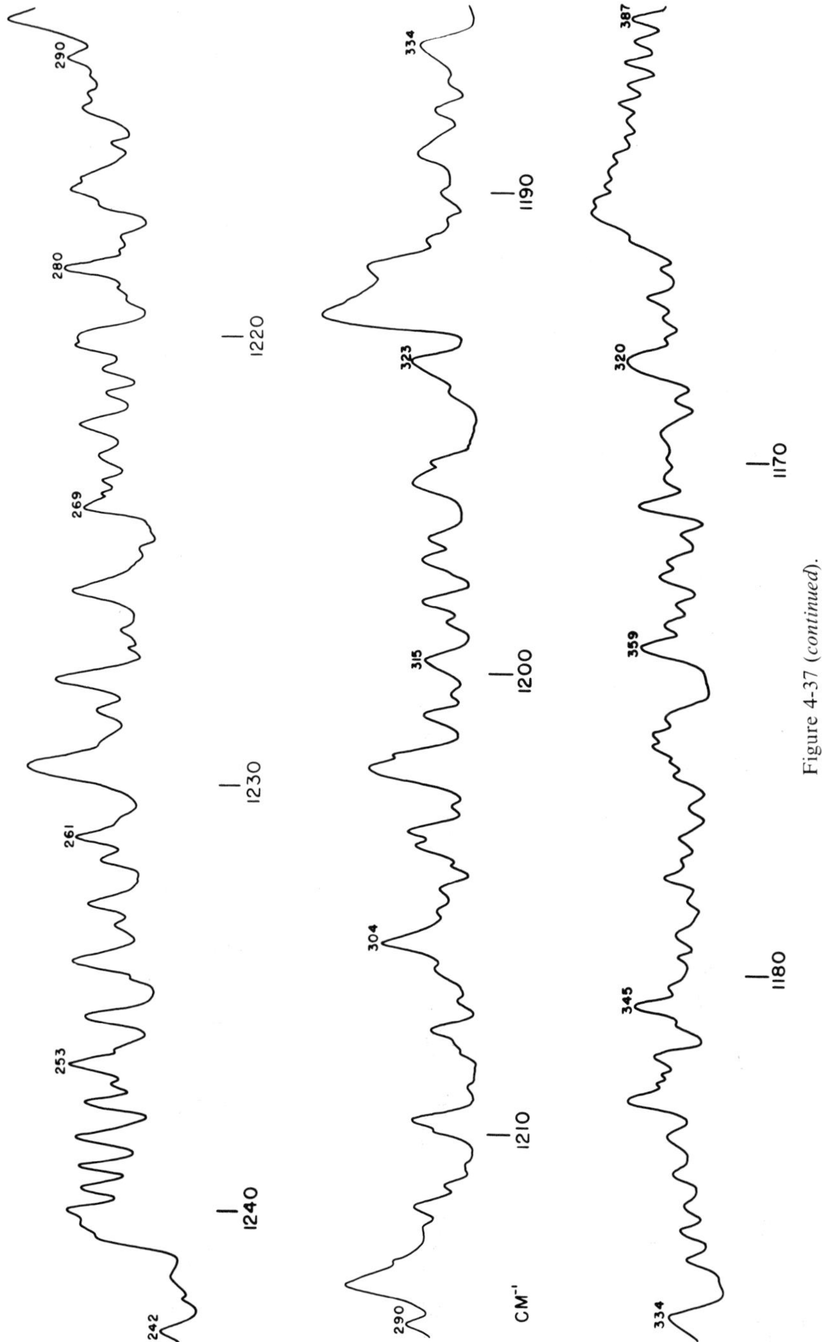

Figure 4-37 *(continued)*.

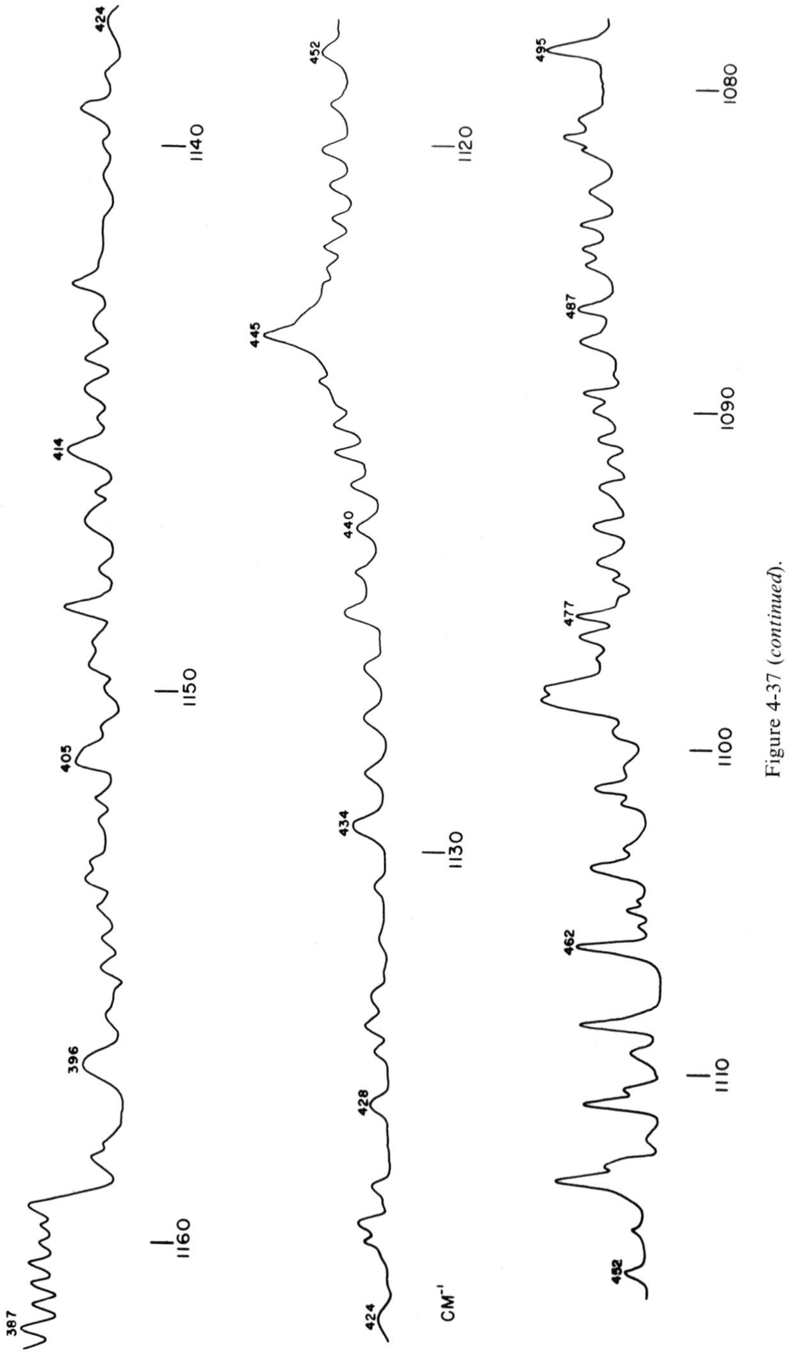

Figure 4-37 (*continued*).

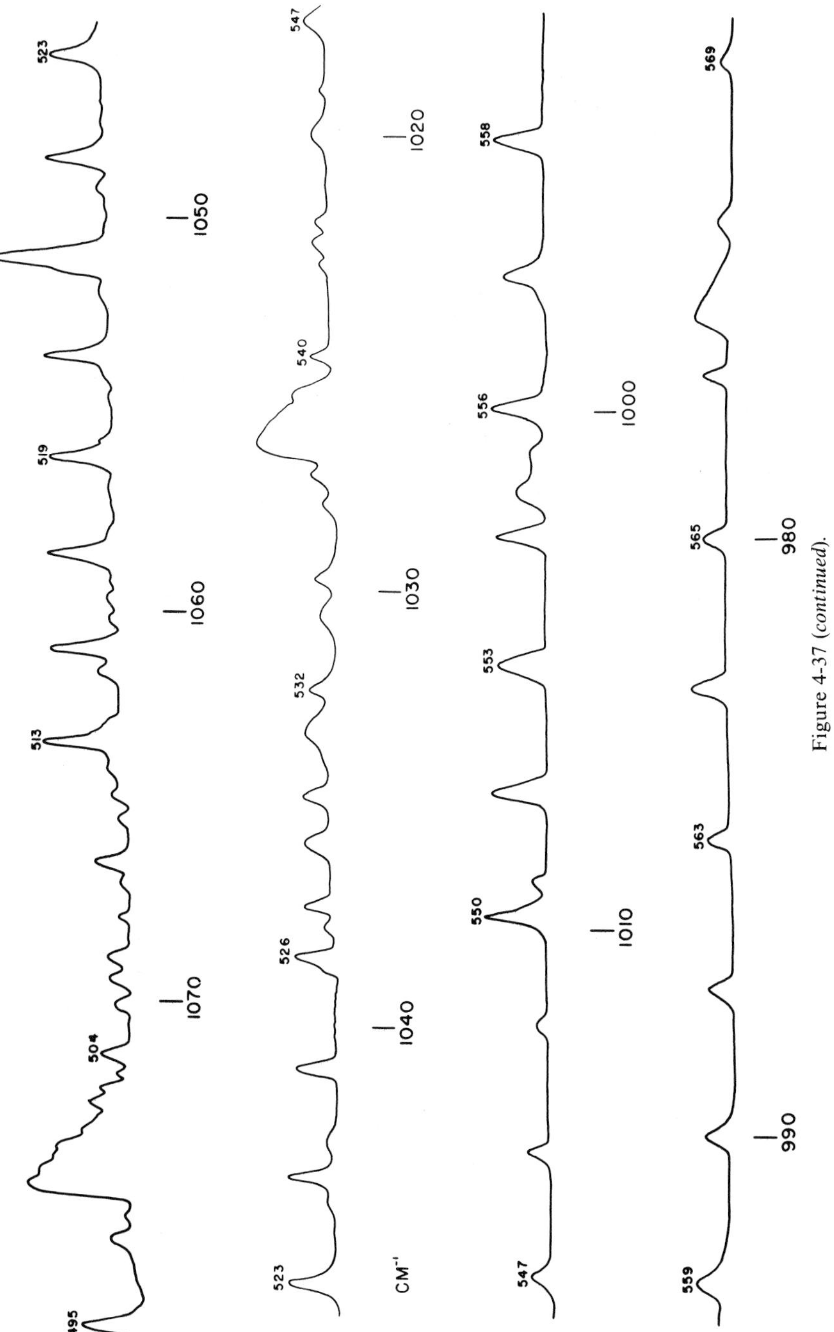

Figure 4-37 (*continued*).

about an axis passing through the carbon and oxygen atoms. Such a symmetry would classify it as belonging to the C_{2v} group. Strictly, the molecule should be considered as an asymmetric rotor. However, two of the moments of inertia are large and nearly equal, while the third is small ($I_A < I_B \approx I_C$), so that the molecule has a spectrum similar to that of a symmetric rotor.

The number of fundamentals can be calculated from the formula $3N - 6$ as follows: $3N - 6 = 3 \times 4 - 6 = 6$.

Three of the six fundamentals give A bands, since the change in dipole moment for these three is along the axis of the least moment of inertia. Two of the fundamentals have B-band contours, since the change in the dipole moment is along the axis of the intermediate moment of inertia I_B. The last fundamental is a C band, since the change in the dipole moment is along the axis of the largest moment of inertia. All are infrared active.

The fundamentals may be presented as shown in Figure 4-38. It should be noted that the fundamentals ν_1, ν_2, and ν_3 are all parallel a_1 vibrations; the fundamentals ν_4 and ν_5 are b_1 vibrations; and ν_6 is a b_2 vibration.

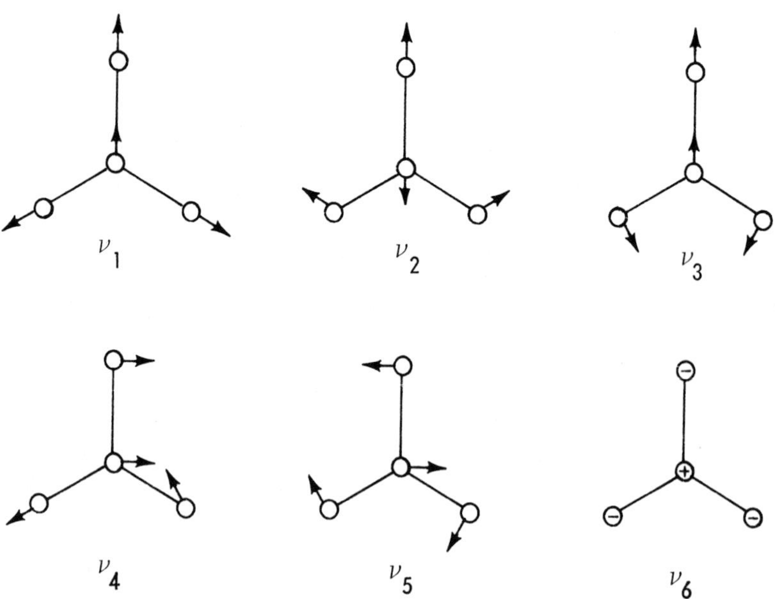

Figure 4-38. The fundamentals of H_2CO.

It can be shown that H_2CO is not a pyramidal molecule with one plane of symmetry by observing the 3:1 intensity alternation of the fine-line structure of some of its bands, which indicates that there are two planes of symmetry in the molecule. The 3:1 intensity alternation is due to the fact that the normal state energy levels with odd K have three times the statistical weight of states with even K. Therefore, subbands originating from *even-K* energy levels will be *weak*, while those from *odd-K* levels will be *strong*.

Earlier, the concepts of planes of symmetry and reflection were described for various symmetry groups. If the model of H_2CO is studied, it can be shown that the ν_1, ν_2, and ν_3 vibrations are symmetric with respect to both symmetry planes in H_2CO. The ν_4 and ν_5 vibrations are symmetric with respect to the plane of symmetry in which the molecule lies and antisymmetric with respect to the second plane. The ν_6 vibration is antisymmetric with respect to the first reflection plane and symmetric with respect to the second, so that its symmetry is opposite to that of ν_4 and ν_5.

In the discussion that follows, it will be shown that ν_3 and ν_5 are coupled by a rotational resonance interaction which changes the rotational structure of ν_3 from that expected for an A band. The rotational structure of the ν_5 band is strongly perturbed by a rotational resonance interaction with the fundamental ν_6, and so the line structure of ν_5 is determined by the interaction between ν_5 and ν_6. This interaction between ν_5 and ν_6 is of the Coriolis resonance type.

We shall also discuss a perturbation which occurs between the combination band $\nu_2 + \nu_5$ and the fundamental ν_4. This is a Fermi resonance interaction and causes the combination band to be displaced from its normal position.

Some of the fine-line structure of the bands for H_2CO can be analyzed by using the theory developed for symmetric rotors, but other structures require the asymmetric rotor theory. Generally, rotational levels identified with small values of K require consideration of the molecule as an asymmetric rotor.

We can explain how the influence of asymmetry will manifest itself in the H_2CO spectrum as follows. For a symmetric rotor each energy level, except $K = 0$, is doubly degenerate. If a molecule is asymmetric, this degeneracy may not exist, and consequently the rotational line structure may show two lines in place of the single line predicted for the corresponding doubly degenerate level of the symmetric rotor. Since H_2CO is nearly symmetric, the splitting of the degeneracy is seen only for values of $K < 3$. Thus, any analysis of lines of $K < 3$ must be done by means of an asymmetric rotor calculation.

In the following discussion, the relationship $I_A < I_B < I_C$ will be assumed for the three moments of inertia of H_2CO, and we will further assume that I_A is the moment of inertia about the C═O axis, I_B the moment about the axis in the plane of the molecule and perpendicular to the C═O axis, and I_C the moment of inertia about the axis perpendicular to the plane of the molecule and to the C═O symmetry axis. If we assume H_2CO to be a symmetric rotor, then $I_B = I_C$.

Further, if we suppose that the symmetric rotor approximation is valid, then for the parallel vibrations (ν_1, ν_2, ν_3) of H_2CO the selection rules for the transitions will be

$$\left.\begin{aligned} \Delta K &= 0 \\ \Delta J &= 0, \pm 1 \end{aligned}\right\} \text{for } K \neq 0 \text{ and } J \geq K$$

$$\left.\begin{aligned} \Delta K &= 0 \\ \Delta J &= \pm 1 \end{aligned}\right\} \text{for } K = 0$$

For the perpendicular vibrations (ν_4, ν_5, ν_6), where the oscillating moment is perpendicular to the symmetry axis, the selection rules are

$$\left.\begin{aligned} \Delta K &= \pm 1 \\ \Delta J &= 0, \pm 1 \end{aligned}\right\} \text{for } J \geq K$$

The term value equation for these symmetric rotor considerations will be similar to equation (4-47), and will be

$$F = AJ(J + 1) + [A - \tfrac{1}{2}(B + C)]K^2$$

where

$$A = \frac{h}{8\pi^2 c I_A} > B = \frac{h}{8\pi^2 c I_B} \approx C = \frac{h}{8\pi^2 c I_C}$$

For parallel vibrations, if A and $\frac{1}{2}(B + C)$ for both the upper and lower state are small, the subbands will nearly coincide, and the total band will appear to have distinct P-, Q-, and R-branches.

For perpendicular bands the subbands do not coincide and, if $A \gg \frac{1}{2}(B + C)$, then the Q-branches stand out rather prominently against the weaker P- and R-lines of the various subbands.

4.16A. The 3-4 μ Region*

A number of overlapping bands are present in the 3–4 μ region, which contains two fundamentals (ν_1 and ν_4) and five overtone and combination bands.

* Throughout the ensuing vibrational analysis of H_2CO, the reader should refer to Figure 4-37, which was taken in part from the doctoral thesis of H. H. Blau (Ohio State University), and in part from the work of Blau and Nielsen [11]. The small numerals associated with some of the peaks refer to tabular data contained in the original thesis.

The fundamental ν_4 is centered at 2843.41 cm^{-1} and stands out in the absorption pattern of this region. Although it is a B band which is hidden by an overlapping band on the low-frequency side, the Q-branch structure and the associated R-branch structure stand out on the high-frequency side.

From the subband lines, such as R_{R_3}, R_{R_4}, etc., we can calculate some of the subband centers and the upper state $\frac{1}{2}(B' + C')$ values, assuming the molecule to be a symmetric rotor. First, however, a value for the lower state $\frac{1}{2}(B'' + C'')$ must be obtained from other data so that the upper and lower state terms may be separated from each other, thus yielding a value for $\frac{1}{2}(B' + C')$. The expression relating the lines of the subband to the terms $\frac{1}{2}(B' + C')$ and $\frac{1}{2}(B'' + C'')$ is

$$R_{R_J} = \nu_0^{\text{sub}} + (B' + B'') + \frac{3(B' + C')}{2} - \frac{(B'' + C'')J}{2} + \frac{(B' - B'' + C' - C'')J^2}{2} \quad (4\text{-}55)$$

A term $[A' - \frac{1}{2}(B' + C')] - [A'' - \frac{1}{2}(B'' + C'')]K^2$ has been omitted from this expression, since it may be considered to be negligibly small. If various values of R_{R_J} are now substituted into equation (4-55) and a value for $\frac{1}{2}(B'' + C'')$ derived from other data is employed, ν_0^{sub} may be calculated.

Having obtained ν_0^{sub}, we can then calculate ν_0 and A' from the expression

$$\nu_0^{\text{sub}} = \nu_0 + A' - \tfrac{1}{2}(B' + C') \pm 2[A' - \tfrac{1}{2}(B' + C')]K + [A' - \tfrac{1}{2}(B' + C')] - [A'' - \tfrac{1}{2}(B'' + C'')]K^2 \quad (4\text{-}56)$$

In this equation the plus sign applies to $\Delta K = +1$ transitions and the minus sign to $\Delta K = -1$ transitions.

On the high-frequency side of ν_4 is the combination band $\nu_2 + \nu_5$. The two bands overlap, so that again only R_R and R_Q lines can be identified. The $\nu_2 + \nu_5$ combination band can be analyzed in the manner described for the ν_4 band; A', $\frac{1}{2}(B' + C')$, and ν_0 values can be determined, assuming that the symmetric rotor model is applicable. The $\nu_2 + \nu_5$ band is quite intense, in fact about one-fourth as intense as the fundamental ν_4. This suggests that some sort of interaction is occurring; however, since no anomalous spacing of the ν_0^{sub} positions is found, the interaction is probably not a Coriolis resonance.

If the observed values of ν_2 and ν_5 are used and anharmonicity is neglected, the calculated position of the $\nu_2 + \nu_5$ band center is

2993.19 cm^{-1}. The observed value of 3003.29 cm^{-1} is somewhat higher, which suggests that Fermi resonance exists between the combination band and ν_4. The choice of ν_4 is based on the fact that it is the nearest band of the same vibrational species.

The fundamental ν_1 is also in the 3–4 μ region. It is overlapped by the lines of the fundamental ν_4, but it appears to be at 2766.39 cm^{-1}. It is an A band and closely resembles a parallel band of a symmetric rotor. Such a band usually has a strong but unresolved Q-branch, and since the strong line at 2766.39 cm^{-1} resembles a Q-branch of this type, it is assigned as the band center for the ν_1 fundamental.

4.16B. The 5.7 μ Region

The band in the 5.7 μ region is a single A band centered at 1746.07 cm^{-1} and is the ν_2 fundamental. It is a parallel type of band of a slightly asymmetric rotor, resembling the parallel band of a symmetric rotor, but has a triplet structure in the P- and R-branches due to the asymmetry of the molecule. The effects of asymmetry are most pronounced for low K and high J values. The triplet structure of the band consists of a strong center line flanked by a weaker component on either side. The different intensities of the three lines can be explained in terms of an intensity alternation. A 3:1 intensity alternation occurs when the subbands originating from odd K states have strong lines and those from even K states have weak lines. The strong line of the triplet consists of the lines from the subbands $K > 3$. The two weak components of the triplet are primarily the $K = 1$ transitions. The splitting of these component lines increases with increasing J, as would be expected from the asymmetry of the molecule. The $K = 2$ lines cannot be resolved.

The $K = 0$ sublevel is not displaced quite as much as the $K = 1$ levels and the lines of this subband should fall between the two components of the $K = 1$ subband. In the band structure, the $K = 0$ lines are part of the strong central line until values of J near 8, at which point the lines begin to split away from the central line. A number of weak lines on either side of the Q-branch, which do not fall within the $K = 1$ doublets, are part of the Q-branch lines of the $K = 0$ or $K = 1$ transitions.

By use of the combination relation

$$R(J - 1) + P(J) = 2\nu_0 + [(B' + C') - (B'' + C'')]J^2 \qquad (4\text{-}57)$$

the term $[(B' + C') - (B'' + C'')]$ and the value of ν_0 can be determined. The calculated value of ν_0 is 1746.07 cm^{-1}, which is the band center of the ν_2 fundamental. The value of $(B'' + C'')$ is known from microwave data, so that the value of $(B' + C')$ can be calculated.

4.16C. The 6–7 μ Region

The fundamental ν_3, which is an A band, appears at 1500.6 cm^{-1}. Water vapor absorption makes it difficult to examine the fine-line structure of this fundamental. The spectrum shown in Figure 4-37 has been measured in a vacuum spectrophotometer, so that the absorption from water appears only at the extreme edge of the R-branch of the formaldehyde band.

The structure of the ν_3 fundamental is quite different from that of ν_2. The Q-branch is partially resolved and the intensity decreases toward the higher-frequency side. This reversal of the situation for the Q-branch implies the following relation for the rotational constant term:

$$\tfrac{1}{2}(B' + C') > \tfrac{1}{2}(B'' + C'')$$

and consequently the R-branch diverges while the P-branch converges. No triplet structure is found in the lines of the band.

The fundamental ν_3 is coupled by rotational resonance to the ν_5 fundamental (and also, to a smaller extent, to ν_6). This interaction is what makes the band structure of ν_3 different from that of a simple A band. Here again we are dealing with a Coriolis resonance.

Using the approximation that the molecule is a symmetric rotor, and again using relation (4-57), we can calculate the term values of ν_0 and $(B' + C') - (B'' + C'')$. The calculated value of ν_0 is 1500.6 cm^{-1}.

4.16D. The 7.5–10 μ Region

The fundamentals ν_5 and ν_6, which are B and C bands, respectively, appear in this region. The two fundamentals overlap in the central portion of the spectrum. Subbands originating from odd K values of normal states are more intense than those originating from even K-states, giving rise to an intensity alternation of 3:1. In the spectrum shown, it is possible to identify lines as originating from even or odd K-states by their intensity. The two fundamentals ν_5 and ν_6 are at 1247.44 cm^{-1} and 1163.49 cm^{-1}, respectively. They interact strongly with a rotational resonance of the Coriolis type, which causes an uneven spacing of some of the lines in the band. The Coriolis effect shows up where the two bands overlap. The Q lines converge strongly, while in the band extremities they diverge abnormally.

The two bands have a similar fine-line structure in the regions where the effects of asymmetry are small, that is, for large K values. Where the effects of asymmetry are important, the selection rules differ for the two types of bands, and the rotational line structure is perturbed for low values of *K*. For example, in the two central Q-branches of ν_5, the lines spread away from the center, leaving the center relatively free of lines; for ν_6, on the other hand, the lines crowd together in the two central Q-branches. This difference helps us to determine that the lower-frequency band is a *C* band and therefore the fundamental ν_6.

4.17. FUNDAMENTAL FREQUENCIES OF THE MOLECULES $CF_2{=}CH_2$, $CF_2{=}CHD$, and $CF_2{=}CD_2$ [14]

The series of molecules $CF_2{=}CH_2$, $CF_2{=}CHD$, and $CF_2{=}CD_2$ is illustrative of how a small change in mass can influence both band positions and band contours. The positions of the fundamentals of these three molecules are given in Figure 4-39. This figure is presented in the form of a correlation chart in which the vibrations of similar species are connected by correlation lines. Symmetric vibrations are connected by solid correlation lines, while asymmetric vibrations are connected by broken lines. For $CF_2{=}CH_2$ and $CF_2{=}CD_2$, the symmetry is C_{2v}. For these molecules, there will be four types of vibrations: a_1, a_2, b_1, and b_2. Of these, the a_2 vibration is not allowed in the infrared but is allowed in the Raman spectrum. The a_1 vibrations for $CF_2{=}CD_2$ are the symmetrical CD_2, CF_2, and C=C stretching and the CD_2 and CF_2 deformation frequencies. The b_1 vibrations are the antisymmetric CD_2 and CF_2 stretching and the rocking of the CF_2 and CD_2 groups. The b_2 vibrations are the in- and out-of-plane wagging of the CF_2 and CD_2 groups.

The molecule $CF_2{=}CHD$ has the symmetry C_s; therefore the normal vibrations are classified as either a' or a''.

If we trace these vibrations in the correlation chart, the following conclusions can be drawn: First, from the assignments of the vibrational frequencies of $CF_2{=}CH_2$ and $CF_2{=}CD_2$, the assignments for the less symmetric $CF_2{=}CHD$ can be verified. Second, the stretching, rocking deformation, and wagging vibrations of the CF_2 group have similar values in all of the three molecules (as would be expected for vibrations of this type) and the masses of the attached group do not greatly influence the band position.

The wavenumbers of the wagging vibrations of the CD_2 and CHD groups differ by 110 cm^{-1}. However, if the band contours are compared, both are quite distinct *C* bands. Thus, in many instances the band contours are more useful for recognizing similar vibrations than are the wavenumbers. Some of the band contours observed for

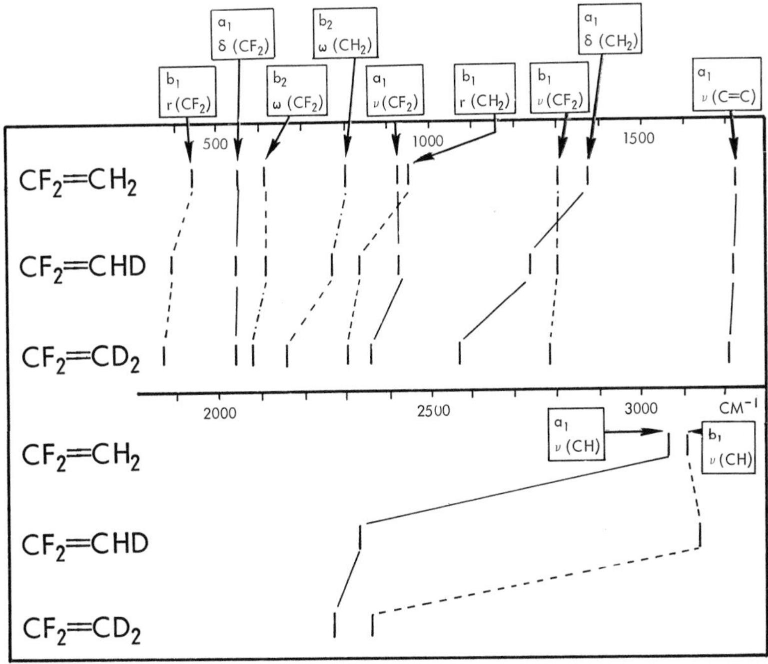

Figure 4-39. Correlation of the fundamentals of $CF_2{=}CH_2$, $CF_2{=}CHD$, and $CF_2{=}CD_2$. [Reproduced with permission from *J. Chem. Phys.* **22**(12): 1983 (1954).]

fundamentals for $CF_2{=}CD_2$ are presented in Figure 4-40. Frequently there are no simple contours that are always found for a distinct vibration. For example, the a_1 vibration of $CF_2{=}CD_2$ will have a band contour that is somewhat related to the contour observed for an a_1 vibration of another molecule. However, the contours can be different enough to make it difficult to identify the vibration as a_1. For comparison, the a_1 vibrations for NF_3 are presented in Figure 4-41. For both $CF_2{=}CD_2$ and NF_3, the P-, Q-, and R-branches of the bands are fairly distinct and the general contours are somewhat similar; however, the differences in the contours are sufficiently large to preclude positive identification of these vibrations as a_1 from this evidence alone.

Perhaps the most distinct type of band is the *C* band associated with a change in the dipole moment along the axis of the least moment of inertia. For many molecules, this band has a distinct Q-branch, with weak and indistinct P- and R-branches. An example

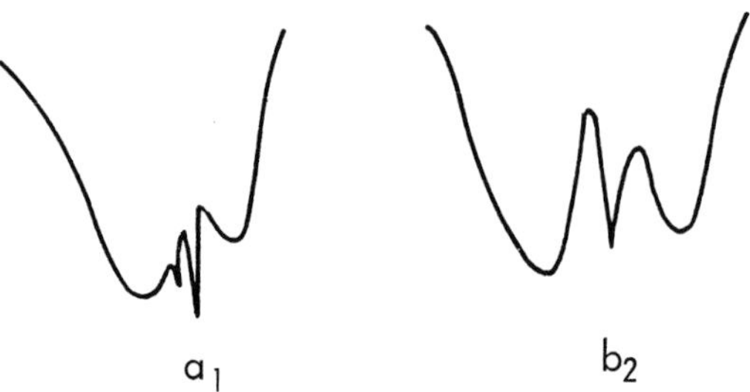

Figure 4-40. Band contours for $CF_2{=}CD_2$ (redrawn from data in Ref. 14).

of such a band is the vibration at 786 cm^{-1} for C_2H_5Cl, shown in Figure 4-42.

4.18. THE INFLUENCE OF ISOTOPIC SUBSTITUTION ON BAND POSITION

If the bond strength in a diatomic molecule may be assumed to remain unchanged when an isotope is substituted for one of the atoms, the ratio of the frequencies of the band position before and after the substitution is given, as has been shown earlier, by the formula

$$\frac{\nu_1}{\nu_2} = \sqrt{\frac{\mu_2}{\mu_1}} \tag{4-58}$$

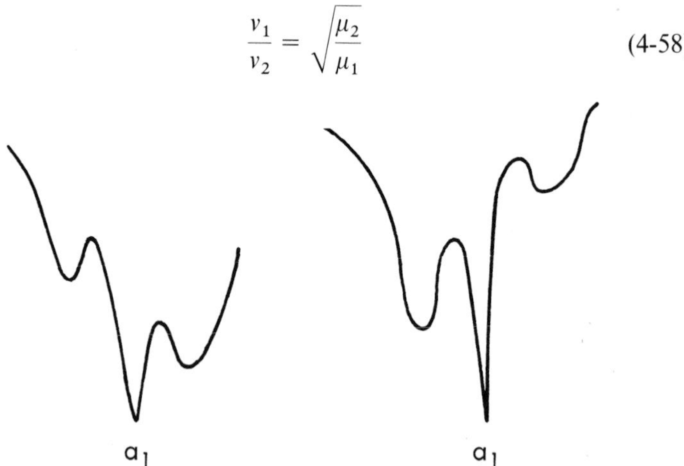

Figure 4-41. Band contours for NF_3 (redrawn from data in Ref. 15).

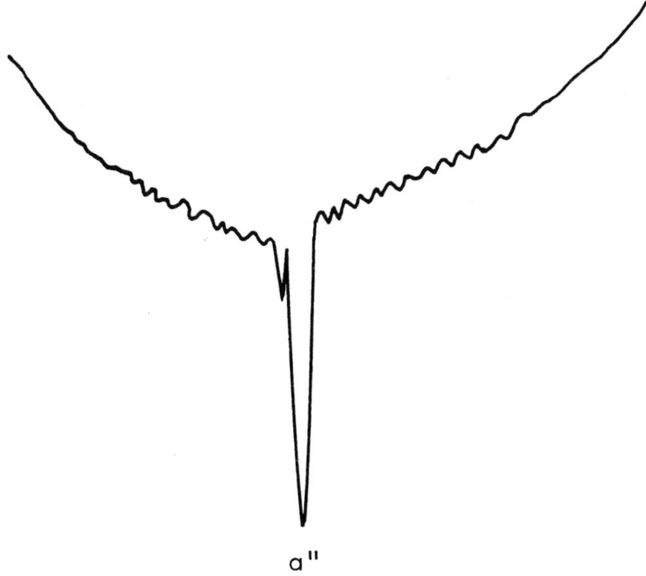

Figure 4-42. A *C*-type band (redrawn from data in Ref. 13).

This formula can be used to calculate the approximate wavenumber shift that occurs when an isotope is substituted in a diatomic molecule. It is only approximately correct for diatomic molecules. For other than diatomic molecules—where the masses of many atoms may be involved in a vibration, and where bending as well as stretching vibrations can occur—the use of this equation is even less satisfactory.

The isotopic shift of a given vibrational frequency will be small if the atoms replaced by the isotopes participate only to a small extent in the vibration, but the shift will be large if these atoms play a major role. Thus, the shift an isotope produces can be used as a measure of the extent to which a particular atom participates in a vibration, provided no factors other than the change in mass are involved in the observed shift. A phenomenon such as Fermi resonance may be enhanced when an isotope is substituted into a molecule, if the shift brings frequencies more nearly into coincidence.

We saw earlier that the frequency ν_{osc} at which a molecule would vibrate if it were a harmonic oscillator can be calculated from the observed position of the fundamental and overtone bands. If this frequency is first calculated for the normal molecule and then for the molecule in which the isotope has been substituted, it is found, as expected, that these harmonic oscillator frequencies obey an equation of the form of (4-58) more closely than the observed frequencies.

For fairly symmetric polyatomic molecules it is possible to obtain an alternative expression for the frequency (or wavenumber) ratio between the normal and the isotopic molecule. For example, for the antisymmetric ν_3 frequencies of H_2O and D_2O this ratio is

$$\left(\frac{\nu_{D_2O}}{\nu_{H_2O}}\right)^2 = \frac{m_O \cdot m_H(m_O + 2m_D \sin^2\alpha)}{m_O \cdot m_D(m_O + 2m_H \sin^2\alpha)} \tag{4-59}$$

where ν_{D_2O} and ν_{H_2O} are the harmonic oscillator frequencies of the vibration of D_2O and H_2O, m_O is the atomic mass of oxygen, m_H is the atomic mass of hydrogen, and α is one-half the apex angle of the molecule.

Other expressions can be derived for the ν_2 and ν_1 vibrations of the H_2O molecule. For other molecules the expressions may be more complex. A rule has been suggested by Teller and Redlich which has been useful in the study of the isotope effect. It will not be discussed here, however, and the reader is referred to the text by Herzberg [16] for a full presentation.

4.19. CONCLUSION

In a brief account, such as presented in this text, one cannot hope to cover the theoretical foundation of Raman and infrared spectroscopy in its entirety. In concluding this chapter, we should like to call to the reader's attention several concepts which, although not discussed in the previous sections, are nonetheless requisite to a complete understanding of the subject.

1. The theoretical prediction of band contours and the separation expected between the P- and R-branches of absorption bands having no fine-line structure are discussed in two important papers [17,18].
2. The calculation of thermodynamic quantities from spectroscopic data is discussed in most advanced texts on thermodynamics.
3. The Wilson *F–G* matrix method for calculating force constants is discussed in most texts on theoretical spectroscopy.
4. The theory of Raman spectroscopy, which is complementary to the theory of infrared spectroscopy, is discussed in a number of advanced texts on spectroscopy.
5. The use of Urey–Bradley force fields has become increasingly popular recently.

In general, the assignment of the vibrational frequencies can be accomplished in many ways. For example, of the following six methods, only the first two have been presented in this text.

1. Determination of infrared and Raman selection rules for the fundamentals.
2. Study of the infrared band contours.
3. Measurement of the depolarization ratios of Raman lines.
4. Study of the selection rule changes that can occur when the spectrum of a molecule is investigated in gaseous and condensed phases.
5. Study of the Raman spectra of single crystals using polarized incident light.
6. Study of single crystals using polarized infrared radiation and a measurement of the absorption spectrum.

All of these techniques are useful and many papers can be found discussing each in detail.

Since this text is intended for readers previously unfamiliar with infrared spectroscopy in general and especially with such theoretical aspects as are presented in this chapter, some general remarks concerning the limitations of theoretical calculations may be in order. Let us consider specifically the problem of determining the structure of a molecule by means of infrared and Raman spectroscopy. It must be pointed out that the structure cannot always be unambiguously assigned, even if experimental conditions are ideal. When experimental conditions are such that vapor state spectra cannot be obtained, further difficulties are encountered in assigning the structure and/or the vibrations of the molecule. For example, selection rules which indicate that an a_2 vibration may not be infrared active may break down in the liquid and/or solid state.

Normal coordinate analysis always involves more force constants than observed frequencies, and a degree of arbitrariness must therefore be introduced. While, in general, one can calculate a set of force constants that are reasonable and will reproduce the observed frequencies, the assignment of a specific frequency to a particular vibration of a molecule is not always possible. For example, on basis of a Urey–Bradley potential energy function the CD_2 deformation vibration in CD_2Cl_2 has been assigned to a band observed at $995\ cm^{-1}$. However, on basis of a general valence field calculation the same vibration has also been assigned to a band observed at $1052\ cm^{-1}$ [22]. Further work on this was reported recently [23]. The multiplicity of force constants for large molecules is given a thorough analysis in recent papers by E. J. O'Reilly [30], and R. G. Parr [31].

It is often possible to assign the fundamentals of one molecule by utilizing force constants calculated for a similar molecule. By combining these data with intensity measurements, knowledge of group frequencies, depolarization ratios measured in the Raman spectrum, vapor band contours, and the position of bands in the various physical states, it is generally possible to make a reasonable vibrational assignment for most small molecules that have some symmetry. The two general approaches, the Urey–Bradley potential field method and the general valence field method, need to be further amplified before vibrational assignments can be made quite accurately. Perhaps the future development in this field will lie in a middle ground between these two approaches.

According to the definition of Fermi resonance given in this chapter, this phenomenon should occur whenever two energy states of the same species are nearly coincident. However, there are many instances of vibrations nearly coincident and of the same species for which no interaction is observed. Clearly, further work must be done to explain this occurrence. Often in solid and/or liquid state spectra interactions occur which cannot be clearly assigned. This phenomenon also requires further clarification.

These are but some of the present limitations that serve to indicate what direction future theoretical work in infrared spectroscopy will have to take. One important concept in spectroscopy, that of lattice vibrations, has not been discussed here. The reader is referred to such theoretical papers as that by Durig and others for an analysis of this topic [32,33].

REFERENCES

1. R. M. Badger, *J. Chem. Phys.* **2**: 128 (1934).
2. W. Gordy, *J. Chem. Phys.* **14**: 304 (1946).
3. A. G. Meister, F. F. Cleveland, and M. J. Murray, *Am. J. Phys.* **11**(5): 239 (1943).
4. J. N. Shearer, T. A. Wiggins, A. Guenther, and D. H. Rank, *J. Chem. Phys.* **25**: 724 (1956).
5. T. A. Wiggins, J. N. Shearer, A. Shull, and D. H. Rank, *J. Chem. Phys.* **22**: 549 (1954).
6. *Infrared Wavenumber Tables*, Butterworths, London (1961).
7. J. S. Garing, H. H. Nielsen, and K. N. Rao, *J. Mol. Spectroscopy* **3**: 496 (1959).
8. E. K. Plyler and E. D. Tidwell, *J. Chem. Phys.* **29**: 829 (1958).
9. H. C. Allen and E. K. Plyler, *J. Chem. Phys.* **26**: 972 (1957).
10. L. H. Jones and R. S. McDowell, *J. Mol. Spectroscopy* **3**: 632 (1959).
11. H. H. Blau and H. H. Nielsen, *J. Mol. Spectroscopy* **1**(2): 124 (1957).
12. D. C. Smith and G. A. Brown, *J. Chem. Phys.* **20**(3): 473 (1952).
13. L. W. Daasch, C. Y. Liang, and J. R. Nielsen, *J. Chem. Phys.* **22**(8): 1293 (1954).
14. W. F. Edgell and C. J. Ultee, *J. Chem. Phys.* **22**(12): 1983 (1954).
15. M. K. Wilson and S. R. Polo, *J. Chem. Phys.* **20**(11): 1716 (1952).

16. G. Herzberg *Infrared and Raman Spectra of Polyatomic Molecules* p. 232, D. Van Nostrand Company, Inc., New York (1945).
17. R. M. Badger and L. R. Zumwalt, *J. Chem. Phys.* **6**: 711 (1938).
18. S. L. Gerhard and D. M. Dennison, *Phys. Rev.* **43**: 197 (1933).
19. E. K. Plyler and H. G. Allen, Jr., *J. Chem. Phys.* **26**(4): 972 (1957).
20. K. N. Rao and E. Polik, *J. Mol. Spectroscopy* **1**: 24 (1957).
21. W. R. Angus, C. R. Bailey, J. B. Hale, C. K. Ingold, A. H. Leckie, C. G. Raisen, J. W. Thompson, and C. L. Wilson, *J. Chem. Soc.* 966 (1936).
22. F. E. Palma and K. Sathianandan, *in* J. R. Ferraro and J. S. Ziomek (eds.), *Developments in Applied Spectroscopy*, Vol. 2, p. 61, Plenum Press, New York (1963).
23. R. S. Dennen, E. A. Piotrowski, and F. F. Cleveland, *J. Chem. Phys.* **49**: 4385 (1968).
24. K. H. Illinger and C. O. Trindle, *J. Chem. Phys.* **48**: 4427 (1968).
25. M. Moshinsky and O. Novaro, *J. Chem. Phys.* **48**: 4162 (1968)
26. K. Ohwada, *Spectrochim. Acta* **24A**: 595 (1968).
27. G. Dellepiane, G. Zerbi, and S. Sandroni, *Chem. Phys.* **48**: 3573 (1968).
28. A. R. H. Cole, G. M. Mohay, and G. A. Osborne, *Spectrochim Acta* **23A**: 909 (1967).
29. J. K. G. Watson, *J. Chem. Phys.* **48**: 4517 (1968).
30. E. J. O'Reilly, *J. Chem. Phys.* **48**: 1086 (1968).
31. R. G. Parr and J. E. Brown, *J. Chem. Phys.* **48**: 49 (1968).
32. J. R. Durig, M. Walker, and F. G. Baglin, *J. Chem. Phys.* **48**: 4675 (1968).
32. A. Bree and R. A. Kydel, *J. Chem. Phys.* **48**: 5319 (1968).

LITERATURE ON VIBRATIONAL ANALYSIS

Dinitromethane: K. Singh, *Spectrochim. Acta* **23A**: 1089 (1967).
Ethylene glycol: W. Savodny, K. Niedenzy, and J. W. Dawson, *Spectrochim. Acta* **23A**: 799 (1967).
Biphenyl: G. Zerbi and S. Sandroni, *Spectrochim. Acta* **24A**: 483 (1968).
Dibromomethane: R. S. Dennen, E. A. Piotrowski, and F. F. Cleveland, *J. Chem. Phys.* **49**: 4385 (1968).
Trimethyamine: P. H. Clippard and R. C. Taylor, *J. Chem. Phys.* **50**: 1472 (1969).
Dibromo- and difluorobenzene: P. N. Gates, K. Radcliffe, and D. Steele, *Spectrochim. Acta* **25A**: 507 (1969).
Ethylbenzene: J. E. Saunders, J. J. Lucier, and J. N. Willis, Jr., *Spectrochim. Acta* **24A**: 2023 (1968).
Diethyl ether: H. Wieser, W. G. Laidlaw, P. J. Kreuger, and H. Fuhrer, *Spectrochim. Acta* **24A**: 1055 (1968).

THEORETICAL BOOKS ON SPECTROSCOPY

Introduction to Molecular Spectroscopy, G. M. Barrow, McGraw-Hill, New York (1962).
Spectroscopy and Molecular Structure, G. W. King, Holt, Rinehart and Winston, Inc., New York (1964).
Spectroscopy and Structure, R. N. Dixon, Methuen & Co., London (1965).
Infrared Spectra of Crystals, S. S. Mitra and P. J. Gielisse, IITRI Press, Chicago (1966).

BOOKS FOR SPECIALIZED STUDIES

Molecular Spectroscopy with Neutrons, H. Boutin and S. Yip, MIT Press, Cambridge, Mass. (1968).
Tables of Wavenumbers, Butterworths, London (1961).
Infrared Spectra of Adsorbed Species, L. H. Little, Academic Press, New York (1966).

CHAPTER 5

The Use of Characteristic Group Frequencies in Structural Analysis

5.1. GENERAL OBJECTIVES

This chapter will be concerned with the use of group frequencies for structural analysis, i.e., the use of infrared spectroscopy to identify the structural groups present in a compound. This type of application of infrared spectroscopy is encountered in many of the laboratories that employ infrared as an analytical tool. Generally speaking, group frequency methods are used when the type of compound under investigation does not lend itself to the theoretical methods developed in Chapter 4, or when the information desired concerning the compound is such as not to require an extensive investigation. Rather than present an extensive listing of group frequencies, we shall attempt to show here how the methods used with group frequencies are developed.

5.2. DEFINITION OF GROUP FREQUENCIES

Group frequencies are vibrations that are associated with certain structural units. For example, the CH_3 group in CH_3CH_2Cl is assumed to have certain characteristic vibrations which will be found in the spectra of many types of compounds in which the CH_3 group appears. This concept of group frequencies may appear to be in conflict with the concept concerning vibrations of molecules developed in the preceding chapter, where we have seen that a molecule vibrates as a unit and that the vibration of one part of the molecule cannot be considered to be isolated from the vibrations of other parts. Nevertheless, the approximate constancy of position of group frequencies has been well established and forms the basis for the structural analysis of compounds.

Not all vibrations are good group frequencies, and some group frequencies which maintain a fairly constant position for most

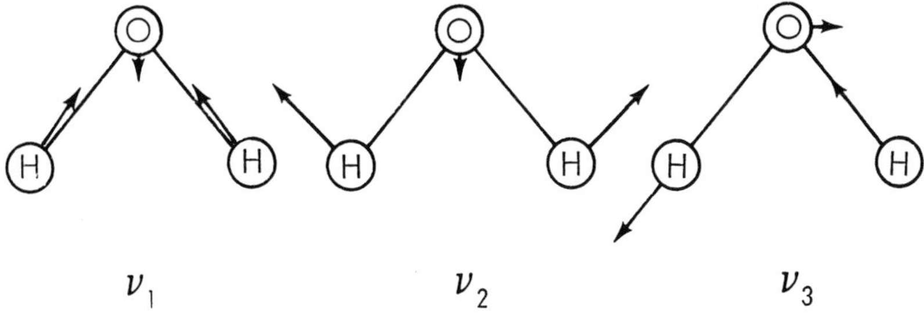

Figure 5-1. The fundamental vibrations of H_2O.

compounds may suddenly appear at a new frequency for certain other compounds.

Group frequencies are described in terms of the motions that the nuclei in a structural group in the molecule undergo during the vibration. For example, a CH_3 group in a molecule can be visualized as having the following vibrations:

1. Symmetric and asymmetric stretch
2. Symmetric and asymmetric scissors (or bend)
3. Twist
4. Wag
5. Rock

It will be recalled that these motions were diagrammed in Chapter 4. In describing the group frequencies of a CH_3 group, one may classify them according to this scheme. However, there is some ambiguity in such a classification.

In a simple molecule such as H_2O, we can describe the three fundamental frequencies as a symmetric stretch (ν_1), a bend (ν_2), and an asymmetric stretch (ν_3). These could be diagrammed as shown in Figure 5-1.

Diagramming vibrations in this manner implies that the frequency ν_1 depends entirely on a stretching force constant k_1, while the bending fundamental ν_2 depends entirely on a bending force constant k_2. There is no contribution of k_2 to the force constant that determines ν_1, and no contribution from k_1 to the vibration ν_2.

This concept of isolated stretching and bending vibrations cannot be applied to many molecules. For example, in Figure 4-27 the vibrations for the series CH_3CF_3, CH_3CF_2Cl, CH_3CFCl_2, and CH_3CCl_3 were presented. It was shown in that section that if the

TABLE 5-I. Fundamental Vibrations of C_3H_8*

CH_2 stretching	$\nu_1(a_1)$, $\nu_{22}(b_2)$
CH_3 asymmetric stretching	$\nu_2(a_1)$, $\nu_{10}(a_2)$, $\nu_{15}(b_1)$, $\nu_{23}(b_2)$
CH_3 symmetric stretching	$\nu_3(a_1)$, $\nu_{16}(b_1)$
CH_2 scissors	$\nu_4(a_1)$
CH_3 asymmetric scissors	$\nu_5(a_1)$, $\nu_{11}(a_2)$, $\nu_{17}(b_1)$, $\nu_{24}(b_2)$
CH_3 symmetric scissors	$\nu_6(a_1)$, $\nu_{18}(b_1)$
CH_2 rocking	$\nu_{19}(b_1)$, $\nu_{25}(b_2)$
CH_3 rocking	$\nu_7(a_1)$, $\nu_{12}(a_2)$, $\nu_{20}(b_1)$, $\nu_{26}(b_2)$
C—C stretching	$\nu_8(a_1)$, $\nu_{21}(b_1)$
C—C—C bending	$\nu_9(a_1)$
CH_2 twisting	$\nu_{13}(a_2)$
CH_3 twisting	$\nu_{14}(a_2)$, $\nu_{27}(b_2)$

* These are only qualitative descriptions of the vibrations of propane, since some of the modes of vibration cannot be clearly distinguished. For a discussion of which modes are uniquely assigned, see H. L. McMurry and V. Thornton, *J. Chem. Phys.* **19**(8): 1014 (1951).

species of vibration are traced through this series, some of the vibrations change their identity and progressively assume the character of other vibrations. For example, a CH_3 rocking vibration in the first member of the series becomes a CCl stretching frequency in the final member. For the molecules intermediate in the series, this vibration must be assumed to have the character of a rock and stretch and should not properly be considered as only one or the other. Other vibrations in this series, such as the CH_3 stretching and scissors, retain their character throughout.

Vibrations which retain their identity can be described as "group frequencies." They can be assumed to be associated with fixed groups in the molecule and they appear in fairly constant regions in a spectrum. When a vibration can be associated with a certain structural group, this is usually noted in the listing of the vibrational analysis of the molecule (for example, see Table 5-I).

Let us examine the vibrational analysis of a molecule to illustrate how group frequencies are related to the observed fundamentals, overtones, or combination bands.

5.3. GROUP FREQUENCIES AND THE VIBRATIONAL ANALYSIS OF A MOLECULE

The molecule C_3H_8 (propane) has 11 atoms and can therefore be expected to have 27 fundamentals. Of these, nine are a_1 vibrations, five a_2, seven b_1, and six b_2. The fundamentals are related to the types of motion occurring in the molecule as shown in Table 5-I. For example, it can be seen that vibrations described as CH_3 and CH_2 stretching and bending exist. Four fundamentals can be

TABLE 5-II. Vibrational Assignments of C_3H_8*

Vibration	Infrared	Raman
ν_{22}	2980 M	
ν_{15}, ν_{23}	2968 S	
ν_1, ν_{10}		2967 M
ν_2	2960 M	2946 S
$2\nu_4, 2\nu_5$		2920 S
ν_3		2903 S
$2\nu_5$		2871 VW
ν_{16}	2885 M	
$\nu_6+\nu_{24}$		2761 VW
$2\nu_{18}$		2725 M
$\nu_{20}+\nu_{21}$	1936 W	
$\nu_8+\nu_{21}$	1730 W	
$\nu_{24}, \nu_4, \nu_{17}$	1470 VS, 1468 S, 1465 S	
ν_{11}, ν_5		1451 S
ν_{18}	1375 S	
ν_6	1370 M	
$\nu_{20}+\nu_{14}$	1338 M	
ν_{12}		1278 W
ν_{25}	1179 M	
ν_7, ν_{19}	1152 M	1152 W
ν_{20}	1053 M	1054 M
ν_{21}	922 M	
ν_8	870 W	867 S
ν_{26}	748 S	

* See also McMurry and Thornton [6] for later work.

described as CH_3 asymmetric stretching vibrations, four others can be described as asymmetric scissors, etc. The observed infrared bands and the vibrational assignments for this molecule are given in Table 5-II.

There are a number of points concerning group frequencies that can be illustrated by use of the data in Tables 5-I and 5-II. First, the designations of vibrations given in Table 5-I correspond to group frequencies, although not all of these would be considered "good" group frequencies since their positions are not constant for a great number of molecules. Second, it can be noted that although there are 27 fundamentals, there are only 12 designations of group frequencies, since many of the fundamentals correspond to the same designation. Moreover, if an instrument of only moderate resolving power is used, not all fundamentals will be separated from one another. Some fundamentals are Raman active only.

The net result is that the infrared spectrum of C_3H_8 does not have as many bands as expected, and most of the observed bands can be assigned to characteristic group frequencies. The CH stretch is assigned to four bands at 2980, 2968, 2960, and 2885 cm^{-1}, the CH_3 and CH_2 scissors give rise to bands at 1470, 1468, 1465, 1375, and 1370 cm^{-1}. Under medium resolution and in the liquid state, some of these closely spaced bands would not be resolved and would give only one strong band. Thus, in the infrared spectrum of this compound two strong bands would be seen in the CH stretching and two in the CH scissors regions. These could be designated as characteristic group frequencies of the CH_3 and CH_2 groups. Examination of a large number of alkanes of this type indicates that most alkanes give similar bands in these regions, and therefore both the CH stretch and scissors are considered good group frequencies. The CH rocking and twisting and C—C vibrations do not retain as constant a position in alkanes as the stretch and scissors vibrations, and therefore are not considered good group frequencies.

It should be noted also that usually only fundamentals make good group frequencies. The combination and overtone bands generally have medium to weak intensity and cannot always be described as being related to the motion of some isolated group.

In order to describe group frequencies in the following sections, the structural unit which appears to be responsible for the band will be utilized. Such groups as CH_3, CH_2, CH, OH, or NH, or rings such as cyclopropane, benzene, etc., can be described as having characteristic group frequencies, as can also structural units such as the isopropyl and *t*-butyl groups. Finally, combinations of structural units such as the amide group ($-\overset{\overset{\displaystyle O}{\|}}{C}-NH_2$) have characteristic group frequencies.

5.3A. The Constancy of Spectral Positions of Group Frequencies

The number of vibrations as well as the positions of the bands is determined by the symmetry of a molecule, the masses of the atoms making up the molecule, and the force constants of the bonds between atoms. It is thus surprising that group frequencies are found and that they retain their spectral position in a series of molecules where the symmetry and mass factors are changing. To explain the relative constancy of spectral positions of group frequencies we must examine the influence of the following four factors:

1. Masses of the atoms of a molecule
2. Force constants of bonds between atoms
3. Symmetry of the molecule
4. Interaction of vibrations (Fermi and Coriolis interactions)

When these factors remain reasonably constant for related vibrations in a series of compounds, a characteristic group frequency will be found. The character of the vibration should remain more or less constant throughout the series of compounds. Examples of systems meeting these conditions are presented in the following sections.

5.3B. Group Frequencies of M—H Structural Units

Structural groups having a hydrogen atom show characteristic group frequencies. This is because the mass of the hydrogen atom is usually quite small compared to the group to which it is attached. The hydrogen atom may be considered as vibrating against an infinite mass and the vibration is therefore not mass sensitive.

The hydrogen is also a terminal atom; therefore, there is only one bond. This reduces the number of force constants that will determine the vibration. Finally, hydrogen is monovalent, so that the force constant is similar for various M atoms. M—H structural units, therefore show characteristic group frequencies unless the vibration is perturbed by another. As a general rule, whenever a terminal group has atoms that differ in mass, a characteristic group frequency is likely to exist. We shall see examples of this below.

5.3C. Terminal Structural Groups such as C═O and Multiple-Bond Groups

The structural unit C═O has an excellent group frequency, which is described as a stretching vibration. Since the C═O group is a terminal group, only the carbon is involved in a second chemical bond. This reduces the number of force constants determining the spectral position of the vibration. The C═O stretching vibration usually appears in a frequency range that is relatively free of other vibrations. This reduces the possibility that a vibration of the same species could interact with the C═O vibration. For example, in many carbonyl compounds the double bond of the C═O has a force constant different from those of such structural units as C—O, C—C, C—H, etc.; only structural units such as C═C have force constants of magnitudes similar to that of the C═O group. The C═C vibration could interact with the C═O if it were of the same species, but generally it is not.

A group such as —C≡N also has good group frequencies since it retains a fairly constant environment in a series of related molecules and probably has a vibration which does not couple with others. Other structural groups with multiple bonds, such as >C═C< or —C≡C—, have fairly good group frequencies because they usually occur as isolated groups in a molecule and do not couple with other groups. They are usually more sensitive to the substituents surrounding them since they are not terminal groups.

5.3D. Group Frequencies of Units Where the Mass Factor Is Important

If one atom has a mass very much different from the masses of the atoms in the remainder of the molecule, then the structural groups with this atom will have characteristic group frequencies. An example would be a unit such as —C—As—, i.e., the case where isolated arsenic atoms are found in a hydrocarbon. The C—As stretching frequency is fairly distinct since the vibration is not likely to couple with another. Other examples are C—Cl, C—Br, C—metal, etc. A C—N group would be expected to give less satisfactory group frequencies since the masses of C and N are similar and both are involved in other chemical bonds.

5.3E. Ring and Other Group Vibrations

We shall also find good group frequencies for rings, such as those in benzene compounds. In most instances the vibration involves a CH group, so that again the small mass of hydrogen is the prime factor which makes these vibrations retain fairly constant spectral positions.

Groups such as $-\overset{\overset{\displaystyle O}{\|}}{C}-NH_2$ also have distinctive group frequencies. Analysis of these group frequencies indicates that they are partly due to a combination of NH and C—N or C=O vibrations. Again, the force constants and mass factors involved in such a vibration remain fairly constant for the compounds in which the vibration is a good group frequency.

5.3F. The Influence of Symmetry on Group Frequencies

The symmetry of a molecule can determine the infrared activity of vibrations as well as their position, so that it is important to understand the symmetry factor. For example, if we compare the group frequencies of CH_3CH_3 to those of CH_3CH_2Cl, we find that only five fundamentals of the former are found in the infrared while the latter compound has 13. The CH stretching and scissors vibrations, which can still be considered good group frequencies, are represented by only two of each type in CH_3CH_3 but by four of each type in CH_3CH_2Cl. Finally, the C—C stretch is not infrared active for CH_3CH_3, while it is for CH_3CH_2Cl. If we refer to Figure 4-26, it is possible to compare the species of vibration of CH_3CH_3 and CH_3CH_2Cl. The introduction of the large mass of Cl in place of an H atom changes the character of some of the vibrations. They can no longer be described as being due completely to a distinct group; that is, they have assumed, in part, the character of other groups. We therefore anticipate less characteristic frequencies for these vibrations and expect that they will be shifted from their normal position. Even the CH_3 scissors vibrations, which are considered good group frequencies, are not as well defined in CH_3CH_2Cl as they are in CH_3CH_3.

5.3G. Summary

The factors which determine the number and position of absorption bands should be kept in mind in the discussion of particular group frequencies in the following sections. They are as follows: the masses of the atoms of the molecule, the force constants acting between atoms, the symmetry of the molecule, and the perturbation of vibrations. Their effects should be considered when attempting to correlate frequency positions with such concepts as electronegativity, resonance, and hydrogen bonding. Whenever such correlations are found for vibrations it is probable that the mass, force constant, or coupling factors are either constant or absent. For example, in correlating spectra of amines an empirical rule such as the following has been suggested [2]:

$$\nu_{\text{sym}} = 345.5 + 0.876\nu_{\text{asym}}$$

where ν_{sym} and ν_{asym} are the frequencies of the symmetric and asymmetric stretching modes of the NH_2 groups. All but one of the above-named factors must remain constant in a series of compounds, or all the factors must vary linearly, for such a linear relation to hold. Even such a phenomenon as hydrogen bonding would need to be similar for all the compounds in the series.

Another example of how correlations between vibrations of molecules and some physical parameters may be suggested is illustrated by the series CH_3F, CH_3Cl, CH_3Br, and CH_3I. The position of the symmetric scissors vibration depends on the halide, and it is tempting to relate this shift to some parameter such as, for instance, the mass of the halide. However, the actual position of this vibration is determined to a great extent by Fermi resonance with a neighboring vibration of the same species [1; p.315].

The general technique of assigning new group frequencies starts with the vibrational analysis of small symmetric molecules containing the structural group in question and proceeds by extending the vibrational assignments to larger asymmetric molecules. If certain vibrations retain fairly constant spectral positions, they can be considered good group frequencies. If the larger asymmetric molecule can be synthesized containing an isotope, the influence of the isotopic substitution on band position can be used to verify the assignment of vibrations to certain groups since, presumably, the vibrations that are shifted are those related to the structural group that contains the new isotope. For example, a CH_2 vibration can be shifted if deuterium is substituted for the hydrogen. However, it is possible in some cases of isotopic substitution that the observed shift is due merely to an uncoupling of vibrations and not entirely to the mass change in the molecule. Raman spectra can also be used in assigning group frequencies.

Sutherland and his co-workers have suggested an interesting method of assigning group frequencies in related compounds [4]. For example, in a homologous series of monosubstituted amides where each succeeding member is formed from the previous member by the addition of a CH_2 unit it is possible to trace the group frequencies by use of the following four criteria:

1. Position of the band in the spectra
2. Intensity of the band
3. Width of the band
4. Behavior of the band under dilution in nonpolar solvents.

Bands which are similar in these four respects are considered related. Satisfactory results were obtained when these criteria were applied, and the results were compared to data obtained by deuteration and other studies.

Having completed the definition of group frequencies and the discussion of the factors that determine them, we can now begin to discuss some known group frequencies. Before we begin, however, it may be of value to hint at some errors that beginners frequently make in using group frequencies for structural analysis.

1. The physical state of a compound when its infrared spectrum is measured can determine the position and number of infrared bands. Therefore it is important that the physical state be listed when the group frequency is reported.
2. Group frequency tables list only group frequency bands. Other bands can appear in the same or in other spectral regions.
3. Group frequencies of one structural unit can appear in the same region as those of another unit, thus making correlation of spectra to structural analysis difficult in some instances.
4. Intensity of bands is a relative factor. For example, CH vibrations in a compound containing only CH groups can appear quite strong, but when other atoms are present, the intensity of these bands may be much lower.
5. The more similar the compounds in a series, the more group frequencies will be found. For example, for *n*-alkanes the position of the CH stretch and scissors will be located at a more exact position than for other molecules. In addition, the rocking and wagging vibrations can be used as group frequencies. On the other hand, in comparing two such different alkanes as CH_3CH_3 and $(CH_3)_4C$ it is not surprising that only very few vibrations are similar. Actually, for these two compounds, only

the CH stretch and scissors appear in fairly similar regions. Also, in comparing CH_3F and CH_3CH_3, only the stretch and scissors of the CH_3 group are found in similar spectral regions. In a comparison of CH_3F and a large branched-chain alkane, only a few group frequencies will be common to both.

5.4. GROUP FREQUENCIES OF THE CH_3, CH_2, AND CH STRUCTURAL UNITS

It is convenient to discuss the group frequencies of the CH_3, CH_2, and CH structural units together since they can be described in terms of the same general motions.

5.4A. Alkanes Containing the CH_3 and CH_2 Units

We have previously discussed the vibrational analysis of C_3H_8 (see Tables 5-I and 5-II) and have shown that, while there are 27 fundamentals, the infrared spectrum shows fewer bands. The characteristic group frequencies can be classified into 12 designations (see Table 5-I), and of these 12 designations only a few of the vibrations are "good" group frequencies. Generally, only the motions of the CH groups of alkanes give rise to good group frequencies, and in this section we shall limit ourselves to these vibrations.

Let us examine the vibrations for CH_3CH_3, $CH_3CH_2CH_3$, and $CH_3CH_2CH_2CH_3$. These are presented in Table 5-III. While in some respects CH_3CH_3 is unique, it is possible, by examining the spectra of these three compounds, to predict which characteristic group frequencies will be found for normal paraffin hydrocarbons. For example, the CH stretching vibrations for these molecules appear between 3000 and 2850 cm^{-1}. From two to four medium to strong bands will be found. If both CH_3 and CH_2 structural units are present in a molecule, four stretching vibrations can be described (an asymmetric and a symmetric one for each of these structural units). The possibility of finding all four in a single spectrum depends on their relative position, their infrared activity, and the resolution of the spectrophotometer.

There may be other bands in the CH stretching region, due to overtones or combination bands, so that one should not expect to find only four distinct bands in this region. The spectra of alkane molecules also depend on the physical state of the sample.

The scissors vibrations of the CH_3 and CH_2 groups also appear in a fairly constant region. Generally, the asymmetric CH_3 scissors is very near the CH_2 scissors, so that it is not possible to distinguish between these two vibrations.

Finally, it can be seen that the CH_3 and CH_2 rocking vibrations appear between 1200 and 700 cm^{-1}. We shall see later that these are

TABLE 5-III. Vibrations for CH_3CH_3, $CH_3CH_2CH_3$, and $CH_3CH_2CH_2CH_3$

Compound	CH_3 stretch		CH_2 stretch		CH_3 scissors		CH_2 scissors	CH_3 rock	CH_2 rock	CH_2 wag	Ref.
	Asym.	Sym.	Asym.	Sym.	Asym.	Sym.	Sym.				
CH_3CH_3 (gas)	2996M	2954S 2915S			1472S	1379W		1190 (Raman only) 822			5
$CH_3CH_2CH_3$ (gas)	2968S 2960M	2885M	2980M		1470S 1465S	1375M 1370M	1468S	870 1155M 923	1053M 748S	1336	1 (p. 361) and 6*
$CH_3CH_2CH_2CH_3$ (gas)	2960S	2860W	~2940	~2870	1466S	1390S	1466S	970S	734M	1340W	7

* We have reassigned some of the vibrations reported in Reference 1 according to the assignments in Reference 6.

TABLE 5-IV. Characteristic Group Frequencies for Normal Paraffin Hydrocarbons (CCl_4 and CS_2 Solutions) (Similar Group Frequencies Are Found for Liquid and Solid State Spectra)

Group frequency designation	Approximate position (cm^{-1})
CH_3 asymmetric stretch	2960
CH_3 symmetric stretch	2870
CH_2 asymmetric stretch	2925
CH_2 symmetric stretch	2855
CH_3 asymmetric scissors	1470
CH_3 symmetric scissors	1380
CH_2 symmetric scissors	1470
CH_2 wag	1305
CH_2 rock	720
CH_3 in-plane rock	1140–1130

often weak and that their position is so variable for complex hydrocarbons that they cannot be considered good group frequencies.

If the factors discussed above are now applied to the analysis of spectra of normal paraffin hydrocarbons, it is possible to give some general spectral positions for characteristic group frequencies. These are listed in Table 5-IV, which presents data obtained from CCl_4 and CS_2 solutions in the proper regions. Since much of the work in infrared laboratories is on samples in solution or in the liquid and solid states, many of the group frequency assignments are reported for these types of samples. Generally, the absorption bands recorded in the gaseous state are at higher frequencies than those observed in solution or in the liquid or solid states.

The intensity of the group frequencies listed in Table 5-IV is determined by the relative number of each type of group. For example, in $CH_3CH_2CH_2CH_2CH_3$ the four stretching vibrations will range from strong to very strong. The two scissors vibrations (CH_2 scissors not being distinguished from the CH_3 asymmetric scissors) at 1470 and 1380 cm^{-1} will be of strong and medium intensity, respectively. The CH_2 wag and rock will be fairly weak in intensity. As the chain length of the hydrocarbon is increased, thus introducing more CH_2 groups, the intensity of the CH_2 vibrations will increase, so that the CH_2 wag near 1305 cm^{-1} and the rock near 720 cm^{-1} will have medium to strong intensity. The 1470 cm^{-1} band will also increase in intensity since it is partly due to a CH_2 group.

Although the group frequency vibrations are given in Table 5-IV for solution spectra, similar bands are found in liquid and solid state spectra of normal alkanes. For example, the spectrum of *n*-hexane in the liquid state is shown in Figure 5-2 (located in pocket on inside back cover). This spectrum was obtained with a sodium chloride prism, so that the bands in the CH stretching region are not completely resolved. It can be seen that two CH stretching vibrations appear in the 3000–2800 cm^{-1} region. The 1470 and 1380 cm^{-1} scissors vibrations are quite distinct as is the 720 cm^{-1} CH_2 rock. The CH_2 wag is less distinct but does appear near 1305 cm^{-1}. Similar group frequencies would be found for other normal alkanes.

Solid state spectra of normal alkanes have similar vibrations to those found in the liquid state, and group frequency assignments are similar to those discussed above. In some instances, bands are split because of the symmetry factors present in the solid state. Also, there is the possibility that some of the motions of extended CH_2 chains will give distinct bands. These facts have been used in structural assignments but are of limited applicability and will not be discussed here.

The CH_2 rock near 720 cm^{-1} is fairly distinct and can be utilized in recognizing long-chain paraffins. Generally, correlation charts show that four or more CH_2 groups are required in a chain to give a distinct band near 720 cm^{-1}, since in shorter chains the band position is quite variable. For example, if we examine the band assignments for C_4H_{10} given in Table 5-III, we find that the CH_2 rock is near 734 cm^{-1}.

The CH_3 rock and CH_2 wags and twists are usually not considered good group frequencies since, although there generally is a band near 1305 cm^{-1} in normal alkanes that is a CH_2 wagging vibration, they are weak and difficult to assign.

We can summarize the group frequency assignments for normal alkanes by stating that the CH stretching and scissors vibrations and the CH_2 rock near 720 cm^{-1} are considered good group frequencies.

5.4B. Branched-Chain Hydrocarbons

Branched-chain hydrocarbons have CH stretching and scissors vibrations for CH_3 and CH_2 groups similar to those found for normal alkanes. The characteristic group frequencies for branched-chain hydrocarbons are listed in Table 5-V. It is important to point out that the CH stretching region can have more or fewer than the four bands predicted by the general group frequency assignments and that the number and intensity of these bands can possibly be related to the branching in the hydrocarbon chain.

Branched-chain hydrocarbons may contain tertiary CH groups. The group frequencies for this CH unit are not as well defined as those of CH_3 and CH_2 groups and are generally weak since the number of tertiary CH groups is usually much less than that of CH_3

TABLE 5-V. Characteristic Group Frequencies for Branched-Chain Hydrocarbons

Group frequency designation	Approximate position (cm^{-1})
CH_3 asymmetric stretch	2960
CH_3 symmetric stretch	2870
CH_2 asymmetric stretch	2925
CH_2 symmetric stretch	2855
CH_3 asymmetric scissors	1470
CH_3 symmetric scissors	1380
CH_2 symmetric scissors	1470
CH stretch	2890
CH bend	1340
—$C(CH_3)_2$ symmetric scissors	1385 and 1375
—$C(CH_3)_3$ symmetric scissors	1395 and 1365
—$C(CH_3)_2$ skeletal vibration	1170, 1155, and 840–790
—$C(CH_3)_3$ skeletal vibration	1250 and 1250–1200

and CH_2 groups in branched-chain hydrocarbons. In correlation tables the tertiary CH stretch is usually listed as appearing in the spectral range from 2880 to 2900 cm^{-1}. It is a weak band and may be masked by bands due to the CH_3 and CH_2 groups. A weak band near 1340 cm^{-1} also seems to be characteristic of the tertiary CH group and is listed in Table 5-V as a characteristic group frequency. In the Raman spectrum the tertiary CH group has a band at 1333 cm^{-1}.

Branched-chain hydrocarbons may contain the isopropyl group —$CH(CH_3)_2$ or the *t*-butyl group —$C(CH_3)_3$. These groupings have characteristic group frequencies. The isopropyl group shows a splitting of the symmetric CH_3 scissors near 1380 cm^{-1} into two bands positioned near 1385 and 1375 cm^{-1}. For the *t*-butyl group this band is also split, with the two bands appearing near 1395 and 1365 cm^{-1}. In addition to the splitting of the symmetric scissors, the isopropyl group shows two characteristic bands near 1170 and 1155 cm^{-1} and the *t*-butyl group has two bands near 1250 and 1250–1200 cm^{-1}. A band in the 840–790 cm^{-1} region for isopropyl groups has limited application.

All these group frequencies have medium to strong intensity if the vibrational group is present in the molecule in equivalent numbers to other groups. All of them are listed in Table 5-V.

5.4C. CH_3 Groups Adjacent to Atoms Other Than Saturated Carbon

There are characteristic group frequencies for CH_3 groups adjacent to atoms other than carbon and for the CH_3 group adjacent to a carbonyl group ($-\overset{\overset{\displaystyle O}{\|}}{C}-$). To illustrate the possible frequencies

TABLE 5-VI. Vibrations of Molecules with CH_3 Groups

CH_3 stretching			CH_3 scissors		CH_3 rocks				Molecule	Ref.
Asym.	Sym.	Other	Asym.	Sym.						
A. CH_3—C(—)—										
3009	2954		1454	1389			1040		CH_3CN (gas)	8
2996	2954	2915	1472	1379		1190		821	CH_3CH_3 (gas)	5
3010	2943		1434	1377			1002		CH_3CCl_3 (gas)	9
3012 2983	2940 2890		1452	1383			1080		CH_3CH_2Cl (gas)	10
3012	2941	2873	1445	1383			1094	1058	CH_3CHCl_2 (gas)	10
2957	2872 2901		1464	1381 1364			971		H_3C—C(CH_3)(—HC=CHCH_2—)—CH_2 (ring: HC=CHCH$_2$ / H$_3$C—C——CH$_2$ / CH$_3$) (liquid)	11
2958	2882		1473 1433	1383			1019 1078		$(C_2H_5)_2B_2H_4$ (gas)	12
2968 2960	2885	2725	1470 1465	1375 1370		1152	1053	748	$CH_3CH_2CH_3$ (gas)	6

CH_3 stretching			CH_3 scissors		CH_3 rocks				Molecule	Ref.
Asym.	Sym.	Other	Asym.	Sym.						
2962	2876		1455	1370	1280				$(CH_3)_4C$ (gas)	13, 19
2994				1379					$CH_3C{\equiv}CH$ (gas)	14
2975		2500	1468	1380			1043		$CH_3C{\equiv}CCH_3$ (gas)	14
2960 2916	2852	2942 2884	1472	1399			996		$CH_3CH{=}CH_2$	14
3035	2975		1443	1408			970		CH_3CF_3 (gas)	15
B. $CH_3-C(=O)-$										
3008	2960	2925	1440 1425	1361					$CH_3-C(=O)-CH_3$ (liquid)	16
			1415	1356			1020		$CH_3C(=O)CH{=}C(O)CH_3$, chelated: $O\ldots Cu-O$	17
C. $CH_3-C(=O)-O-$										
3048 2997	2961		1445 1401	1340		1068	990		$CH_3-C(=O)-OH$ (gas) (monomer)	18

TABLE 5-VI (continued)

CH$_3$ stretching			CH$_3$ scissors		CH$_3$ rocks				Molecule	Ref.
Asym.	Sym.	Other	Asym.	Sym.						
3035 (CH$_3$O) 3012 (l) (CH$_3$C) 2970 (CH$_3$C)	2966 (CH$_3$O) 2942 (CH$_3$C) (R)		1469 1450 (CH$_3$O) 1427 (l) 1412 (l) (CH$_3$C)	1440 (CH$_3$O) 1375 (CH$_3$C)	1248 1187 (CH$_3$O) (R) 1169 (CH$_3$C)				$CH_3-C(=O)-O-CH_3$ (gas)	20
D. CH$_3$—O										
2960 2925	2875		1511 1492	1484	1247	1196			B(OCH$_3$)$_3$	21
2955	2875		1500	1500	1150–1205				DB(OCH$_3$)$_2$	21
2955	2875		1493	1493	1150–1205				HB(OCH$_3$)$_2$	21
2977	2844		1477 1415	1455	1233	1116			CH$_3$OH (gas)	22
3093	2993	2954 2887 2830	1466	1443		1190	1156		$CH_3-O-N=C(H)(Cl)$ (gas)	23
2986	2914		1466*	1466*		1180			CH_3-O-CH_3	1 (p. 354)

CH_3 stretching			CH_3 scissors		CH_3 rocks				Molecule	Ref.
Asym.	Sym.	Other	Asym.	Sym.						
3012 3045	2969	2943	1465 1454	1445		1230	1168	1032 1371	CH_3—O—C(H)=O (gas)	20
3025 (CH_3O) 3012 (CH_3C) 2970 (CH_3C)	2956 (CH_3O) 2942 (CH_3C)		1469 (CH_3O) 1427 (CH_3C) 1450 (CH_3O) 1412 (CH_3C)	1440 (CH_3O) 1375 (CH_3C)		1248 (CH_3O)	1169 (CH_3C) 1187 (CH_3O)	980 (CH_3C)	CH_3—O—C(=O)—CH_3	20
E. CH_3—N										
2960	2918		1466	1420		1172			$Ag^+(CH_3—N{=}NO_2)^-$ (solid)	24
			1464 1441 1425	1391					CH_3—N=NO(OCH_3) (liquid)	24
2953 (R)	2916 (R)		1482 (R)	1432 (R)		1164 (R)			$NH_4^+(CH_3—N{=}NO_2)^-$ (liquid)	24
3014	2966		1459	1410			1041		CH_3NC (gas)	25
2994	2951		1453	1377		1181	1107		$CH_3N{=}C{=}O$	14

TABLE 5-VI (continued)

CH_3 stretching			CH_3 scissors		CH_3 rocks				Molecule	Ref.
Asym.	Sym.	Other	Asym.	Sym.						
3030		2596 2391	1430				1013		$CH_3—N{=}N—CH_3$	1 (p. 359)
			1470 1450	1395					$(CH_3)_2N—C({=}O)—H$	26
			1421 1400	1383 1343					$H_3C—C{\equiv}\overset{+}{N}—CH_3$	27
2967	2822 2777		1466	1402		1272	1183 1104		$(CH_3)_3N$ (gas)	28
	2940		1445 (CH_3N, CH_3C)	1413 (CH_3N) 1373 (CH_3C)		1159 (CH_3N)	1040 (CH_3C)	987 (CH_3C)	$CH_3—C({=}O)—N(H)—CH_3$	29
	2940		1455 (CH_3N) 1418 (CH_3C)	1375 (CH_3C) 1418 (CH_3N)		1165 (CH_3N)			$Cd(CH_3—C({=}O)—N(H)—CH_3)Cl_2$	30
F. CH_3X										
3006	2965		1471	1475		1196			CH_3F (gas)	31
3042	2966		1455	1355		1015			CH_3Cl (gas)	31

CH_3 stretching			CH_3 scissors		CH_3 rocks				Molecule	Ref.
Asym.	Sym.	Other	Asym.	Sym.						
3056	2972		1445	1305		952			CH_3Br (gas)	31
3060	2970		1440	1252		880			CH_3I (gas)	31
G. CH_3—Si										
2975	2915		1405	1258				818 841	$(CH_3)_2SiCl_2$ (liquid)	32
2969	2911		1454 1414	1248				848 857	$(CH_3)_3SiCl$ (liquid)	33
2957			1430	1285		1253			$(CH_3)_4Si$	19
H. CH_3—S										
			1460 1420 1441	1335 1325 1304			1037 1024	975 905	$(CH_3)_2S$ (gas)	34
3012	2907		1430	1300				940	CH_3—S—NO (gas)	35
2973	2908		1455 1440 1419 1405	1319 1304			1016 1006 929 915		$(CH_3)_2SO$ (gas)	36

TABLE 5-VI (continued)

CH_3 stretching			CH_3 scissors		CH_3 rocks				Molecule	Ref.
Asym.	Sym.	Other	Asym.	Sym.						
I. CH_3—P										
3000	2938		1450	1346			1017 977		CH_3PH_2 (gas)	37
2970	2850 2920		1417 1430	1310 1298			960 1067 947		$(CH_3)_3P$ (gas)	38
			1420 1437	1340 1292 1305			950	872 866	$(CH_3)_3PO$	38
J. CH_3—B										
2958	2841		1441	1326 1316			1064	976	$(CH_3)_2B_2H_4$ (gas)	39
2950	2849		1443	1328 1321			1055	975	$(CH_3)_2B_2D_4$ (gas)	39
2924 2958	2841		1437	1324 1312			1017	935	$(CH_3)_2B_2H_2(CH_3)_2$ (gas)	40
2941	2857		1424	1319			971		$CH_3B_2H_5$ (gas)	41
2958	2857		1433	1319			976	952	$CH_3B_2D_5$ (gas)	41

K. CH_3—Metal

CH_3 stretching			CH_3 scissors		CH_3 rocks				Molecule	Ref.
Asym.	Sym.	Other	Asym.	Sym.						
2982	2917		1442	1244				830	$Ge(CH_3)_4$	13
2979	2915		1465	1205				776	$Sn(CH_3)_4$	13
2999	2918		1462	1169				767	$Pb(CH_3)_4$	13

(R) = Raman.
* Perturbed vibrations.

that will be found for molecules which have these types of CH_3 groups we shall refer to the vibrations observed for small and symmetric molecules. A listing of these types of molecules is given in Table 5-VI. Again, we should call attention to the fact that as we search for generalizations concerning group frequencies we may be ignoring accidental vibrational perturbations or other factors. Thus, any conclusions that are drawn are general and subject to further verification.

CH_3 ***Stretching Vibrations.*** If we examine the frequencies listed for various molecules in Table 5-VI, we shall find that only a few generalizations can be made concerning CH_3 stretching frequencies. There is some indication that an electronegative element adjacent to the CH_3 group can raise the CH symmetric stretching frequency. For example, in a molecule such as $CH_3—O—\overset{\overset{\displaystyle O}{\|}}{C}—CH_3$ the $CH_3—O—$ group has the CH_3 symmetric stretch at 2966 cm^{-1}, while the $CH_3—\overset{\overset{\displaystyle O}{\|}}{C}—$ group has it at 2942 cm^{-1}. The asymmetric stretch for these groups also appears at a slightly higher frequency. Another example, is the series $(CH_3)_4M$, where M is C, Si, Ge, Sn, or Pb. We see that the asymmetric stretch moves from 2962 cm^{-1} for the first member of the series to 2999 cm^{-1} for the last member. The symmetric stretch moves from 2876 to 2919 cm^{-1} from the first member to the second and then remains at a fairly constant position.

It has been shown that for compounds having methyl groups attached directly to aromatic rings a series of four bands in the CH stretching region shifts to low wavenumber positions while for aromatic methyl esters the bands appear at their highest position [[67]].

The structural groupings $CH_3—O$ and $CH_3—N$ have a stretching frequency in the 2830–2760 cm^{-1} region. For example, ethers have a band of medium intensity in the 2830–2815 cm^{-1} region, while amines have a band in the 2820–2760 cm^{-1} region. These are listed as characteristic group frequencies in Table 5-VII. A summary chart of stretching vibrations is presented in Figure 5-3. This work was done using an LiF prism to obtain the high resolution required to separate stretching frequencies.

To generalize, it appears that both the symmetric and the asymmetric stretching frequencies are slightly shifted from their position in alkanes, depending on the atom adjacent to the CH group.

CH_3 ***Scissors Vibrations.*** There appear to be several generalizations that can be made concerning the group frequencies of the CH_3

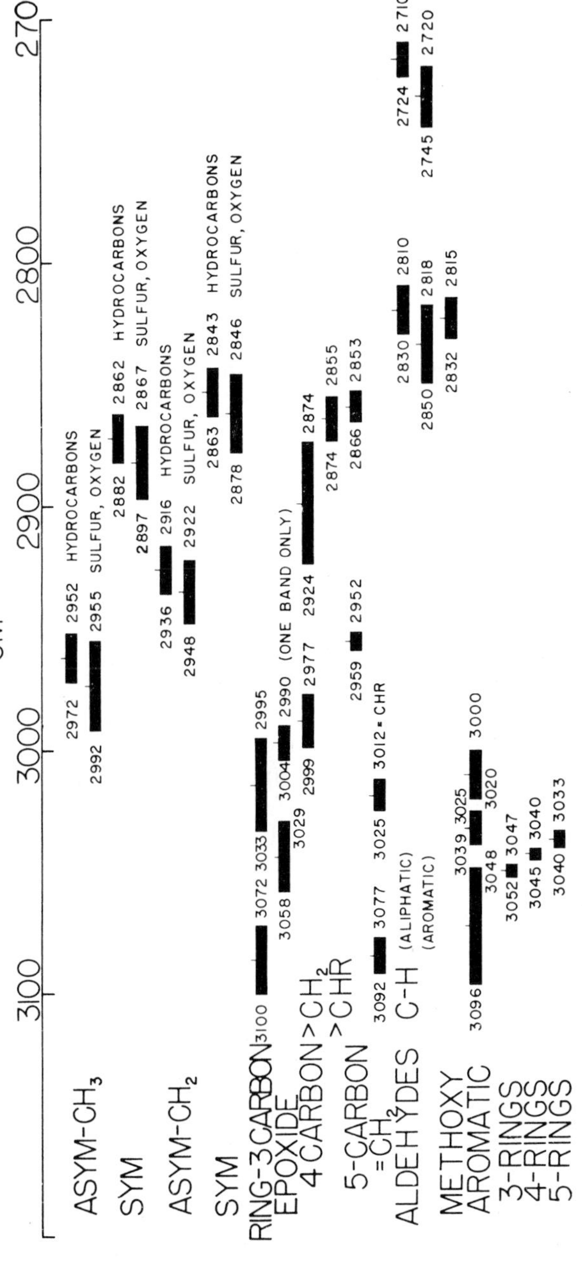

Figure 5-3. CH stretching frequencies. [Based on a private communication from Dr. Stephen Wiberley, Rensselaer Polytechnic Institute, Troy, N.Y.]

scissors vibration if we examine this vibration for the series of compounds listed in Table 5-VI.

The $CH_3-\overset{O}{\overset{\|}{C}}-$ *Group*. Let us first examine the vibrations for a CH_3 group adjacent to a carbonyl group.

The frequency position of the symmetric scissors when a CH_3 group is adjacent to a saturated carbon atom is near 1380 cm^{-1}. We have previously shown that this is a very constant position for most types of alkanes. When the methyl group is adjacent to a carbonyl group, for instance, in ketones having structures such as $CH_3-\overset{O}{\overset{\|}{C}}-C-$, the symmetric scissors moves to a lower value near 1360–1355 cm^{-1}. In Table 5-VI several examples of vibrational assignments of the symmetric scissors in this range can be found, and a sufficient number of ketones have been investigated to make this a fairly certain group frequency assignment (see footnote 1 to Table 5-VII). However, it appears that this group frequency assignment loses some of its validity for structures such as $CH_3-\overset{O}{\overset{\|}{C}}-O$, where the carbonyl is in turn adjacent to an oxygen atom. Here the symmetric scissors is found at 1375 cm^{-1}. For methyl acetate, two symmetric scissors are found, due, respectively, to the $CH_3-\overset{O}{\overset{\|}{C}}-$ and CH_3-O- groups, the latter appearing at 1440 cm^{-1} and the former at 1375 cm^{-1}.

We have listed the position of the symmetric scissors in the 1360–1355 cm^{-1} range as a characteristic group frequency of methyl groups in ketone structures in Table 5-VII.

The asymmetric scissors of CH_3 groups in $CH_3-\overset{O}{\overset{\|}{C}}-$ structures appears to be at a lower frequency than that found in alkanes (1460 cm^{-1}). While this has not been investigated for a large number of compounds it is listed in Table 5-VII as a characteristic group frequency.

A number of acetylacetonates have been investigated, and for these the CH_3 asymmetric scissors appears in the 1415–1380 cm^{-1} range while the symmetric scissors appears in the 1390–1350 cm^{-1} range. These too are listed as characteristic group frequencies in Table 5-VII.

TABLE 5-VII. Characteristic Group Frequencies for the CH_3 Unit

Structural group	Vibration	Positions	Remarks
CH_3—M (M = metal)	Asymmetric stretch Symmetric stretch	Dependent to a small extent on M	
CH_3—O (ethers)	Stretching	2830–2815[1]	
CH_3—N (amines and imines)	Stretching	2820–2760[1]	
CH_3—C(=O)— (acetylacetonates)	Asymmetric scissors Symmetric scissors	1415–1380 1390–1350	
CH_3—C(=O)— (ketones)	Asymmetric scissors Symmetric scissors	1450–1400 1360–1355[1]	
CH_3—C(=O)—O	Asymmetric scissors Symmetric scissors	1450–1400 1400–1340	
CH_3—O—	Symmetric scissors of CH_3 group	1470–1430[4] 1445–1430[3] (esters)	Asymmetric and symmetric scissors may coincide
CH_3—N	Symmetric scissors	1440–1390	
CH_3—N— (amides)	Asymmetric scissors Symmetric scissors	1500–1450 1420–1405[2]	
CH_3—M (M = metal)	Asymmetric scissors Symmetric scissors	1450–1400 1400–1150	
CH_3—B	Asymmetric scissors Symmetric scissors	1460–1405[1] 1320–1280[1]	
CH_3—S	Symmetric scissors	1325–1300	
CH_3—Si	Symmetric scissors	1260[1]	
$(CH_3)_2Si$—	Symmetric scissors	1260[1]	
$(CH_3)_3Si$—	Symmetric scissors	1250[1]	
CH_3—P	Symmetric scissors	1320–1280[1]	
CH_3—Ge	Symmetric scissors	Near 1250	
CH_3—Sn	Symmetric scissors	Near 1205	

TABLE 5-VII (continued)

Structural group	Vibration	Positions	Remarks
CH_3—Pb	Symmetric scissors	Near 1170	
CH_3X (X = halogen)	Symmetric scissors	Shifted depending on halide (1500–1250)	

(1) Assignments are reported in *Introduction to Practical Infrared Spectroscopy*, A. D. Cross, Butterworths Scientific Publications, London, 1961.

(2) Assignment reported by M. Beer, H. B. Kesseler, and G. B. B. M. Sutherland, *J. Chem. Phys.* **29**(5): 1097 (1958).

(3) Assignment reported by A. R. Katritzky, A. M. Monro, J. A. T. Beard, D. P. Dearnaley, and N. J. Earl, *J. Chem. Soc.* 2182 (1958).

(4) The CH_3—O symmetric scissors in anisole has been reported at 1300 cm^{-1} with fairly strong evidence that this is a correct assignment [see: J. H. S. Green, *Spectrochim. Acta* **18**: 39 (1962).]

The CH_3—O— *Group.* The symmetric scissors of the methyl group adjacent to oxygen atoms in structures like CH_3—O—C$\lessgtr$ appears at a higher value than the 1380 cm^{-1} found for a CH_3—C$\equiv$ group. For CH_3OH the symmetric scissors is at 1455 and 1430 cm^{-1} while in CH_3OCH_3 it is at 1466 cm^{-1}. In methyl acetate it is, as noted above, at 1440 cm^{-1}. For methyl esters a band near 1440 cm^{-1} has been suggested as a characteristic group frequency (see footnotes 3 and 4, Table 5-VII). The shifting to higher frequencies of the symmetric vibration in CH_3—O— units may cause the band to coincide with the asymmetric scissors near 1460 cm^{-1}, so that it is probable that in liquid or solid state spectra only a single band will be observed in this region. We shall see several examples of this in later sections. Even in the vapor state under higher resolution dimethyl ether has only a single band, which is assigned to the asymmetric and symmetric scissors. In Table 5-VI, the compounds $HB(OCH_3)_2$ and $B(OCH_3)_3$ also have only a single band.

The CH_3—N$\lessgtr$ *Group.* For the structural group CH_3—N$\lessgtr$, where a methyl group is adjacent to a nitrogen atom, the symmetric scissors appears to shift to higher frequencies, as it did for the CH_3—O— group. The general range 1440–1390 cm^{-1} can be listed as the region where this vibration appears. Sutherland and his coworkers [4] have suggested that in many amides where this group is present the CH_3—N$\lessgtr$ group has a symmetric scissors near 1408 cm^{-1}. The assignments for the molecule CH_3—C(=O)—N(CH_3)—H given in

Table 5-VI furnish an excellent example of how the symmetric scissors of a methyl group in $CH_3—N\lt$ appears at a higher frequency than that of the methyl group in $CH_3—C\equiv$, both groups being in the same molecule. Again, that generalizations concerning group frequency positions are not always accurate is illustrated by the molecules CH_3CN and CH_3NC. The position of the symmetric scissors is at 1390 and 1410 cm^{-1}, respectively. The high position in CH_3CN would not be expected from the assignment suggested for the $CH_3—C$ group.

For molecules having several CH_3 groups, the scissors vibration may result in more than two bands near 1400 cm^{-1}. For example, the spectrum of $(CH_3)_3P$ shows bands at 1430, 1417, 1310, and 1298 cm^{-1}, all ascribable to the asymmetric and symmetric scissors of the CH_3 group.

The $CH_3—M$ *Group.* In general, the position of the symmetric scissors of CH_3 groups adjacent to atoms other than carbon is quite variable, in contrast to the asymmetric scissors. Examples listed in Tables 5-VI include such groups as $CH_3—Si$, where the symmetric scissors can appear near 1260 cm^{-1}, $CH_3—P$, where it is near 1300 cm^{-1}, and $CH_3—S$, where it is also near 1300 cm^{-1}. Characteristic group frequencies can be listed for all of these groups and are given in Table 5-VII; however, the frequency positions listed should be considered as approximate since the molecular environment surrounding the structural group can change them significantly. Perhaps the series $(CH_3)_4M$, where M is C, Si, Ge, Sn, or Pb, furnishes the best illustration of how the symmetric scissors moves progressively to lower frequencies as the central atom changes from C to Pb (see Table 5-VI).

The position of the symmetric scissors of a methyl group in a structure such as CH_3X, where X is a halogen, ranges from a high of 1475 cm^{-1} in CH_3F to a low of 1250 cm^{-1} in CH_3I.

Summary. In contrast to the symmetric scissors, the asymmetric scissors for most of the compounds listed in Table 5-VI appear in a fairly constant range between 1450 and 1400 cm^{-1}. The conclusion is inevitable that the symmetric scissors vibration is much more sensitive to the atom adjacent to the CH_3 group than is the asymmetric vibration. For the methyl halides it has been shown that this sensitivity results in part from a perturbation of the symmetric scissors vibration by another vibration of the same species [1; p.315]. Bellamy has suggested that the frequency shift of the symmetric scissors can be correlated with the electronegativity of the atom adjacent to the CH_3 groups [14].

The CH_3 rocking frequencies are not considered good group

TABLE 5-VIII. Vibrations of Molecules with CH_2 Groups

Compound	Physical state	CH_2 stretching		CH_2 scissors	CH_2 wag	CH_2 twist	CH_2 rock	Ref.
		Asym.	Sym.					
A. $-CH_2-C(=O)-$								
$CH_3CH_2-C(=O)-CH_2CH_3$	Solution			1414				26
$COOHCH_2CH_2COOH$	Solid			1422	1205	1180	803	46
B. CH_2 in Rings								
(cyclopentane ring)	Gas	2965 2944	2868 2876	1462 1487 1453	1283 1206 1258	1020 1104	1030 896 717 617	42
(cyclobutanone ring, β, α, α; =O)	Gas	3000(β) 2978(α)	2978(β) 2933(α)	1479(β) 1470(α) 1402(α)	1242 1209	1402(α) 1332(β) 1242(α) 1209(α)	1073(α) 735(β)	43
$CH_2-C{=}O$ / O O / $O{=}C-CH_2$ (glycolide ring)	Solid			1435	1335	1275	997	44

Compound	Physical state	CH_2 stretching		CH_2 scissors	CH_2 wag	CH_2 twist	CH_2 rock	Ref.
		Asym.	Sym.					
(ring)N=O	Liquid	2972Sh	2952 2884	1490 1460	1302 1282 1219	1302 1282 1219	1302 1282 1219 992	45
CH_2—C=O / NH NH / O=C—CH_2	Solid			1468	1340	1249	998	44
$Cl_2Pt(CH_2)_2CH_2$	Solid	3025 2954	2948Sh	1414M	1255	1165	1087VS 948M	47
$(C_3H_6)PtCl_2(py)_2$	Solid	2992W 2938M	2917M	1437Sh	1238W	1217M	1038S 976W	47
CH_2—CH_2 / O	Gas	3079S 3063S 3005VS		1490M 1470M	1153M 1120M	1143M	821S 807S	48
CH_2—O / CH_2—O PCl	In $CHCl_3$			Near 1480				49

TABLE 5-VIII (continued)

Compound	Physical state	CH_2 stretching		CH_2 scissors	CH_2 wag	CH_2 twist	CH_2 rock	Ref.
		Asym.	Sym.					
C. —CH_2—C								
$NCCH_2CH_2CN$	Liquid	2985 (gauche)	2965 (gauche)	1425 (gauche)	1338 (gauche) 1270 (*trans*)	1197 1230	962 (gauche) 951(R) (*trans*)	50
$CH_3CH_2CH_3$	Gas	2980		1468		1179 1152		6
$NCCH_2CN$	Gas			1422	1322	1222(*l*)	936(*l*)	51
CH_3CH_2Cl				1450	1280(R)	1245(R)		10
D. —CH_2—O								
CH_2—O—NO_2 / CH_2 / CH_2—O—NO_2				1430				26
Ethyl esters	In $CHCl_3$			1475–1460	1378–1366		873–838	52

Compound	Physical state	CH_2 stretching		CH_2 scissors	CH_2 wag	CH_2 twist	CH_2 rock	Ref.
		Asym.	Sym.					
E. CH_2—N								
$CH_3-C(=O)-N(H)-CH_2-CH_2-CH_3$				1439 (CH_2N)				4
CH_2N_2	Gas	3115	3074 3066	1416				58
$B[N(C_4H_9)_2]_3$	Solid			1466				53
F. —CH_2—NO_2								
$NaCH_2NO_2$	Solid	2920MW	2847W	1443W	985MS		1185W	54
G. CH_2X_2								
CH_2F_2	Gas	3015	2949	1508	1435	1262	1176	55, 83
CH_2Cl_2	Liquid	3045	2990	1424	1266	1157	899	56, 83
CH_2Br_2	Gas	3065	2988	1388	1190	1089	754	1 (p. 348), 83
CH_2I_2	Gas	3049	2967	1351	1107	1035	717	57, 83

TABLE 5-VIII (continued)

Compound	Physical state	CH_2 stretching		CH_2 scissors	CH_2 wag	CH_2 twist	CH_2 rock	Ref.
		Asym.	Sym.					
H. $NH_2CH_2CH_2NH_2$								
$NH_2CH_2CH_2NH_2$	Solution	2930 2917	2891 2858	1469 1456	1305 1298	1360	775 761	59
$Pt(NH_2CH_2CH_2NH_2)_2Cl_2$	Solid			1454M	1395W 1373M	1311M 1275M	897M	60
$Pt(NH_2CH_2CH_2NH_2)PtCl_4$	Solid			1472W 1459	1391M	1281VW 1318W	897W	60
$Pt(NH_2CH_2CH_2NH_2)Cl_2$	Solid	2945	2883	1467M 1453W	1366M	1309W	873W	60
$Cu(NH_2CH_2CH_2NH_2)_2PtCl_4$	Solid			1475W 1463W	1376M	1282M	889W	60
$Ni(NH_2CH_2CH_2NH_2)_2PtCl_4$		2947	2897	1463	1399VW 1373W	1282M	882VW	60
$Pd(NH_2CH_2CH_2NH_2)_2Cl_2$		2960	2875	1458	1395VW 1372M	1298W 1280	900W	60
$Pd(NH_2CH_2CH_2NH_2)Cl_2$		2969	2914	1475M 1460M	1369M	1307	880W	60
$Pd(NH_2CH_2CH_2NH_2)_2PtCl_4$				1475W 1464W	1373M	1317W	896W	60

Compound	Physical state	CH_2 stretching		CH_2 scissors	CH_2 wag	CH_2 twist	CH_2 rock	Ref.
		Asym.	Sym.					
I. —CH_2—M—								
$(C_4H_9)_2SnCl_2$	Crystal	2965 2930	2875	1468 1462 1440	756W 710WSh		874S	61
$(C_2H_5)_2P \cdot BH_2$	Solid			1422S	1242M 1260M			62
$(C_2H_5)_2P \cdot BCl_2$	Solid			1412MS	1244W			62
$(C_2H_5)_2P \cdot BBr_2$	Solid			1410VS	1239MW			62
$(C_2H_5)_2P \cdot BI_2$	Solid			1410S	1250MW			62
$(C_2H_5)_2Zn$	Liquid			1415W or 1465	1225			63
$(C_2H_5)_4Sn$	Liquid	3005 2900S	2950SSh	1428S	1235M		660VS 673VS	63
$(C_2H_5)_2Hg$	Liquid	3100 2930VS	2850SSh	1430	1332	1305W	672S 671S	63
CH_2B in ethyldiboranes		2925	2845	1435	1291 1245	1291 1245	826–755	64
$(C_2H_5)_2Cd$	Liquid	3010– 2840		1421M	1226M		663M 670M	65

The symbols Sh, M, W, S, V indicate shoulder, medium, weak, strong, and very, respectively. (*l*) = liquid; (R) = Raman.

frequencies. For many of the molecules listed in Table 5-VI the rocking frequencies appear as medium to strong bands in a region from 1300 to 800 cm^{-1}. The number and type of CH_3 rocking frequencies will be determined by the symmetry of the molecule. A comparison of a series of somewhat similar symmetry such as $(CH_3)_4M$, where M is C, Si, Ge, Sn, or Pb, shows that the rocking vibration moves to a progressively lower position as M is changed from C to Pb.

While some trends seem to exist for the rocking vibrations of related molecules, the variation in position is too great to allow them to be generalized into a group frequency assignment.

5.4D. CH_2 Groups Adjacent to Atoms Other Than Saturated Carbon

In a previous section we have seen that a CH_2 structural group in saturated alkanes has characteristic group frequencies. In this section, we shall consider CH_2 groups in structures where an atom other than a saturated carbon is adjacent to the CH_2 as, for example, in compounds such as $NH_2CH_2CH_2NH_2$ and CH_2Cl_2. For the present we shall not discuss structures like $CH_2{=}CR_1R_2$, where the CH_2 is a terminal group.

To find characteristic group frequencies for the CH_2 group, we can utilize the vibrational analyses reported for molecules containing this group. In Table 5-VIII, the vibrational analyses of a number of these molecules are presented. Only vibrations which may be characteristic group frequencies are listed. A CH_2 structural unit can have a symmetric and an asymmetric stretch, a symmetric scissors, and wags, twists, and rocks. These are the general classifications listed in Table 5-VIII.

The $-CH_2-\overset{\overset{\displaystyle O}{\|}}{C}-$ *Group.* Let us consider first the influence of a carbonyl group adjacent to a CH_2 group, as for example in ketones with the structure $-CH_2-\overset{\overset{\displaystyle O}{\|}}{C}-$.

The asymmetric and symmetric stretching frequencies of CH_2 do not appear to be shifted to any great extent by the presence of the carbonyl group.

The symmetric scissors of the $-CH_2-\overset{\overset{\displaystyle O}{\|}}{C}-$ group shifts from 1465 cm^{-1} to the 1435–1405 cm^{-1} range. This range for $-CH_2-\overset{\overset{\displaystyle O}{\|}}{C}-$

scissors has been reported as a characteristic group frequency by other workers (see footnote 4, Table 5-IX).

The $-CH_2-\overset{\displaystyle O}{\overset{\|}{C}}-$ group may be part of a ring structure, as in the following compounds:

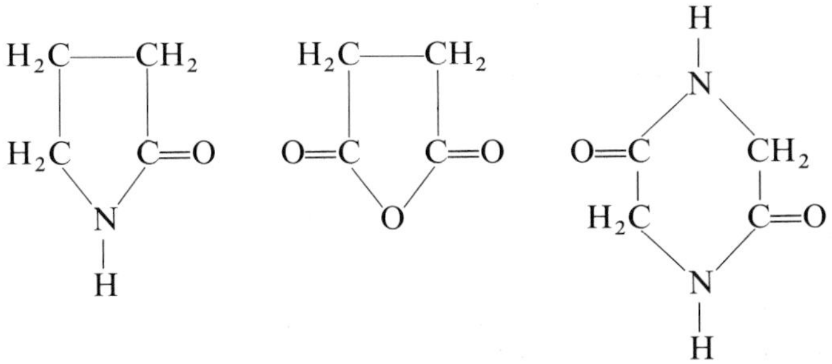

```
                                                       H
                                                       |
H2C——CH2          H2C——CH2                             N
 |     |           |     |                           /   \
H2C   C=O         O=C   C=O                      O=C       CH2
  \   /              \  /                          |        |
    N                 O                           H2C      C=O
    |                                                \     /
    H                                                   N
                                                        |
                                                        H
```

For these molecules several scissors vibrations appear higher than 1435 cm^{-1}. It is possible that the presence of the nitrogen group or the ring structure is influencing the position of the CH_2 scissors. Cyclobutanone has its CH_2 scissors at 1479 and 1470 cm^{-1}, both of which are higher values than are found for the $-CH_2-\overset{\displaystyle O}{\overset{\|}{C}}-C\lessgtr$ structure in acyclic systems. We list a general range of 1475–1425 cm^{-1} for this group in ring systems.

A wagging vibration for the group $-CH_2COOH$ is also listed as a group frequency in Table 5-IX.

The $-CH_2-O-$ *Group.* It is difficult to find examples of molecules having $-CH_2-O-$ groups that do not also have CH_3 groups. Some ring structures such as ethylene oxide appear to have a CH_2 scissors higher than 1470 cm^{-1}, the position for this group in alkanes. However, there is also a second vibration which can be described as a CH_2 scissors. For example in ethylene oxide, bands at 1490 and 1470 cm^{-1} can be described as scissors. The structure

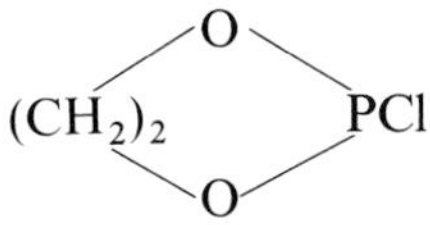

```
        O
      /   \
(CH2)2      PCl
      \   /
        O
```

has a scissors at 1480 cm^{-1}. It is therefore necessary to list the $-CH_2-O-$ group in ring systems as a group frequency distinct from that of the same group in nonring systems.

TABLE 5-IX. Characteristic Group Frequencies of the CH_2 Group Adjacent to Polar Groups

Structural unit	Vibration	Spectral range (cm^{-1})
Cyclopropane derivatives	Ring	1020–1000[1]
Cyclobutane derivatives	Ring	1000–960[1]
Cyclohexane derivatives	Ring	1005–925 and 1055–1000[1]
$—CH_2—C(=O)—$ (small ring systems)	Scissors (several bands)	1475–1425
$—CH_2—C(=O)—$ (acyclic systems)	Scissors	1435–1405[4]
$—CH_2—NO_2$	Scissors	1425–1415[2]
$NCCH_2CH_2CN$	Scissors	1425
$NCCH_2CN$	Scissors	1420
$—CH_2—N$ (amides)	Scissors	1440[3]
$—CH_2—O—$ (small ring systems)	Scissors (several bands)	1500–1470
$—CH_2—O—$ (acyclic systems)	Scissors	1470–1435
$—CH_2—O—$ (esters)	Scissors	1475–1460[6]
$—CH_2—S—$	Scissors	1415
$—CH_2—$Metal	Scissors	1465–1400 (depending on metal)
$NH_2CH_2CH_2NH_2$ (in metal complexes)	Scissors (2 bands) Wag	1480–1450 1400–1350
$—(CH_2)_{(x \geq 4)}O—$		742–734
$—CH_2—C(=O)OH$	CH_2 wag	1200[5]
$—CH_2X_2$ (X = halogen)	Scissors	1435–1385

[1] Assignments from L. J. Bellamy, *The Infrared Spectra of Complex Molecules*, John Wiley & Sons, Inc., New York (1958).

[2] Assignment reported by F. Erkstein, P. Gluzinski, W. Sobotka, and T. Urbanski, *J. Chem. Soc.* 1370 (1961).

In a compound such as

$$\begin{array}{l} CH_2\text{—}O\text{—}NO_2 \\ | \\ CH_2 \\ | \\ CH_2\text{—}O\text{—}NO_2 \end{array}$$

a band near 1430 cm^{-1} is probably the CH_2 scissors, and in general, the —CH_2—O— vibrations in acyclic structures appear near 1430 cm^{-1}.

A band for the structural unit —$(CH_2)_{(x \geq 4)}$—O— in the 742–734 cm^{-1} region is related to the —$(CH_2)_{(x \geq 4)}$—C— band near 720 cm^{-1} in alkanes. It is listed as a characteristic group frequency in Table 5-IX.

The —CH_2—N *Group and Other* —CH_2—M *Groups.* The group —CH_2—N in amines and amides appears to have symmetric and asymmetric stretching vibrations in the expected positions.

The scissors vibration appears in the 1450–1405 cm^{-1} range. In compounds such as $CH_3\text{—}\overset{\overset{\large O}{\|}}{C}\text{—}\overset{\overset{\large H}{|}}{N}\text{—}CH_2\text{—}CH_3$ the scissors is near 1440 cm^{-1}. This has been reported as a characteristic frequency for amides by Sutherland and his co-workers [4], and is listed in Table 5-IX as such. In ethylenediamine the scissors is at 1460 cm^{-1} while in ethylenediamine complexes with such metals as Pt, Ni, Cu, and Pd it appears in the 1480–1450 cm^{-1} range.

In —CH_2NO_2 structures, the scissors is in the 1425–1415 cm^{-1} range and is listed as a characteristic group frequency (see footnote to Table 5-IX). Examining the spectra of compounds where the CH_2 group is adjacent to such elements as Si, B, and Sn suggests that the scissors vibration does shift slightly from the 1465 cm^{-1} position. For example, in dibutyltin dichloride the CH_2 scissors is near 1453 cm^{-1}. In $(C_2H_5)_2PBH_2$ it is at 1412 cm^{-1}, while in $(C_2H_5)_2Hg$ it is at 1430 cm^{-1}.

---

(3) Assignment reported by M. Beer, H. B. Kesseler, and G. B. B. M. Sutherland, *J. Chem. Phys.* **29**(5): 1097 (1958).

(4) Assignments similar to those reported by B. Nolin and R. N. Jones, *J. Am. Chem. Soc.* **75**: 5626 (1953).

(5) Assignment reported by T. Shimanouchi, M. Tsuboi, T. Takenishi, and N. Iwata, *Spectrochim. Acta* **16**: 1328 (1960).

(6) A. R. Katritzky, J. M. Lagowski, and J. A. T. Beard, *Spectrochim. Acta* **16**: 954 (1960).

In the compounds CH_2X_2, where X is a halogen, the scissors of the CH_2 group moves from 1435 cm^{-1} in CH_2F_2 to 1388 cm^{-1} in CH_2Br_2.

Other Frequencies of the —CH_2— Group. The wagging, twisting, and rocking frequencies are quite sensitive to the groups adjacent to the CH_2 group. For example, the rocking frequency in CH_2Cl_2 is at 899 cm^{-1}, while the same vibration in CH_2Br_2 is at 807 cm^{-1}. For a series of related compounds such as ethylenediamine complexes of Pt, Cu, and Pd, the wagging, twisting, and rocking frequencies appear in similar regions. Even here, however, some shifting of these bands is noted. For example, for $Pd(en)_2Cl_2$* the rock is at 900 cm^{-1}, while for $Pd(en)Cl_2$ bands at 868 and 821 cm^{-1} are identified as rocking frequencies. The wagging vibrations for these two compounds are near 1380 and 1353 cm^{-1}, respectively.

Metal atoms adjacent to the CH_2 group also shift the scissors vibration slightly from the normal 1465 cm^{-1} position.

In general the CH_2 scissors is therefore a good group frequency, only slightly perturbed by its environment.

5.4E. Saturated Ring Systems

The CH_2 group can be present in a ring structure as for example in cyclopropane or cyclobutane. It is possible for ring deformation modes to be characteristic group frequencies for these types of ring compounds. Cyclopropane has at least two vibrations which can be described as ring deformation modes. A fairly constant band in the 1020–1000 cm^{-1} region found for many cyclopropane derivatives appears to be a characteristic group frequency. It is probably due to a ring vibration, which for cyclopropane is found at a frequency near 1025 cm^{-1}.

A CH_2 scissors vibration can also be visualized for CH_2 ring systems, and usually one or more bands are found near 1465 cm^{-1} which are characteristic group frequencies of the CH_2 group.

A small ring such as cyclopropane can have the CH stretching vibrations appear higher than the usual positions expected for the CH_2 group in alkane structures. For example, in cyclopropane a CH stretch near 3040 cm^{-1} is observed. Small-ring systems appear to have CH_2 scissors vibrations shifted slightly from the expected position. However, since more than one vibration can be described as a scissors of the CH_2 group, this correlation appears not as satisfactory as others.

Characteristic group frequencies (near 1000 cm^{-1}) have been suggested for cyclobutane and cyclohexane derivatives, and these are so listed in Table 5-IX. Vibrational analysis of small ring compounds is discussed in Reference (84).

* en = $NH_2CH_2CH_2NH_2$.

TABLE 5-X. Groups with Characteristic Frequencies in the Same Regions as the Characteristic Group Frequencies of the CH_3 and CH_2 Groups

Group (or compound)	Spectral region (cm^{-1})
A. 3000–2800 cm^{-1}	
—OH (hydrogen-bonded) e.g., carboxylic acid dimers, enolized β-diketones, tropolones	3300–2500
—NH (hydrogen-bonded) e.g., in amine salts	2900–2300
B. 1500–1175 cm^{-1}	
α, α-dihalo nitro	1340–1325
α-halo nitro	1355–1340
α, β-unsaturated nitro	1360–1335
Tertiary nitro	1360–1340
Primary, secondary nitro	1385–1360
Nitramines	1300–1260
N—N=O	1500–1440
Pyridine derivatives	1300–1250
—N—C(=S)—N—	1430–1130
R—O—SO_2—R′	1420–1330
$(RO)_2SO_2$	1440–1350
RSO_2Cl	1375–1340
RSO_2N—	1370–1300
CF_3—CF_2—	1365–1325
CF_3 (attached to aryl)	1330–1310
—P—C_6H_5	1450–1435
—P—CH_3	1320–1280
—P=O	1350–1175
Si—C_6H_5	1430–1425
—B—C_6H_5	1440–1430
—B—O—	1350–1310
—B—N—	1380–1330
HCO_3^-	1420–1400
NH_4^+	1485–1390
N_3^-	1375–1175
NO_3^-	1410–1340
NO_2^+	1410–1370
=CH_2	1460
—B—O—	1445
	1425
Aromatics	1500–1400

TABLE 5-XI. Groups Interfering with the CH_3 and CH_2 Group Frequencies

Vibration	Normal position	Groups absorbing in the same region			
$C{-}(CH_3)_2$	1175–1165 1170–1140	$La(NO_3)_3$	1140;	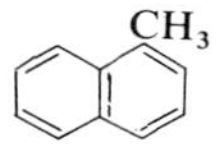 CH_3	1171;
		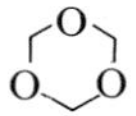 O O O	1168;	$CH_2{=}CHCH{=}CHCH{=}CH_2$	1166;
		$(CH_3)_2C{=}C(CH_3)_2$	1167;	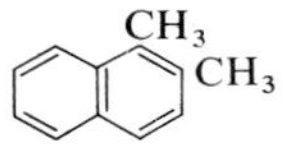 CH_3 CH_3	1153;
		$CH_3CH_2CH(NO_2)CH_3$	1144;	$CH_2{=}CHCH{=}CHCH{=}CHCH{=}CH_2$	1140;
		 N CH_3	1149		
$C{-}(CH_3)_2$	840–790	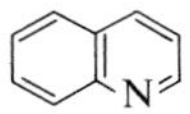 N HCl	795;	$Pb(NO_3)_2$	800;
		$CH_3CH{=}C(CH_3)CH_2CH_3$	823;	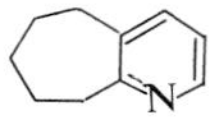 N	803;
		$Cu(NO_3)_2 \cdot 6H_2O$	835;	$Pb(NO_3)_2$	805;
		$CH_3CH_2CH(NO_2)CH_3$	845;	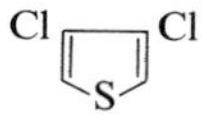 Cl S Cl	836;
		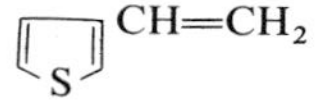 S $CH{=}CH_2$	829;	CH_3CH_2OH	802;

Vibration	Normal position	Groups absorbing in the same region			
		$(CH_3)_3CNO_2$	801;	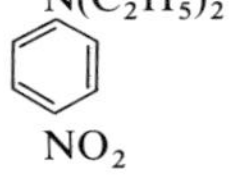 $N(C_2H_5)_2$ / NO_2	810
C—$(CH_3)_3$	1255–1245 1250–1200	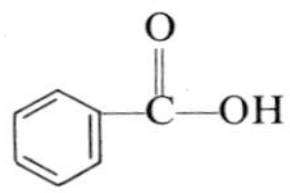 O=C—OH	1250;	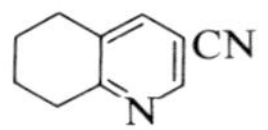 CN / N	1248; 1240;
		CH_3NO_2	1211;	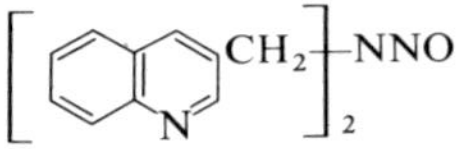 [N CH_2]$_2$—NNO	1245;
		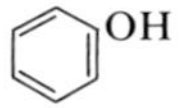 OH	1235;	CH_3—CF_3	1230;
		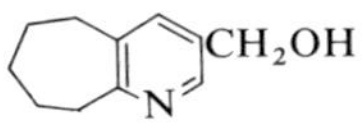 N / CH_2OH	1227;	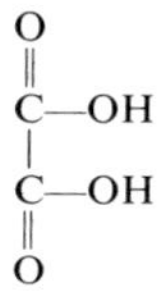 O=C—OH / C—OH=O	1224;

TABLE 5-XI (continued)

Vibration	Normal position	Groups absorbing in the same region			
		$CH_2(C\equiv N)_2$	1220;	$CH_3-C(=O)-O-CH_3$	1218;
		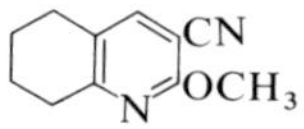 CN, OCH_3, N	1217;	$CH_3-O-C(=O)-H$	1214;
		$CClF_3$	1210;	$CH_3-C(=O)-O-CH_3$	1204;
		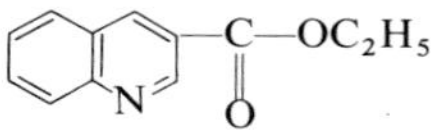 N, $-C(=O)-OC_2H_5$	1202;		1205;
		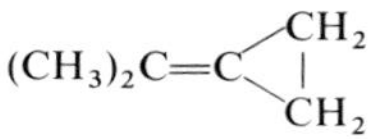 $(CH_3)_2C=C(CH_2)(CH_2)$	1239;	$C_6H_5NO_2$	1247;
		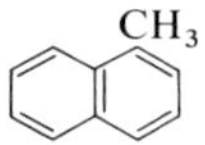 CH_3	1216;	$CH_3CH=C(CH_3)CH_2CH_3$	1211;
		$CH_3(CH_2)_4OH$	1232		

Vibration	Normal position	Groups absorbing in the same region			
C—CH_3	1470–1435	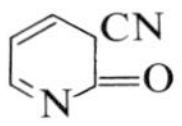 CN, N, =O	1465;	$Pr(NO_3)_3$	1455;
		$(C_6H_5)_3SiCl$	1464;	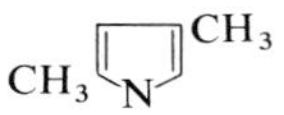 CH_3, N, CH_3	1448;
		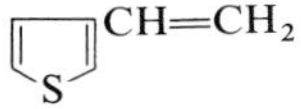 $CH{=}CH_2$, S	1438;	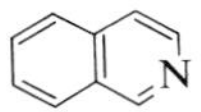 N	1445;
			1452		
C—CH_3	1385–1370	$Ce(NH_4)_2(NO_3)_6$	1385;	KNO_3	1380;
		$C_6H_5SiHCl_2$	1379;	$LiNO_3$	1370;
		$MgNO_3$	1365		
C—$(CH_3)_2$	1385–1380 1370–1365	Same as above and including			
		$(C_6H_5)_3SiCl$	1376;	CH_3NO_2	1379

TABLE 5-XI (continued)

Vibration	Normal position	Groups absorbing in the same region			
C—$(CH_3)_3$	1395–1385 1365	Same as above and including		CH_3 N CH_3	1392;
		O, CH_3, NO_2, N—C_3H_7-*n*	1389;	$CH_3\overset{NO_2}{C}HCH_3$	1396;
		$(CH_3)_2CH$, $CH(CH_3)_2$, $CH(CH_3)_2$	1365;	CH_3, CH_3	1365;
		O, CH_3, NO_2, N—C_3H_7-*n*	1365;	$CH_3CH_2C(CH_3)_2CH_2CH_3$	1365

Vibration	Normal position	Groups absorbing in the same region			
$—CH_3$	2975–2950 2885–2860	CH_2I_2	2967;	$CH_2{=}CHCH{=}CHCH{=}CH_2$	2877;
		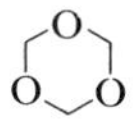 O O O (ring structure)	2869		
$—CH_2—$	2940–2915 2870–2845	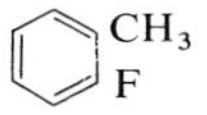 (benzene ring with CH_3, F)	2934;	$(CH_3)_2C{=}C(CH_3)_2$	2865;
		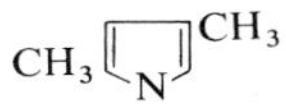 CH_3 (ring with N) CH_3	2870;	$(CH_3)_3CNO_2$	2857

5.4F. Groups Having Absorption Bands Which Interfere with Group Frequency Assignments of CH_3 and CH_2 Groups

The preceding sections on group frequencies of CH_3 and CH_2 groups may have given the impression that it is always possible to identify the CH_3 and CH_2 structural units from their group frequencies. In the laboratory, where the unknowns to be identified may be impure, it is not always possible to be absolutely certain of the presence of the CH_3 and CH_2 groups. Certainly observation of only a single characteristic group frequency cannot be considered sufficient evidence that a certain group is present since many groups can give bands in the same region. This is especially true of group frequencies other than those of the CH_3 and CH_2 groups since most group frequencies are far less distinct than these.

It will be the purpose of this section to list the group frequencies that can interfere with the group frequencies of the CH_3 and CH_2 groups and to examine the spectra of complex compounds for possible identification of the frequencies. A partial listing of compounds that have interfering bands which are not group frequencies also will be given. It should be kept in mind that, when one group frequency cannot be identified because it is obscured by other bands, other group frequencies of the same group may be observable and these may still permit an identification of the structural group. An experienced interpreter, therefore, uses all the frequencies associated with a particular group in correlating infrared absorption bands with structure.

Structural groups having absorption bands which interfere with the group frequency assignments of CH_3 and CH_2 units can be classified into two types. The first type includes those having characteristic group frequencies in some of the same regions as the CH_3 and CH_2 groups. The second type includes specific compounds having an absorption band not generally classified as a group frequency in the same region as the group frequencies of the CH_3 and CH_2 groups. Examples of the first type can be easily predicted. Compounds of the second type, however, are difficult to list, since only by observing their spectra can the interference be detected.

In Table 5-X the first type of interfering groups is listed. Let us first examine the stretching vibrations. The frequency range 3000–2800 cm^{-1} is listed as the region where CH_3 and CH_2 stretching vibrations are found. This rather broad region has been chosen to allow for the shifts which can occur for the CH_3 and CH_2 stretching vibrations under all possible conditions.

From the listing in Table 5-X, we can see that hydrogen-bonded OH and NH groups can have bands in the CH stretching region. Fortunately, the OH and NH groups usually have other distinct vibrations that can be used to recognize their presence. Some examples of this will be given in the next section.

The CH_3 and CH_2 scissors vibrations can appear in a region from 1500 to 1175 cm^{-1}. The list of groups having interfering frequencies in the scissors region is therefore rather extensive. We have seen previously that the CH_3 and CH_2 scissors are considered good group frequencies. In spite of the large number of groups which may interfere with the assignment of CH_3 and CH_2 scissors, these frequencies can still be recognized in the spectra of many complex compounds. Band intensity and sharpness and the presence of other related group frequencies can normally be used to distinguish the CH_3 and CH_2 scissors from most interfering groups.

No attempt will be made to list the group frequencies interfering with the rocking, twisting, and wagging vibrations of CH_3 and CH_2 groups since they absorb in too broad a region and are not considered good group frequencies.

In Table 5-XI a brief listing is given of compounds having absorption bands in the regions occupied by the characteristic group frequencies of the CH_3 and CH_2 groups. These vibrations are not characteristic group frequencies for the compounds listed, although they may be indicative of general classes of compounds absorbing these regions. This table is given merely to illustrate that characteristic group frequencies can often be obscured by bands of other compounds.

5.4G. The Spectra of Compounds with CH_3 and CH_2 Groups

A large number of spectra could be presented in this section, but to do so would only emphasize the fact that in many compounds the characteristic group frequencies of the CH_3 and CH_2 groups are merely a few of many bands, sometimes difficult to identify, but rarely obscured completely. In general they are recognizable and very useful in structural identification.

A few selected spectra will suffice to illustrate this point. Figures 5-4, 5-5, and 5-6 show the 1500–1300 cm^{-1} scissors region for a series of sulfides, thiols, alcohols, and nitro compounds. We expect to find bands near 1465 and 1380 cm^{-1} representing the scissors of the CH_3 and CH_2 groups. In addition, if isopropyl or *t*-butyl groups are present, the 1380 cm^{-1} band should be split. Upon examination of the spectra shown in these three figures we do indeed find that, in general, the group frequencies of the CH_3 and CH_2 groups can be identified and that they do appear near the frequencies expected for them. However, we also note that their intensity relative to each other and to other bands often differs. Moreover, the position and number of absorption bands differ for related compounds. Only for

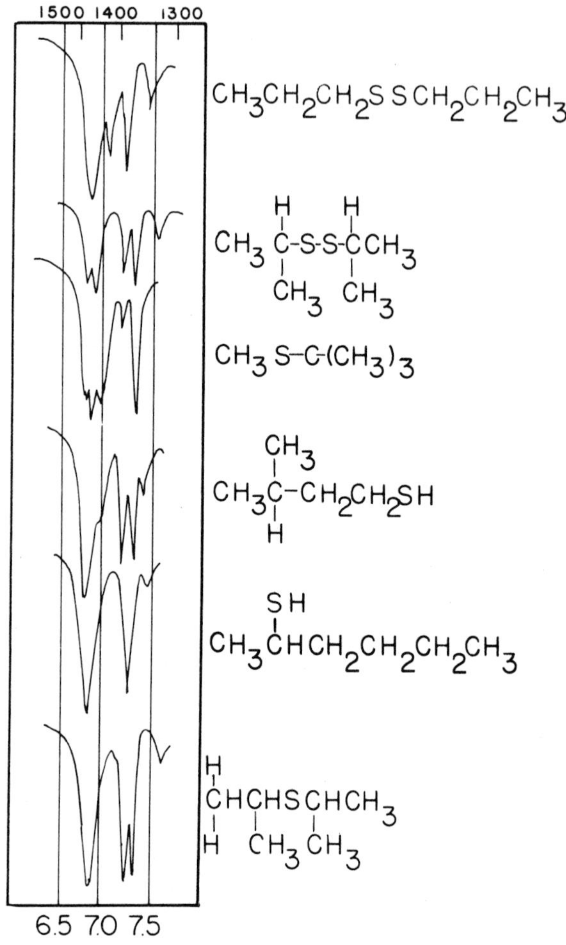

Figure 5-4. The 1500–1300 cm^{-1} region for a series of sulfides and thiols.

the long-chain compounds do the CH_3 and CH_2 vibrations appear distinct. The series of compounds whose spectra are given in Figures 5-4, 5-5, and 5-6 were chosen so that vibrations other than those of CH_3 and CH_2 groups would appear. It is therefore not surprising that in these spectra it is difficult to separate the CH_3 and CH_2 group frequencies from other bands. In many spectra one will find much the same problem in identifying all the group frequencies. In

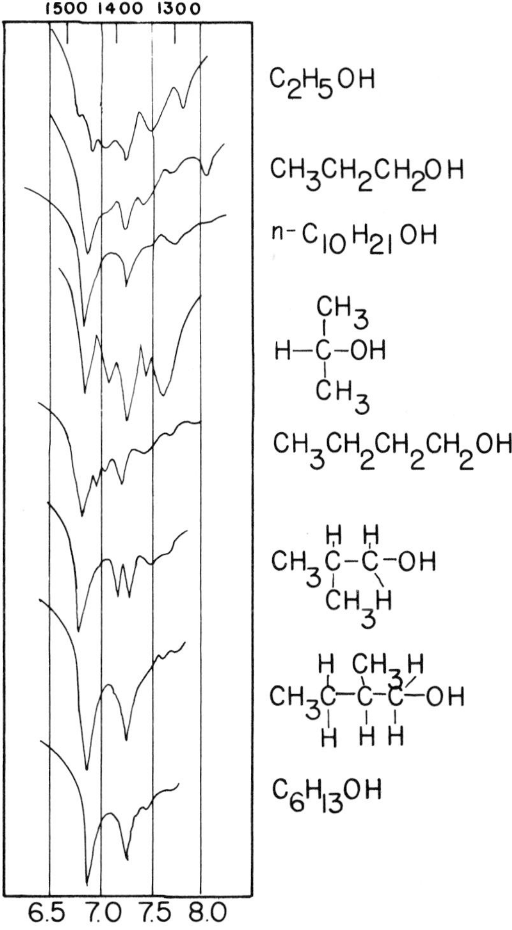

Figure 5-5. The 1500–1300 cm^{-1} region for a series of alcohols.

many instances, unless supplementary data which indicate the type of groups present in a compound are available, it is difficult to assign the structure of the compound completely from the observed bands.

It should be noted that there is a resemblance of band contours for a series of compounds, such as alcohols, in a region such as that of the CH_3 and CH_2 scissors vibrations. This kind of knowledge is often used by an experienced interpreter to identify possible groups.

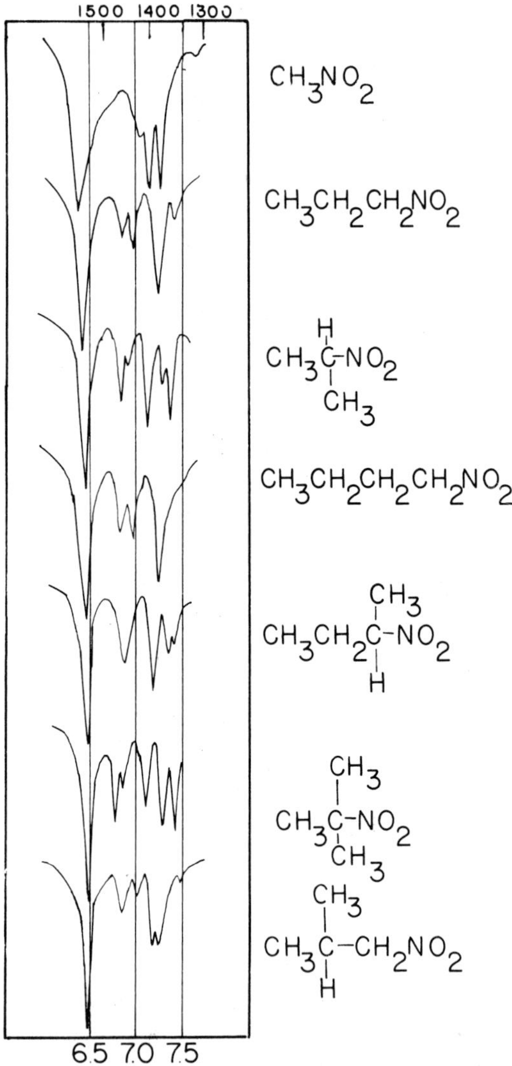

Figure 5-6. The 1500–1300 cm^{-1} region for a series of nitro compounds.

It is difficult to systematize this type of knowledge so that a beginning interpreter of spectra can utilize it, and yet it is often useful.

The spectra illustrated in Figures 5-4, 5-5, and 5-6 are representative of the influence of negative groups on the CH_3 and CH_2 group frequencies.

The CH_3 and CH_2 groups can also be present in olefinic and aromatic compounds and these will be discussed next.

5.4H. CH_3 and CH_2 Group Frequencies in Alkenes

If we examine the spectra of simple olefinic compounds containing CH_3 and CH_2 groups, we shall find that the group frequencies of the CH_3 and CH_2 groups can usually be identified.

The CH stretching of the structural unit =CH is in the 3040–3010 cm^{-1} range, which is higher than the CH stretching vibrations of the CH_3 and CH_2 groups.

The CH_3 and CH_2 scissors vibrations near 1465 and 1380 cm^{-1} are not perturbed to any great extent when the CH_3 or CH_2 group is adjacent to a double bond. For example, in Figure 5-7 the spectrum of 3-methyl-cis-2-hexene is presented. The two scissors vibrations at 1465 and 1380 cm^{-1} are quite distinct.

A system such as $CH_2{=}CR_1R_2$ has a characteristic group frequency of the $=CH_2$ group in the 1420–1410 cm^{-1} region which cannot always be distinguished from the 1465 cm^{-1} scissors of the CH_3 and CH_2 groups. For the compound $CH_2{=}CHCH_2Cl$, which has no CH_3 group, the 1420 cm^{-1} band of the $=CH_2$ group is easily recognized (see Figure 5-8). The spectrum of

$$\begin{array}{l}CH_2{=}CHCH_2CH_2\overset{}{C}HCH_3\\ \qquad\qquad\qquad\qquad\ \ |\\ \qquad\qquad\qquad\qquad\ CH_3\end{array}$$

is shown in Figure 5-9. The strong band at 1465 cm^{-1} represents the CH_3 and CH_2 scissors while that at 1420 cm^{-1} is due to the $=CH_2$ group.

In the spectra of many liquid olefins containing the $=CH_2$ group, the 1420 cm^{-1} band is often obscured by the strong 1465 cm^{-1} band. Possibly if the solution spectra of these olefins were determined, the 1420 cm^{-1} band would be discernible.

Dienes such as 1,3-butadiene have a strong 1420 cm^{-1} band. Dienes containing CH_3 groups often have only a single band near 1420 cm^{-1} instead of the two bands expected at 1465 and 1420 cm^{-1}. Some dienes have a series of three or four maxima in the 1500–1400 cm^{-1} range, making it difficult to recognize the presence of CH_3 and CH_2 groups in these dienes.

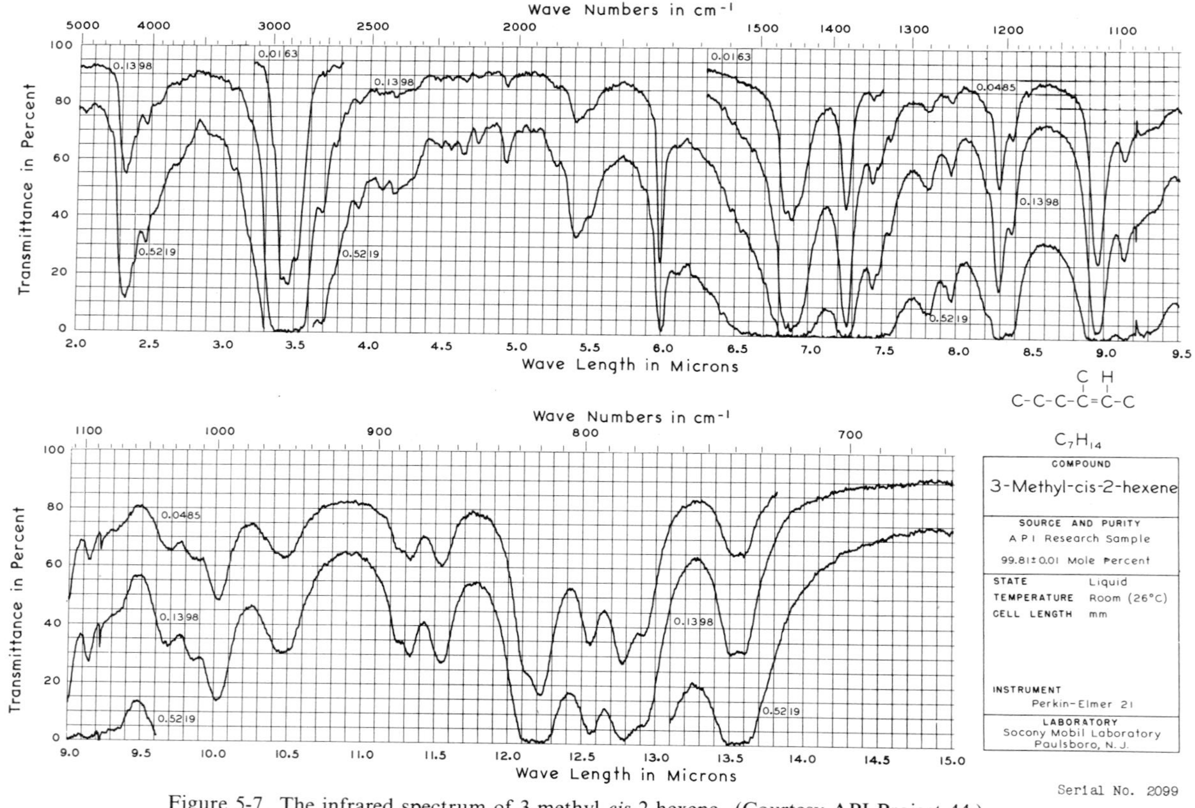

Figure 5-7. The infrared spectrum of 3-methyl-*cis*-2-hexene. (Courtesy API Project 44.)

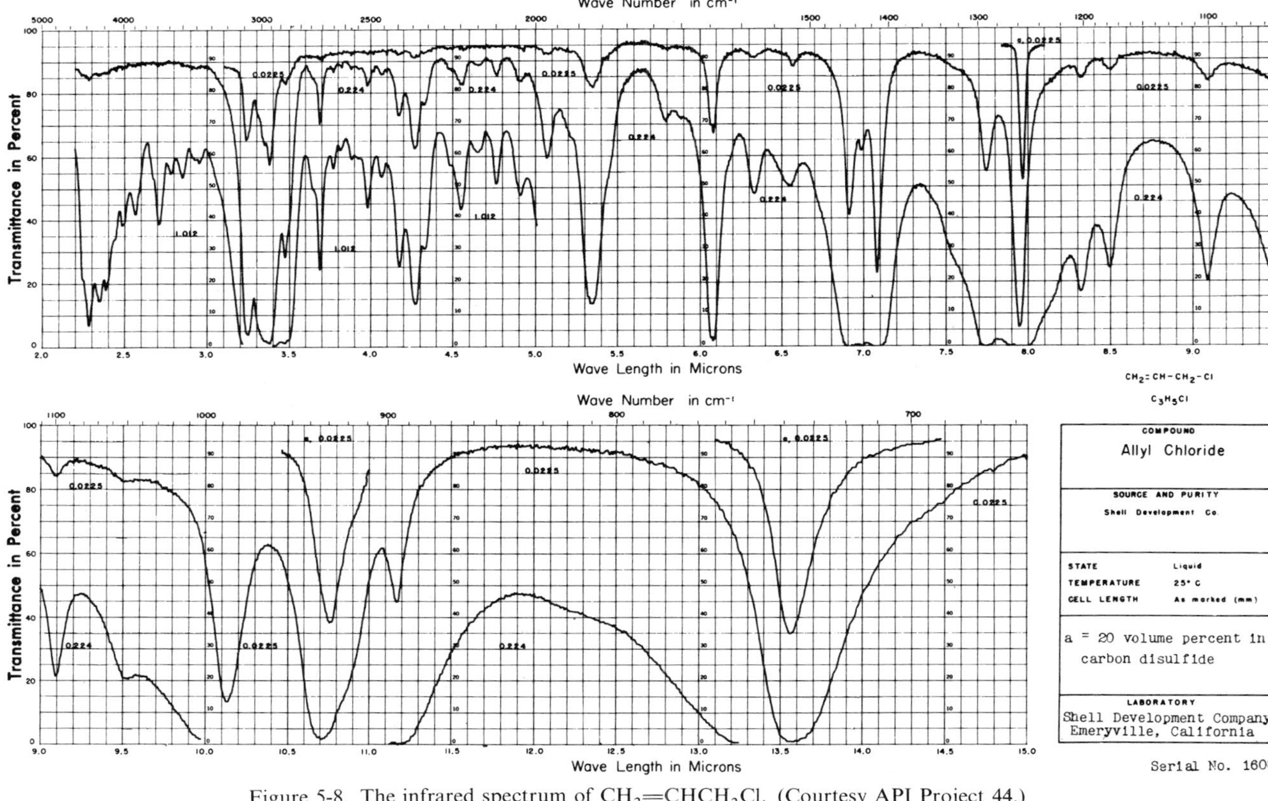

Figure 5-8. The infrared spectrum of $CH_2{=}CHCH_2Cl$. (Courtesy API Project 44.)

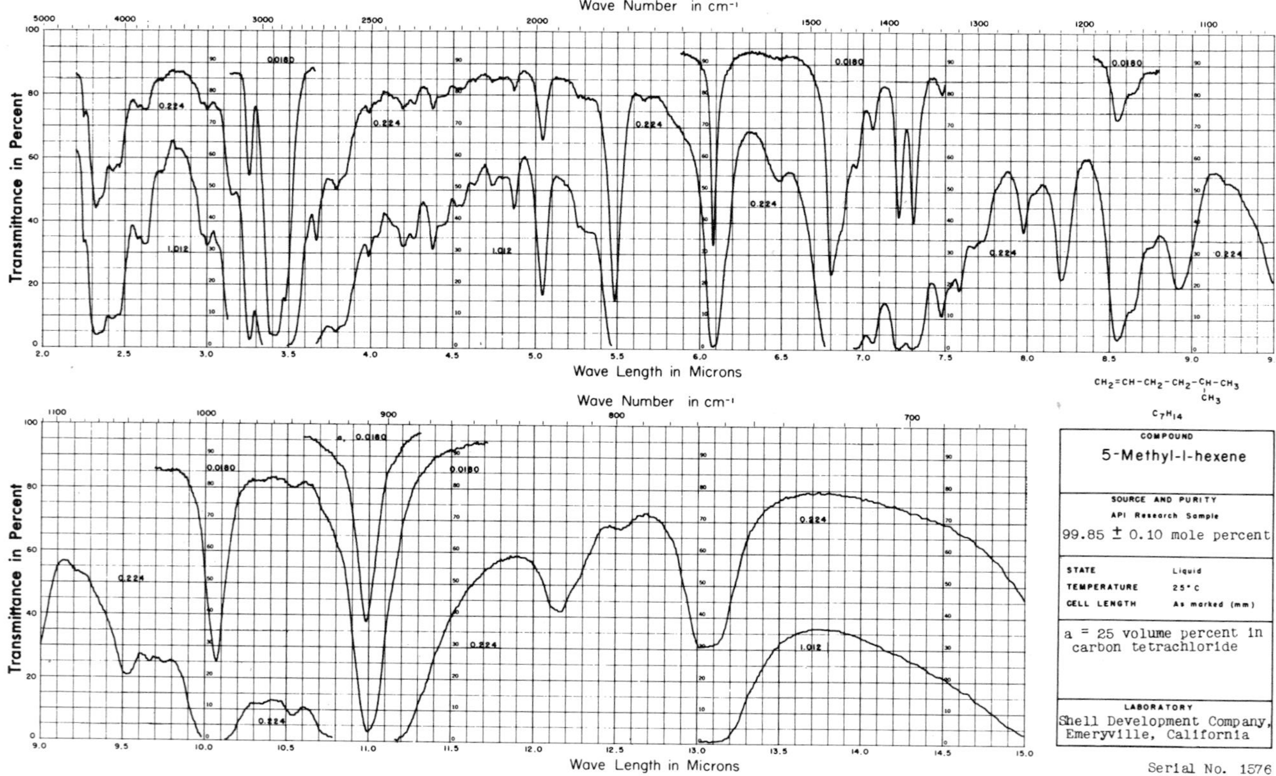

Figure 5-9. The infrared spectrum of $CH_2{=}CHCH_2CH_2CH(CH_3)_2$. (Courtesy API Project 44.)

The presence of a chlorine atom adjacent to the double bond does not appear to shift or obscure the group frequencies of the CH_3, CH_2, and $=CH_2$ vibrations in the 1500–1350 cm^{-1} region. In some instances enhancement of band intensity occurs when halogens are present in the molecule.

In compounds having —O—, $—CH_3$, $—CH_2$, and C=C groups, the group frequencies of the CH_3, CH_2, and $=CH_2$ vibrations may be difficult to identify. In part, the difficulty of identification may be due to the method of sample preparation. For example, an ethylenic acid may be a solid and the spectrum of the material may have broad bands, making it difficult to assign vibrations. The spectrum of

$$\begin{array}{l} CH_2{=}CCOOH \\ \qquad\quad | \\ \qquad\quad CH_2COOH \end{array}$$

has one broad band in the 2850 cm^{-1} and another in the 1440 cm^{-1} region, making it difficult to separate the group frequencies of the $=CH_2$ and CH_2 groups. The presence of an oxygen atom can also change the relative intensity of bands, thus making it more difficult to identify the group frequencies. To illustrate some of the variations which occur in intensity and position of group frequencies in olefinic compounds containing oxygen atoms, a brief listing of the vibrations observed for some of these compounds is given in Table 5-XII.

The structural units which may have group frequencies appearing in the same region as those of the CH_3 and CH_2 groups were listed in Table 5-X. Alkenes having these groupings would of course be expected to have spectra in which the CH_3 and CH_2 groups would be difficult to identify.

5.4I. CH_3 and CH_2 Group Frequencies in Aromatic Compounds

The characteristic group frequencies for CH_3 and CH_2 structural units when these units are attached to aromatic ring systems remain fairly constant in position and generally are not obscured by aromatic ring vibrations. However, an aromatic ring vibration near 1500 cm^{-1} can sometimes obscure the 1465 cm^{-1} scissors vibration of the CH_3 and CH_2 groups for some molecules, and for solid aromatic compounds, where sample preparation is difficult, the bands may be broad and may overlap.

In simple compounds such as toluene or ethylbenzene the CH_3 and CH_2 scissors and stretching vibrations can be easily distinguished, especially if the spectrum of the sample is determined in solution. The spectra of the pure liquids have somewhat broader bands but

TABLE 5-XII. The CH_3 and CH_2 Group Frequencies in Olefins Containing Oxygen Atoms

Compound	State	Prism	CH_3 and CH_2 stretching vibrations (cm^{-1})	CH_3 and CH_2 scissors and other vibrations (cm^{-1})
$H_2C{=}CHOCH_2CH_2CH_2CH_3$	Liquid	NaCl	3150W 2890S	1460M, 1375M
$CH_3CH{=}CHC({=}O)OH$	Solid	NaCl	Broad band near 3000	1450S, 1440Sh, 1375M
$CH_2{=}CHCH_2OH$	Liquid	NaCl		1430S
$CH_2{=}C(CH_2COOH){-}COOH$	Solid	NaCl	Broad band centered near 2850	Broad band at 1440 and sharp band at 1320
$CH_2{=}CHOCH(CH_3)_2$	Liquid	NaCl	3150W 2950S (split)	1460S (split) 1370S 1340S 1320S
$CH_3CH_2{-}O{-}CH{=}CH_2$	Liquid	NaCl	3150W 2975S 2850M	1470W 1445W 1380S

the vibrations are still distinguishable (for examples, see Figures 5-20 to 5-25). If isopropyl or *t*-butyl groups are substituted on the benzene ring, the expected splitting of the symmetric scissors of the CH_3 group is also observable. The hexa-substituted benzene ring, where the substitution is methyl or ethyl and the sample is determined in KBr, makes the scissors vibrations of the CH_3 and CH_2 groups quite broad. The spectrum of the ethyl compound is given in Figure 5-10. Table 5-XIII lists the group frequencies observed for a number of compounds having aromatic, CH_3, and CH_2 structural units, and includes the compounds under discussion in this section.

Polyaromatic ring systems, such as the compound

$$C_6H_5CHCH_3C_6H_5,$$

generally have spectra in which the 1500 cm^{-1} aromatic ring vibration is distinct from the 1465 and 1380 cm^{-1} scissors of the CH_3 group, although for some of these compounds, listed in Table 5-XIII, a band near 1400 cm^{-1} is also present. The spectrum shown in Figure 5-11 for a compound of this type shows six bands in the 1500–1350 cm^{-1} region.

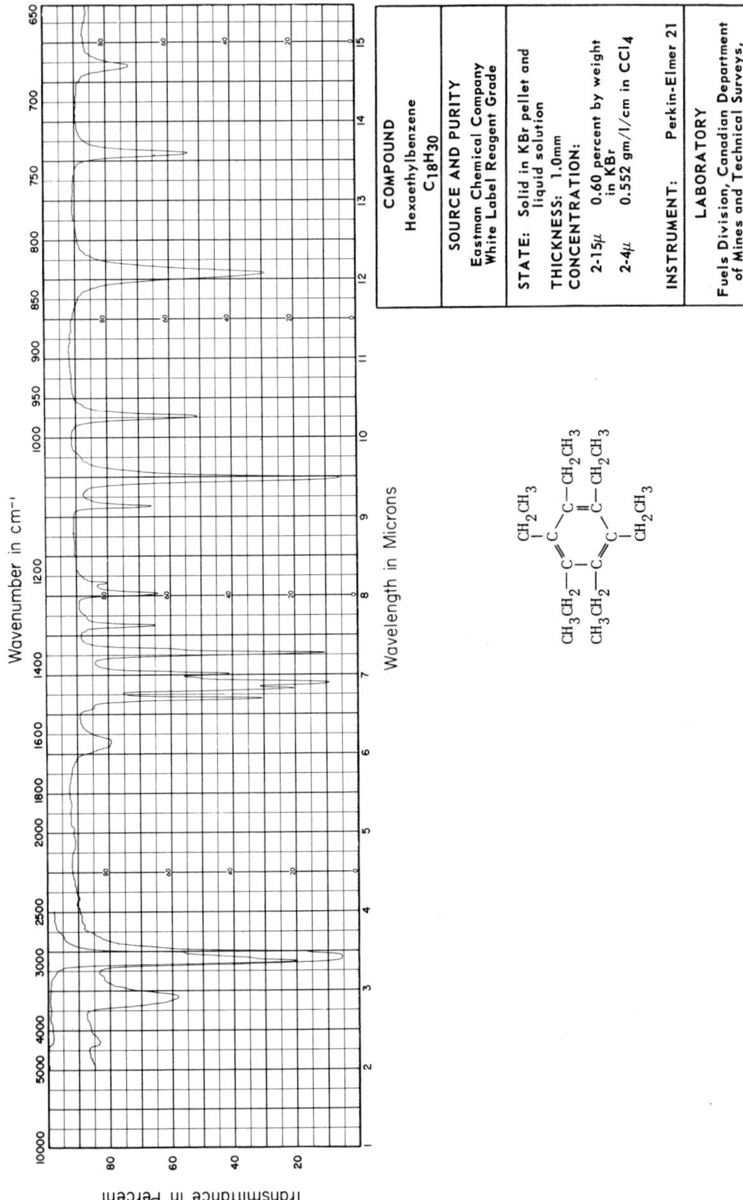

Figure 5-10. The infrared spectrum of hexaethylbenzene. (Courtesy API Project 44.)

TABLE 5-XIII. Vibrations of Aromatic Compounds

Explanation: The structural formula of the compound is listed in column 1 together with the physical state of the sample. The symbol *l* is used for liquid state spectra, *l′* for solution spectra (where CS_2 and CCl_4 were used), and *s* for solid state spectra in a KBr pellet. The intensity of the band is listed after the spectrum position of the vibration. The symbols used are S, M, W, Sh, Sp, St, V, B, indicating strong, medium, weak, shoulder, sharp, split band, very, and broad, respectively.

An asterisk on the spectral position number indicates that a second group frequency is interfering with the group frequency assignment. If the interfering group is known, it is listed below the vibration.

Bands in parentheses are group frequencies associated with the group listed in the column where the parentheses appear but are not the vibrations associated with the motion listed at the top of the columns. Question marks after entries indicate doubtful assignments.

Compound		CH stretching	1465 scissors	1380 scissors	Aromatic ring group frequencies				
					1600	1580	1500	1450	Other
CH_3	(*l*)	3030S 2900S 2850Sh	1455SB	1380M	1600S	1580MS	1480VS		750M 714W 700S
C_2H_5	(*l*)	3030SSp 2960VSSp 2870Sh	1480S 1465Sh 1455S 1440S	1375MW	1600M		1500M		695S 750S
HO—⟨⟩—CH_3	(*l* and *l′*)	3030VS 3300W 2900W 2850W	1480*MS (Ring)	1360VW			1495*S (CH_2)		
⟨⟩—OCH_3 (*l*)	(*l*)	3030MS 2950S 2850MS	1470SSp	1340M	1605S		1500S		

Compound	CH stretching	1465 scissors	1380 scissors	Aromatic ring group frequencies				
				1600	1580	1500	1450	Other
$CH_3-O-C_6H_4-C(=O)-CH_3$ (*l*)		1460S	1350M	1600S	1580M	1505M		830S
$C_6H_5-CH_2-C(=O)-NH_2$ (*l*)		1485*S (Ring)			1585S	1485*S (CH_2)		740M 700M
$C_6H_5-C(=O)-CH_3$ (*l* and *l′*)	3030S 2950Sh 2850Sh	1450S 1430Sh	1355VS (1080M)	1600S	1580M 1550M			760S 690S
$C_6H_5-CH_2NH_2$ (*l*) (*l*)				1620*S (NH_2)		1510S		740M 700M
$C_6H_5-C(=O)-C(CH=O)(CH_3)-C_6H_5$		1450	1390 1370	1600	1580	1500		700
$C_6H_5-CH_2-C(=O)-O-C(=O)-CH_2-C_6H_5$				1600	1590	1495		750 770 700

TABLE 5-XIII (continued)

Compound		CH stretching	1465 scissors	1380 scissors	Aromatic ring group frequencies				
					1600	1580	1500	1450	Other
$CH_3-C_6H_5$	(*l* and *l'*)	2980VS 2900S· 2850Sh	1455S	1370W	1600S		1490VS		725S 695S
$CH_3-C_6H_4-CH_3$	(*l*)	3030S 2950S 2850M	1456*S	1375M	1605M		1530S		795M
$(CH_3)_2CH-C_6H_4-CH(CH_3)_2$	(*l'*)	2940VS 2900M 2850M	1465S	1385M 1365M (1155) (1105)					
$(CH_3)_3C-C_6H_4-C(CH_3)_3$	(*s*)	2940VS 2900M 2850M	1470SSp	1400MS 1360S (1270S) (1200W)	1610W		1515MS		835VS
$C_6(CH_3)_6$ (ring with six CH_3 groups)	(*s*)	2950SB 2850Sh	1450VSB	1390S					

Compound	CH stretching	1465 scissors	1380 scissors	Aromatic ring group frequencies				
				1600	1580	1500	1450	Other
Hexaethylbenzene, $C_6(C_2H_5)_6$ (*s*)	2950SB 3030Sh	1490S ? 1455SSp 1428M	1380VS					
C_6H_5—$C(CH_3)_2$—$(CH_2)_5$—CH_3 (*l* and *l*′)	3030M 2915S 2840M	1450S 1460Sh	1375MS (746S)	1600S		1490S		760M ? 695S
$(C_6H_5)_2CH$—CH_3 (*l* and *l*′)	3030S 2950S 2850M	1450S	1370W	1600S	1580M	1490S		725W 695S
o-NH_2—C_6H_4—$C(=O)$—O—CH_3 (*l*)	3000S			1590* (NH_2)	1560* (NH_2)			

TABLE 5-XIII (continued)

Compound	CH stretching	1465 scissors	1380 scissors	Aromatic ring group frequencies 1600	1580	1500	1450	Other
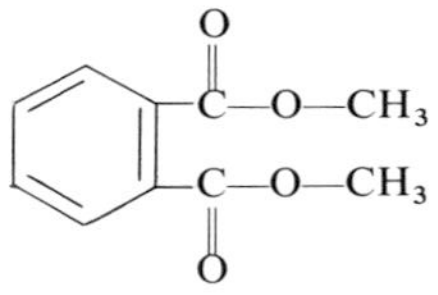 $C_6H_4(C(=O)-O-CH_3)_2$ (ortho)		1415		1600	1590	1490		745
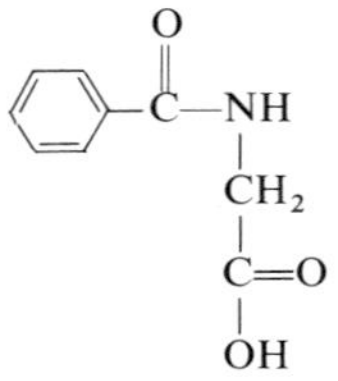 $C_6H_5-C(=O)-NH-CH_2-C(=O)-OH$		1445		1600* (CO)	1505* (NH)			720 690
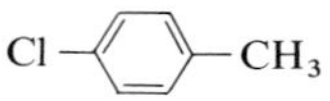 $Cl-C_6H_4-CH_3$ (para) (*l*′)	3000M 2870S 2800M	1450W 1400W	1390W	1605W	1550M ?	1495VS		800S
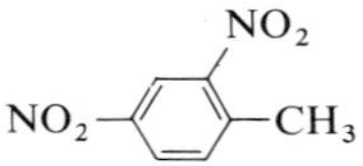 $NO_2-C_6H_3(NO_2)-CH_3$ (*l*′ and *s*)	2990S 2830Sh	1460*Sh (NO_2)	1340SB	1605S	1520*B (NO_2)			912S ? 835S

Compound	CH stretching	1465 scissors	1380 scissors	Aromatic ring group frequencies				
				1600	1580	1500	1450	Other
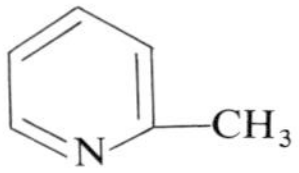 N, CH_3 (*l*)	2950S 2915M 3050Sh 2850Sh	1440VS	1380M	1590S	1570S	1475VS		730S ?
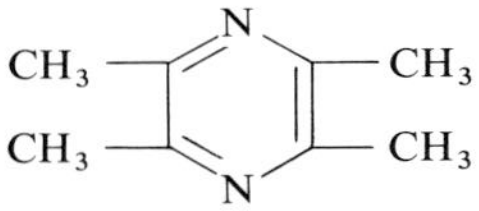 CH_3, CH_3, N, N, CH_3, CH_3 (*l′*)	2950S 3000Sh 2850Sh	1450M 1410S	1355W					
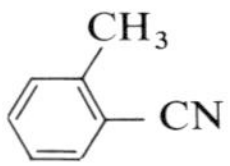 CH_3, CN (*l* and *l′*)	3080S 2940S 2875Sh	1470S	1385M	1600S		1490S		750S
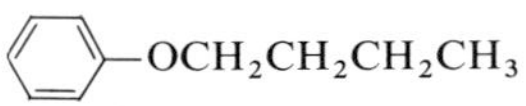 $-OCH_2CH_2CH_2CH_3$ (*l*)	3030M 2950S 2880M	1470S	1390M	1605S		1500S		755S 690S
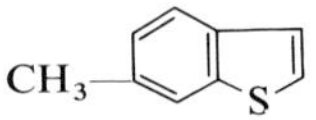 CH_3, S	3030M 2890S 2860Sh	1445SB 1410W	1390W	1610W	1575W	1500M		760SB
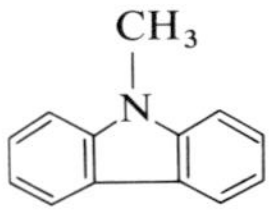 CH_3, N (KI)	3030W 2915W	1465S 1450S 1420S	1355MS	1600S		1490SB 1480S		745VSB

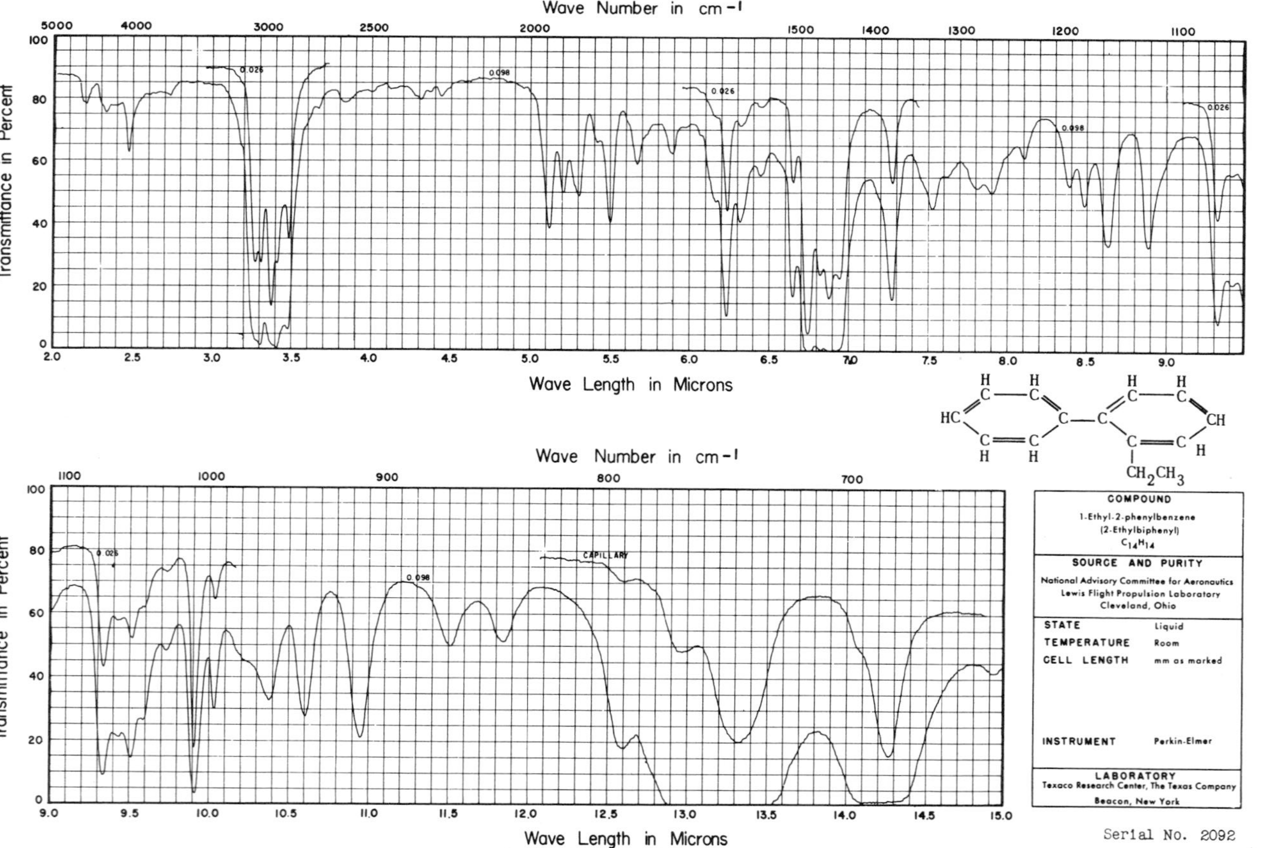

Figure 5-11. The infrared spectrum of 2-ethylbiphenyl. (Courtesy API Project 44.)

The introduction of nitrogen into the ring system can make the assignment of the CH_3 and CH_2 scissors difficult. However, if the spectrum is obtained in solution, usually both scissors vibrations can be seen. Several examples of the frequencies found for this type of compound are given in Table 5-XIII.

The presence of CO, CN, and NO_2 structural units in aromatic systems containing the CH_3 and CH_2 units can still further confuse the identification of the CH_3 and CH_2 group frequencies. Since the type of structure will determine to a large extent whether vibrations will interfere with the normal group frequency assignments of the CH_3 and CH_2 groups, it is not possible to generalize. The groups which have bands that can interfere with the group frequency assignments of the CH_3 and CH_2 groups were listed previously in Table 5-X. The examples given in Table 5-XIII were chosen so as to include most groupings which can interfere with group frequency assignments of CH_3 and CH_2 units. An example in which the 1465 cm^{-1} scissors vibration is obscured by an NO_2 vibration occurs for the compound p-$NO_2C_6H_4CH{=}CHCOOC_2H_5$. The NO_2 vibration near 1530 cm^{-1} obscures the 1465 cm^{-1} band. For many of the complex compounds containing CO, aromatic ring, CH_3, and CH_2 groups, the scissors vibrations of the CH_3 and CH_2 groups still remain distinct vibrations, easily identified. As an illustration, Figure 5-12 gives the spectrum of p-$CH_3OC_6H_4COCH_3$.

Halogens in the ring do not give bands interfering with the assignments of CH_3 and CH_2 group frequencies.

The presence of an OH group on the ring can make the identification of the CH_3 scissors vibrations difficult, as illustrated by the spectrum shown in Figure 5-13. While a broad band near 1450 cm^{-1} can be identified, the 1380 cm^{-1} band is either shifted or absent. In the *para* derivative of this compound, the 1300 cm^{-1} band is present.

The $C{\equiv}N$ group in a ring does not give bands which interfere with the CH_3 scissors vibrations although the scissors vibrations appear weak relative to the other bands in the spectrum.

In summary, we can state that in general the stretching and scissors vibrations of CH_2 and CH_3 groups can be identified, although for some of the aromatic compounds more bands may appear in the 1450–1300 cm^{-1} region than just the two scissors vibrations. In addition, for some compounds, the CH_3 and CH_2 group frequencies may be obscured by vibrations from groups such as NO_2.

5.5. GROUP FREQUENCIES ASSOCIATED WITH ALKENES

The group frequencies associated with alkenes are of two types: one is associated with the $C{=}C$ stretching motion and appears in

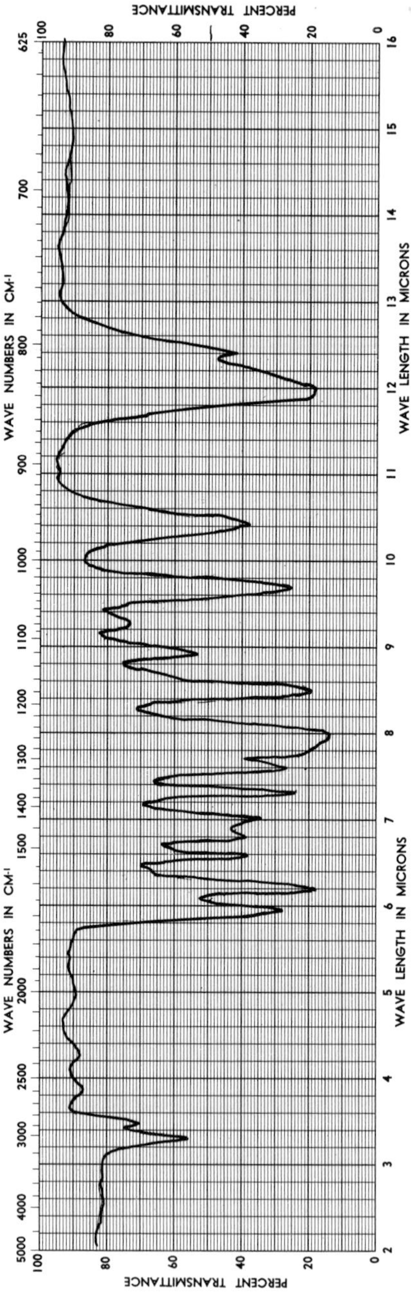

Figure 5-12. The infrared spectrum of p-$CH_3OC_6H_4COCH_3$.

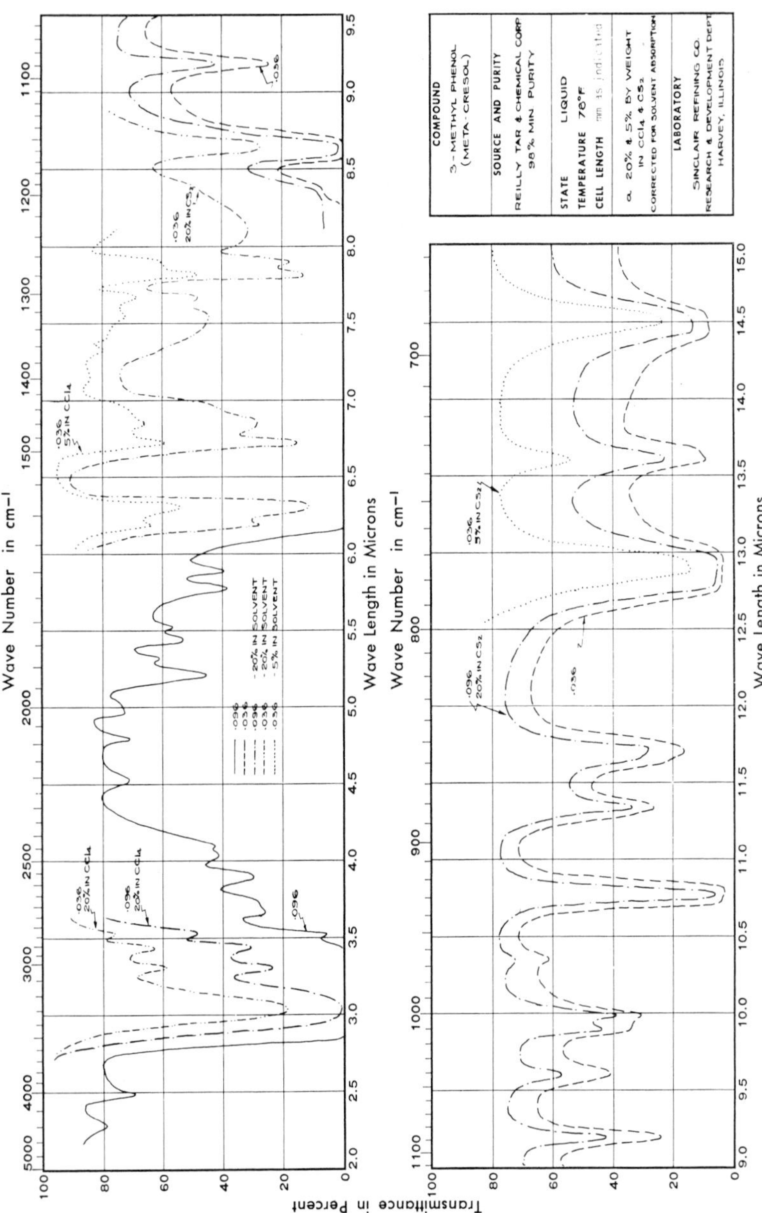

Figure 5-13. The infrared spectrum of *m*-cresol. (Courtesy API Project 44.)

the spectral region from 1680 to 1580 cm^{-1}; the other is associated with the protons attached to the carbon atoms of the double bond. An example of the second type is furnished by a structural unit such as $R_1R_2C{=}CHR_3$, which will have distinct group frequencies associated with the CH stretch and CH deformation frequencies.

5.5A. 1680–1580 cm^{-1} Region for Alkenes

As mentioned above, the C=C stretching frequencies appear in this region. The exact position in each case is determined by the structural groups attached to the C=C unit. The intensity of the

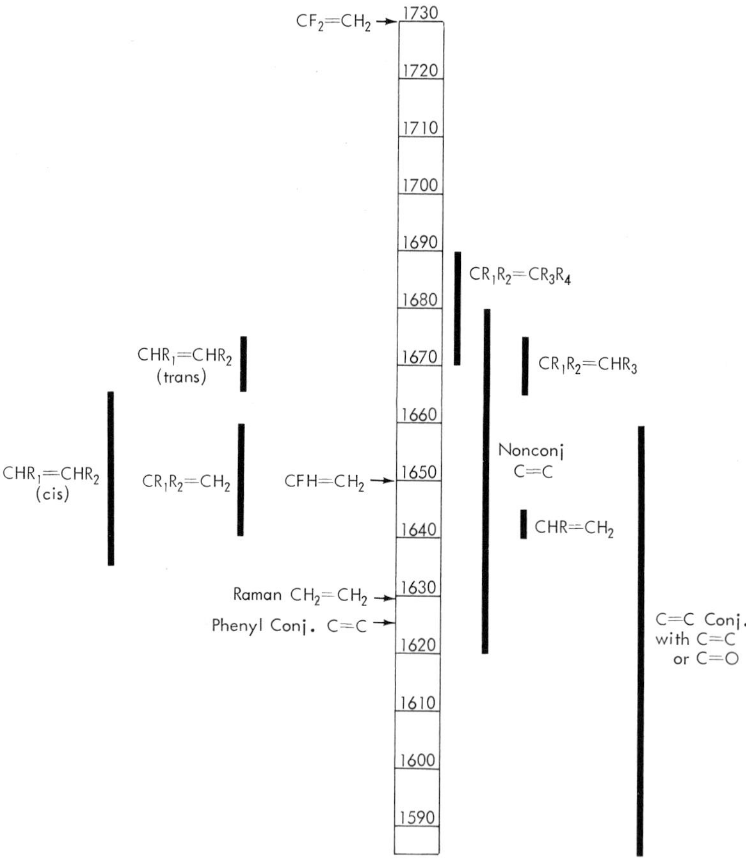

Figure 5-14. Group frequencies of alkenes.

band is also a function of these structural groups. Many compounds that do not have olefinic structure, such as water, alcohols, some alkanes, imines, oximes, and ketones, also have infrared bands in the 1680 to 1580 cm^{-1} region. Because of interference from these bands it is often necessary to look for group frequencies associated with the hydrogen atom in the unit C═C—H in order to establish the presence of the C═C group.

Figure 5-14 has been prepared to summarize the group frequencies associated with the C═C stretch. As can be seen from the diagram, fluorine-substituted compounds absorb at the highest frequencies. Compounds with nonconjugated double bonds absorb between 1680 and 1620 cm^{-1}. In compounds where a phenyl group is conjugated with the double bond a strong band can appear in the region near 1630 cm^{-1}. When the double bond is conjugated with a second double bond, a strong band is found near 1600 cm^{-1}. Ethylene has a strong C═C stretching frequency at 1628 cm^{-1} in its Raman spectrum, but no infrared band appears because the stretching frequency causes no net change in the dipole moment of this symmetrical molecule.

We can summarize some of the factors which influence the position of the C═C stretching frequency as follows:

1. The substitution of hydrogens around the double bond determines the position of the stretching frequency of the double bond.

2. When halogens are substituted for the hydrogens next to the double bond, the stretching frequency shifts to higher values. For example, one fluorine substituted in ethylene will shift the band for ethylene from 1628 to 1650 cm^{-1}, while if a second fluorine atom is substituted on the same atom, the band will appear at 1730 cm^{-1}. The presence of the halogen also enhances the intensity of the band.

3. When the double bond is conjugated with an aromatic ring, the stretching frequency is shifted to lower values compared to a nonconjugated alkene of similar structure. The shift is usually not greater than 30 cm^{-1}.

4. The conjugation of a carbonyl group with the double bond also produces a downward shift in frequency, usually not greater than 30 cm^{-1}.

5. The conjugation of one double bond with a second produces a large downward shift in frequency.

6. Conjugation of the double bond with any of the groups discussed above generally enhances the intensity of the double bond group frequency.

7. For both conjugated and unconjugated dienes two bands can occur in the C═C stretching region.

8. When two double bonds are conjugated with each other, the lower-frequency band of the C=C stretching frequency is the stronger one. Conjugated steroids often show this doublet.

9. The carbonyl stretching frequencies can occur in the region where the double bond stretching frequencies appear, and when both these structural groups are present in a molecule, the double bond frequency may appear as a shoulder on the stronger carbonyl band.

10. The position of the stretching frequency for the double bond in unsaturated tertiary amines is influenced considerably when the amine is converted into a salt.

11. The presence of electronegative atoms in olefinic structures causes shifts in the group frequencies found for the unsubstituted olefin.

12. Aromatic, carbonyl, alkyl, and hydroxyl compounds can have strong bands in the region of the C=C stretch and can therefore interfere with the band assignment of the C=C stretch. In some cases, such as the aromatic compounds, careful sample preparation can remove this interference.

13. When two double bonds occur in a molecule and their structural environments differ, quite frequently they will show different C=C stretching and CH deformation frequencies. Usually the deformation frequency will be more sensitive to structural environment than the C=C stretching frequency.

5.5B. 970–660 cm^{-1} Region for Alkenes

The CH deformation vibration of alkenes gives distinct bands in the 970–660 cm^{-1} region. The positions of these bands are determined by the substitution around the double bond. Five types of groupings can occur around a double bond. The five, and the regions in which they show characteristic group frequencies, are

$R_1R_2C{=}CH_2$	895–885 cm^{-1}
$R_1R_2C{=}CHR_3$	850–790 cm^{-1}
$HR_1C{=}CH_2$	995–985 cm^{-1}
	915–905 cm^{-1}
$R_1HC{=}CHR_2$ (*cis*)	730–665 cm^{-1}
$R_1HC{=}CHR_2$ (*trans*)	980–960 cm^{-1}

The 970–660 cm^{-1} region of the infrared spectrum also shows a large number of nonolefinic group frequencies, and the bands listed above may therefore be obscured. Aromatic compounds form one large group which can have bands in this region.

5.5C. 3100–3000 cm^{-1} Region for the —C═C—H Group

Either one or two bands can exist which may be described as olefinic CH stretching frequencies. When only one R group is attached to the double bond ($RCH{=}CH_2$) two bands are found, at 3040–3010 and 3098–3075 cm^{-1}, respectively. If two R groups are present ($RRC{=}CH_2$), a single band is found at 3079 cm^{-1}. For RRC═CRH there is a single band at 3019 cm^{-1}.

The presence of a halogen in the molecule can shift all of these frequencies. For example, in 1,1-dichloroethylene the stretching frequency is found at 3085 cm^{-1}.

When the double bond is present in combination with aliphatic groups, the ═C—H band can appear as a shoulder on the aliphatic CH stretching band.

5.5D. Interpretation of Spectra of Alkenes

Spectra of some alkenes were presented in an earlier section (see Figures 5-7 and 5-8). Additional alkene spectra are presented in Figures 5-14 to 5-19 (Figures 5-15 and 5-16 are located in the pocket on the inside back cover). We will examine these spectra for characteristic group frequencies, considering first the CH stretching vibrations.

The two chlorine-substituted olefins having ═CH and $=CH_2$ structural units show distinct CH stretching vibrations (see Figures 5-8 and 5-15). The position of the CH stretching vibration is in the expected region, near 3000 cm^{-1}. In the compound 3-chloro-1-propene the alkyl $-CH_2-$ group has a band near 2900 cm^{-1}.

For the alkenes containing $-CH_3$ as well as $-CH_2-$ and $=CH_2$ groups the CH stretching region contains bands indicative of each group. The intensity of each band is determined by the relative number of each type of group. In compounds where two CH_3 groups or one CH_3 and one $-CH_2-$ appear, the olefinic ═CH stretching vibration is less intense than the stretching vibration of the alkane groups (see Figures 5-18 and 5-19). In general, the olefinic CH stretching vibration is rather weak, and one should expect to see this vibration as a shoulder or a sharp weak band on the high-frequency side of the stronger alkane CH stretching bands. In some spectra the olefinic CH stretch may not be resolved from the alkane vibrations (see Figure 5-7). An instrument with better resolution would probably separate the olefinic CH stretch from the alkane CH.

The C═C stretching vibration, which appears in the 1680–1580 cm^{-1} region, can have variable intensity, depending on the compound in which the group is present. For the symmetric compound tetrachloroethylene no C═C vibration is present in the infrared spectrum. Trichloroethylene has a very strong C═C

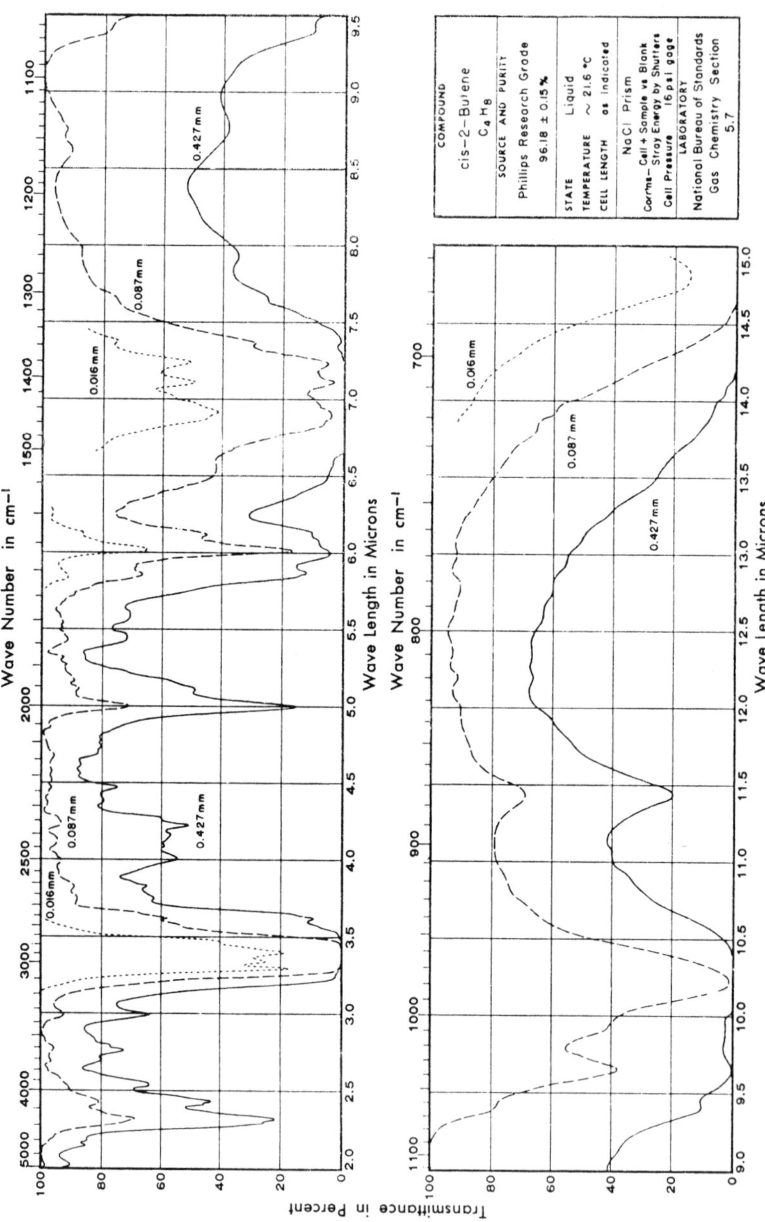

Figure 5-17. The infrared spectrum of *cis*-2-butene. (Courtesy API Project 44.)

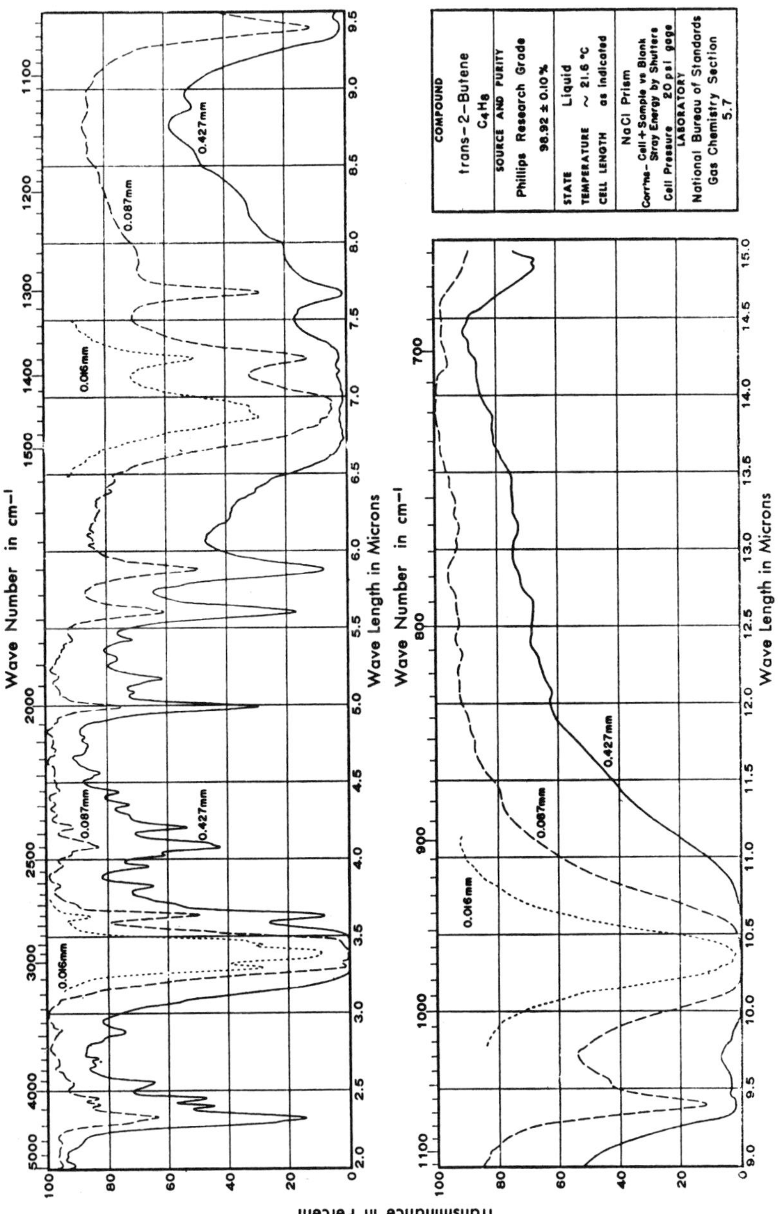

Figure 5-18. The infrared spectrum of *trans*-2-butene. (Courtesy API Project 44.)

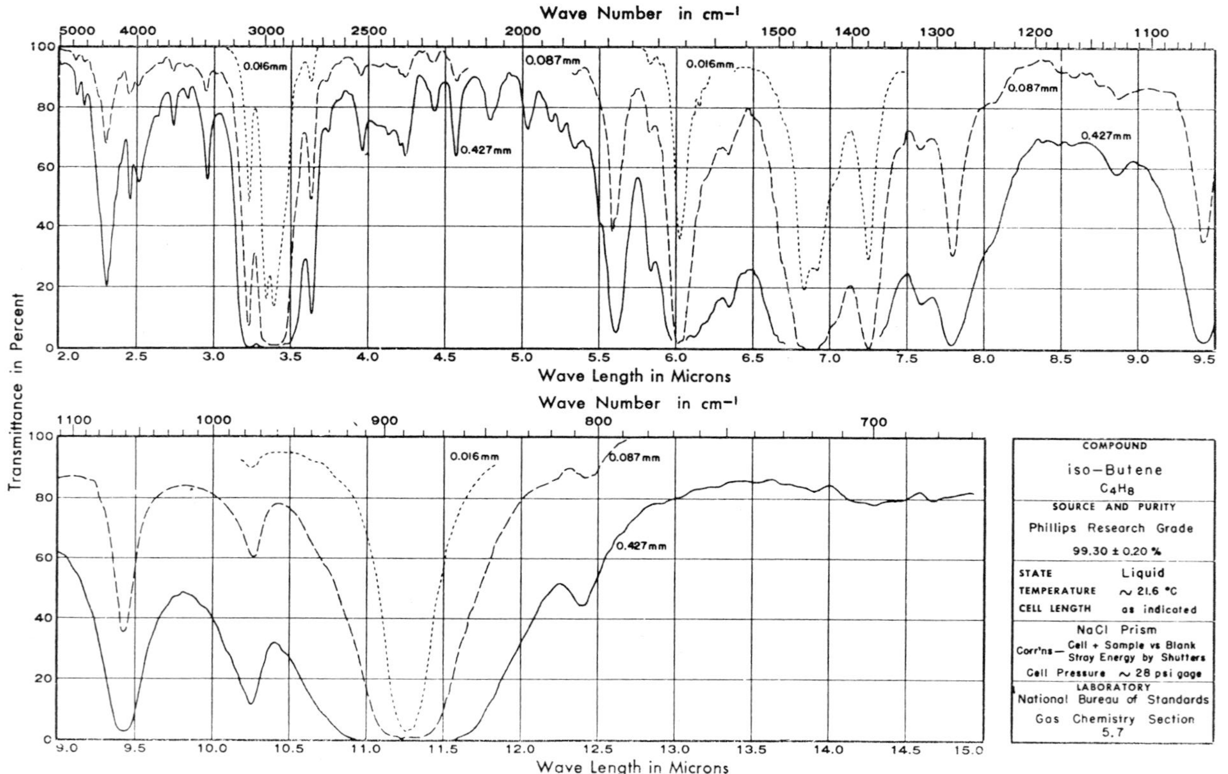

Figure 5-19. The infrared spectrum of 2-methylpropene. (Courtesy API Project 44.)

stretching near 1640 cm^{-1}. The position of the C═C stretching vibration is sensitive to the groups substituted on the double bond, as described in an earlier section. The compound 5-methyl-1-hexene contains the group CH_2═CH—. The spectrum of this compound has the double bond vibration near 1640 cm^{-1}, while in 3-methyl-*cis*-2-hexene, where the RRC═CH— group is present, the vibration is at 1665 cm^{-1}.

The 970–660 cm^{-1} region contains a strong group frequency vibration at a position determined by the structure around the double bond. The sensitivity of this group frequency to structure is exemplified by the positions of the band in *cis* and *trans* alkenes. In *cis*-2-butene the band is near 675 cm^{-1}, while in the *trans* compound it is near 960 cm^{-1}. The olefinic group frequencies in the 970–660 cm^{-1} region can be found in the spectra shown in Figures 5-7, 5-8, 5-9, 5-16, 5-17, 5-18, and 5-19.

5.6. GROUP FREQUENCIES ASSOCIATED WITH ALKYNES

A compound containing the structural unit C≡C may have either a hydrogen atom next to it (i.e., CR≡C—H) or other groups such as CH_3, etc. It is convenient to discuss group frequencies for alkynes in terms of both the C≡C stretch and, if the C≡C—H group is present, the stretching and deformation frequencies of the —C≡C—H group.

The C≡C structural unit has characteristic absorption frequencies in the region from 2260 to 2100 cm^{-1}. The exact position depends on the groups substituted around the triple bond, and the intensity of the band depends, to some extent, on the position of the triple bond in the molecule.

For monosubstituted acetylenic compounds, where the structural unit is RC≡C—H (R represents a group which is not conjugated with the triple bond), a strong band is found for the C≡C stretch in the region from 2140 to 2100 cm^{-1}. When the R group is conjugated with the triple bond, the characteristic stretching frequency becomes more intense and slightly shifted toward lower frequencies. If R is a halogen, the band also may shift slightly toward lower frequencies, but this does not occur with sufficient regularity to state a definite rule.

For the structural unit $R_1C{\equiv}CR_2$, a characteristic C≡C stretching vibration can be found in the 2260–2190 cm^{-1} region. The intensity of the band will be determined by the groups R_1 and R_2. If these groups are identical, the stretching motion of the triple bond causes no net change in the dipole moment of the molecule and therefore no absorption band will be observed. If the two R groups differ only slightly in mass and in the nature of their constituents, then a weak band will be found. A compound such as $C_3H_7—C{\equiv}C(CH_2)_2Cl$ will have only a very weak stretching frequency, since the position of the $C{\equiv}C$ in the molecule is such that the change in dipole moment during a stretching motion is almost zero. The compound n-$C_5H_{11}C{\equiv}CBr$ also has a very weak band in the triple bond stretching region. When the triple bond is near the end of a simple hydrocarbon chain, its intensity will be large since the change in dipole moment for the stretching of the triple bond will be large.

The region from 2260 to 2100 cm^{-1}, where the stretching frequency of the triple bond occurs, can also contain bands due to compounds that do not contain alkyne groups. Compounds absorbing in this region include a number of saturated hydrocarbons, saturated nitro hydrocarbons, fluorocarbons, trioxanes, aromatic ring systems, silicon compounds, thiophenes, etc. Nitriles and isocyanates also have group frequencies in this region.

It would appear from the fact that so many compounds offer interference in this region that it is very difficult to identify the alkyne structure in a completely unknown material. However, many groups which have a frequency that might be mistaken for an alkyne vibration also have other group frequencies which can be used to identify them. Thus, by a process of elimination it may still be possible to identify an alkyne from a single group frequency.

5.6A. Group Frequencies Associated with the Unit C≡C—H

The structural unit C≡C—H shows a CH stretching frequency near 3300 cm^{-1}. The deformation frequencies associated with this unit are not as well defined, although it has been suggested that bands in the 1200–600 cm^{-1} region are due to this vibration. The only other structural units which may have bands near 3300 cm^{-1} are OH and NH groups, and these bands are usually so characteristic in shape that they can be differentiated from the alkyne stretching frequency. In some liquid-state spectra of alkenes and alkanes medium to weak bands can occur as high as 3300 cm^{-1}, and these may be mistaken for alkyne group frequencies. By determining the spectra of these compounds in solution it is possible to examine only the very strong vibrations, which are characteristic group frequencies. Examples of such compounds are

$$CH_3CH_2\underset{\displaystyle CH_3}{\underset{|}{C}}HCH_2CH_2\underset{\displaystyle CH_3}{\underset{|}{C}}HCH_3, \quad (CH_3)_3CCH_2-\overset{\displaystyle O}{\overset{\|}{C}}-NH_2$$

and

$$(CH_3)_2CHCH(CH_3)_2,$$

which have bands of medium intensity near 3300 cm^{-1}. The carbonyl overtone vibration also appears near 3300 cm^{-1}. A number of pyridine derivatives can have fairly strong bands as high as 3330 cm^{-1}.

A band near 1330 cm^{-1} appears to be indicative of the $—C{\equiv}C—CH_2—$ linkage [66].

5.7. CHARACTERISTIC GROUP FREQUENCIES FOR BENZENE DERIVATIVES

The frequencies associated with aromatic rings and the regions in which they occur are as follows:

A. The CH stretch, which is usually quite sharp, near 3030 cm^{-1}.
B. A series of bands related to the substitution on the benzene ring, in the 2000–1660 cm^{-1} region.
C. The skeletal in-plane ring vibrations, near 1600, 1580, and 1500 cm^{-1}.
D. A second series of bands related to the substitution in the benzene ring, in the 1225–950 cm^{-1} region.
E. The CH out-of-plane vibrations, in the 950–650 cm^{-1} region.

Each of these regions will be discussed in detail.

5.7A. Aromatic CH Stretching Frequencies

CH stretching frequencies associated with the benzene ring or other aromatic rings appear near 3030 cm^{-1}. Since this is in the region of the olefinic CH stretch, the aromatic CH vibration can be confused with the olefinic CH unless other aromatic vibrations are identified. The aromatic CH stretch is usually a band of medium to weak intensity. For some aromatic compounds several bands can be found near 3030 cm^{-1}. When a sodium chloride monochromator is used, the aromatic CH stretch may not be resolved from the stronger alkane CH vibrations.

Examining the spectra previously given for aromatic compounds (see Figures 5-11, 5-12, 5-13) as well as the spectra presented in Figures 5-20 to 5-25 (all located in pocket on inside back cover), we can locate the aromatic CH stretch near 3030 cm^{-1}. In benzene (Figure 5-20) two sharp bands are seen even when a sodium chloride monochromator is used. Hexaethylbenzene (Figure 5-10) shows no aromatic CH stretching frequency, as expected. The spectrum of the bicyclic aromatic compound 2-ethylbiphenyl, shown in Figure 5-11, has two strong CH stretching frequencies near 3030 cm^{-1}. In the spectrum shown in Figure 5-13, in which an OH vibration occurs near 3400 cm^{-1}, the 3030 cm^{-1} aromatic vibration is a weak band.

5.7B. Benzene Ring Vibrations in the 2000–1660 cm^{-1} Region

The pattern of bands in this region shown by various substituted benzene ring compounds is illustrated in Figure 5-26. It can be seen that a distinct pattern is found for each type of aromatic substitution. For example, a monosubstituted benzene derivative shows a series of four maxima beginning at about 1880 cm^{-1}. The assignments of these frequencies to combination and overtone bands have been given by Kakiuti [67] and Whiffen [68]. Since for most compounds these are weak bands, a thicker cell or more concentrated solution is used if this region is to be examined carefully.

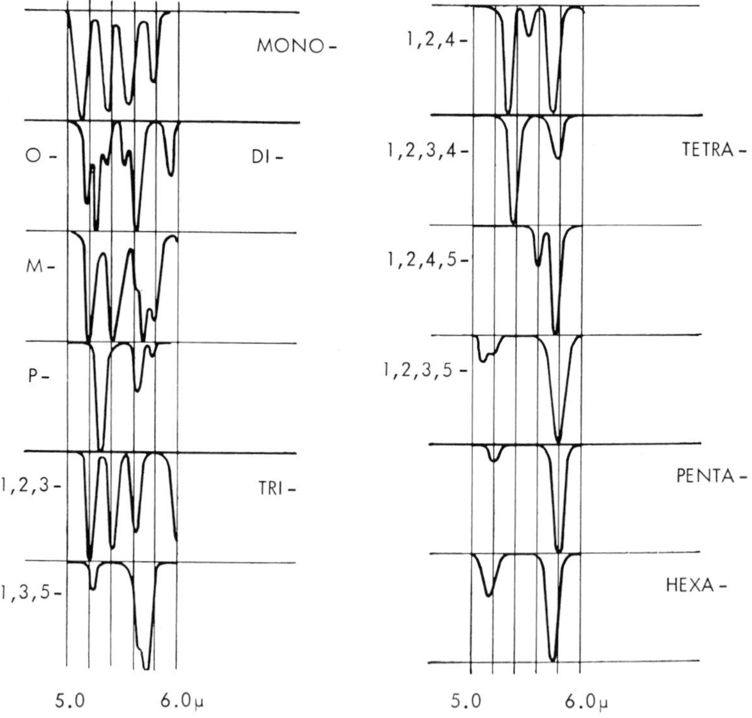

Figure 5-26. Characteristic benzene ring vibrations in the 2000–1600 cm^{-1} region. [Reprinted with permission from *Anal. Chem.* **23**: 709, 1951.]

Substitution of elements and groups such as O, F, and NO_2 in the ring breaks down these patterns to some extent, since such electronegative groups can shift these bands as well as introduce new absorption bands. Moreover, if the compound contains a carbonyl group, the bands may be obscured, since the carbonyl frequency is very strong in this region.

The types of compounds other than benzene derivatives that show medium to strong bands in the 2000–1660 cm^{-1} region include alkenes such as *trans*-$CH_3CH{=}CHCH_2CH_3$ and alkanes such as $(CH_3)_2CHCH(CH_3)CH(CH_3)_2$. In addition, fluorocarbons in saturated rings or straight chains also have a number of strong bands in this region, as do trioxane and thiophene.

In the spectra of aromatic compounds having only alkyl groups as substituents, the patterns of bands expected for mono-, di-, or tri-substitution, etc., are easily recognized. For example, in Figure 5-21, where the spectrum of toluene is presented, the pattern of four bands in the 2000–1660 cm^{-1} region is quite distinct. For other examples, one can examine the spectra in Figures 5-22, 5-23, and 5-24. The hexasubstituted aromatic ring compound shown in Figure 5-10, does not contain the expected two bands. It is possible that a thicker sample would reveal them. The spectrum of the diaromatic compound shown in Figure 5-11 has the pattern of bands associated with both mono- and di-substituted rings, as expected from the structure of the compound.

In the spectrum of acetophenone presented in Figure 5-25, the carbonyl band interferes to some extent with the pattern of bands, but three of the four bands expected for a monosubstituted compound are present.

5.7C. Aromatic Ring Frequencies in the 1600–1450 cm^{-1} Region

One of the best series of group frequencies for recognizing the presence of aromatic ring structures occurs in the 1600–1450 cm^{-1} region. Since this region is near the 1465 cm^{-1} CH_3 scissors vibration, we have discussed some of these bands for aromatic compounds in Section 5.4I (see Figures 5-10, 5-11, and 5-12 and Table 5-XIII). We shall refer to the data presented in that section as well as spectra shown in Figures 5-20 to 5-25.

Generally, correlation tables list up to four bands associated with aromatic rings in the 1600–1450 cm^{-1} region. These occur near 1600, 1585, 1500, and 1450 cm^{-1}. Their intensity is variable, depending on the groups present in the aromatic compound. For aromatic compounds in which a substituent is conjugated with the ring the 1585 cm^{-1} band is usually strong. The exact positions of all four bands depend on the type of aromatic compound and the method of sample preparation for spectral analysis. The possibility of assigning the 1450 cm^{-1} band depends on the absence of alkyl groups, which have interfering bands near 1465 cm^{-1}.

For specific examples of cases that either do or do not permit these vibrations to be identified, let us re-examine the data presented in Table 5-XIII. It should be recalled that the vibrations marked with asterisks are those which are difficult to assign due to interference by other vibrations.

For some of the compounds listed, we find that either the 1600 or the 1585 cm^{-1} band is sufficiently strong to be identified, but that the other is either weak or absent. In general, at least one strong band is found in the 1620–1570 cm^{-1} region. In compounds with NH, NH_2, or NO_2, and in some cases C=C or C=O groups, the 1600 cm^{-1} band can be obscured by vibrations from these groups. Several examples of this interference are listed in Table 5-XIII.

Many nonbenzenoid compounds can have strong bands in the 1600–1450 cm^{-1} region. For example, the alkene

$$CH_2{=}CHCH(CH_3)CH_2CH_3$$

has a strong band at 1592 cm^{-1}. The aromatic compound thiophene has a band near 1590 cm^{-1}. Other types of compounds having medium to strong bands in the 1600–1450 cm^{-1} region include nitroalkanes, pyridines, alkenes, saturated five- and six-member ring compounds, alcohols, and many inorganic compounds.

In examining the spectra of simple aromatic compounds such as those presented in Figures 5-20 to 5-25, one should have little difficulty in assigning at least one of the two high-frequency bands near 1600 and 1585 cm^{-1} as aromatic ring vibrations. In the spectrum of benzene (Figure 5-20) the 1500 cm^{-1} band is quite strong. In the spectrum of acetophenone (Figure 5-25) the 1450 cm^{-1} band is obscured by the CH_3 scissors vibration but the 1600 and 1585 cm^{-1} bands are quite distinct. As expected, the spectrum of hexaethylbenzene (Figure 5-10) has no 1600 or 1585 cm^{-1} vibrations, since there are no hydrogens on the ring. This compound has bands near 1490 cm^{-1} which may be indicative of the hexasubstituted ring. The spectrum of the diaromatic compound 2-ethylbiphenyl shown in Figure 5-11 has a very strong band near 1600 cm^{-1} and weaker bands near 1585 and 1500 cm^{-1}.

5.7D. Benzene Ring Vibrations in the 1225–950 cm^{-1} Region

While a number of characteristic group frequencies of aromatic compounds can appear in the 1225–950 cm^{-1} region, their weak intensity and variability of position make them only fair group frequencies. Also, this region contains a large number of bands in addition to those found for aromatic compounds. Because of these factors, this region is not as useful as the 2000–1660 cm^{-1}, 1600–1450 cm^{-1}, and 950–650 cm^{-1} regions for recognizing aromatic ring frequencies. The many nonaromatic compounds which have strong bands in this region include fluorocarbons in both the five- and six-member ring systems and in straight chains, alkanes, nitroalkanes, alkenes, pyridine derivatives, and polycyclic aromatic compounds.

5.7E. Group Frequencies of Benzene Derivatives in the 950–650 cm^{-1} Region

Table 5-XIV lists the characteristic group frequencies of aromatic compounds in the 950–650 cm^{-1} region. The position of these group frequencies can be related to the substitution on the benzene ring. For example, if the ring is monosubstituted, two strong bands occur in this region, one in the 770–730 cm^{-1} region and the other in the 710–690 cm^{-1} region. If the single substituent group is an alkyl group, the two bands will be found in the region expected and can

TABLE 5-XIV. Aromatic Group Frequencies in the 950–650 cm^{-1} Region

Ring substitution	Position of group frequency (cm^{-1})	
Monosubstituted	770–730S 710–690S	(5 free hydrogens)
1:2 disubstituted	770–735S	(4 free hydrogens)
1:3 disubstituted	900–860M	(1 free hydrogen)
	810–750S	(3 free hydrogens)
	725–680M	(Not always seen)
1:4 and 1:2:3:4 substituted	860–800S	(2 free hydrogens)
1:2:3 trisubstituted	800–770S	(3 free hydrogens)
	720–685M	(Not always seen)
1:2:4 trisubstituted	860–800S	(2 free hydrogens)
	900–860M	(1 free hydrogen)
1:3:5 trisubstituted	900–860M 865–810S 730–675S	(1 free hydrogen)
1:2:3:5, 1:2:4:5, and 1:2:3:4:5 substituted	900–860M	(1 free hydrogen)

be easily identified. If a monosubstituted benzene derivative has polar groups, the two bands may still be recognizable, but in some cases will be obscured or shifted out of the expected range (e.g., benzoic acid). Examination of the data presented in Table 5-XIII for monosubstituted aromatic compounds indicates that in almost every compound listed the two bands are identifiable. Even for diaromatic compounds where only one ring is monosubstituted, the bands can be assigned.

We will not discuss the characteristic group frequencies of aromatic compounds in the 950–650 cm^{-1} region any further, but the reader, as an exercise, may analyze the data presented in Table 5-XIII and Figures 5-11, 5-12, 5-19, 5-20, 5-21, 5-22, 5-23, 5-24, and 5-25 for examples of such group frequencies.

Again it must be pointed out that strong bands from many nonaromatic compounds can occur in the 950–650 cm^{-1} region, so that a positive identification of the aromatic group frequencies cannot always be made. Many alkenes, inorganics, organometallics, and even alkanes may have strong bands in the 950–650 cm^{-1} region.

5.7F. Summary of Benzene Ring Group Frequencies

By way of summarizing our discussion of benzene ring group frequencies, and in particular of some of the groups that interfere with the group frequency assignments of aromatic rings, some suggestions on interpreting spectra of aromatic compounds may be presented:

1. The group frequencies of aromatic compounds near 1600 cm^{-1} are excellent for recognizing the presence of aromatic rings in the spectrum of an unknown compound, although some olefinic nitro, amino, and carbonyl compounds can also have bands there.

2. A compound such as $H_3N^+-C_6H_4-SO_3^-$ can have a strong band near 1600 cm^{-1} that is representative of the NH_3^+ group. This band could thus be easily misidentified as a ring vibration.

3. In a biphenyl derivative where each ring is substituted in a different manner, characteristic group frequencies for each ring should be present in the spectrum.

4. A conjugated aromatic ring normally has a band near 1585 cm^{-1}, while an unconjugated ring has one near 1600 cm^{-1}. However, many aromatic compounds can have bands which appear between these two frequencies. For example,

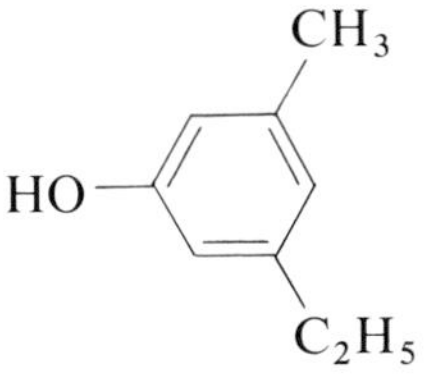

CH_3

HO—

C_2H_5

has a band at 1590 cm^{-1}.

5. In a compound such as (naphthyl)—NH_2, bands indicative of a trisubstituted as well as of a disubstituted benzene ring will be found. For this compound, a band at 780 cm^{-1} is indicative of the trisubstituted ring, while one at 760 cm^{-1} is indicative of the disubstituted ring. In general, polycyclic aromatic compounds show bands similar to those expected for a single aromatic ring having similar substitutions.

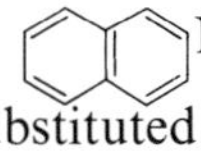

6. A compound such as (phenyl)—CH=CHCO—(phenyl) will show different absorption bands for the two rings, due to the difference in the type of structure adjacent to each ring. Thus, the two bands that appear at 1490 and 1440 cm^{-1} in the spectrum of this compound are due to ring 1 and ring 2, respectively. To distinguish two bands so close to one another, the spectrum must be obtained in a medium where an overlapping of bands will not occur.

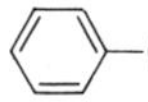

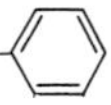

7. Some nonaromatic structures which have absorption bands in the 1600–1450 cm^{-1} region include the NH_2 group, which has a band near 1600 cm^{-1}, the NH_3^+ and NH groups, which have bands near 1500 cm^{-1} (as in (phenyl)—$CONHCH_2COOH$), the CH_3—N(H)—C(=O)— group, which has an absorption band at 1520 cm^{-1}

(as in —N(H)—C(=O)—N(CH$_3$)—CH_3),

and the NO_2 group.

8. Some aromatic nitro compounds have a band near 1575 cm^{-1} which might be confused with the normal group frequency assignment for conjugated rings at 1585 cm^{-1}.

5.8. VIBRATIONS OF PYRIDINE AND ITS DERIVATIVES

We shall only briefly mention the vibrations of pyridine and its derivatives. A brief bibliography of recent work on these compounds is included in the Appendix. Pyridine and its derivatives should have spectra similar to those of benzene and its derivatives since these compounds are structurally related. Pyridines have CH stretching vibrations near 3070–3020 cm^{-1}. The C═C and C═N vibrations appear in the 1660–1590 cm^{-1} and 1500 cm^{-1} regions. Ring vibrations appear near 1200, 1100–1000, 900–650, and 710 cm^{-1}. The assignments for substituted pyridines are similar to the assignments for the corresponding substituted benzenes in the 900–700 cm^{-1} region if it is assumed that pyridine itself is similar to a monosubstituted benzene ring.

A number of workers have studied the influence of electron-attracting and -withdrawing substituents on the position of absorption bands (see reference [1] in the Appendix). An interesting use of solvent effects to identify the carbonyl frequencies of pyridines also has been reported (see reference [32] in the Appendix).

5.9. GROUP FREQUENCIES ASSOCIATED WITH THE C═O STRUCTURAL UNIT

The C═O structural unit has a stretching vibration which is an excellent group frequency. It generally results in a very intense and sharp band, appearing in the 1850–1650 cm^{-1} region for most compounds. In a series of related compounds the spectral range it will appear in is fairly narrow. For example, most aliphatic aldehydes have a C═O stretching frequency in the range from 1750 to 1700

TABLE 5-XV. Carbonyl Frequencies [69]

Compound	In solvents	As liquid	As vapor
Acetophenone	1697–1676	1687	1709
Benzophenone	1671–1651		1681
Cyclohexanone	1726–1700	1713	1742
Dimethylformamide	1696–1661	1666	1716
Acetyl chloride	1810–1798	1805	1821
Methyl acetate	1754–1726		1770
Acetone	1724–1703	1715	1738
Ethylene carbonate	1851–1805*		1867*

* A second band appears in the region, which is not assigned as a carbonyl vibration.

cm^{-1}. The position of the C=O vibration is rather sensitive to the physical state of the compound. For example, if we examine Table 5-XV, we can see that for a number of compounds the C=O vibration appears at its highest frequency for the vapor state, at an intermediate frequency for solution, and at the lowest frequency for the liquid (or solid) state. It has been shown that the concentration of the C=O compound in a solution can also determine the position of the vibration.

Some work has been done on correlating the position of the C=O stretching frequency with the substituents attached to the group. The data presented in Table 5-XVI illustrate the influence of aliphatic substituents on this vibration. It has been shown that for a series of compounds of the general formula

$$(R_1)(R_2)N{-}C_6H_4{-}NH{-}CH(CN){-}C(=O){-}R_3$$

the vibration can appear anywhere in the frequency range 1756–1687 cm^{-1}, depending on the nature of groups R_1, R_2, and R_3.

TABLE 5-XVI. Carbonyl Frequencies in $R_1{-}C(=O){-}R_2$ Compounds [70]

R_1	R_2	Carbonyl frequencies
CH_3	CH_3	1715
CH_3	C_2H_5	1715
CH_3	n-C_3H_7	1710
CH_3	i-C_3H_7	1715
CH_3	t-C_4H_9	1708
i-C_3H_7	i-C_3H_7	1710
t-C_4H_9	t-C_4H_9	1687
n-C_3H_7	n-C_3H_7	1710
CH_3	CF_3	1765*

* Shift due to polar effects of substituents.

For such related aromatic compounds as [4-pyridone] and [2-pyridone], the C=O stretch is found at 1590 cm^{-1} and 1650 cm^{-1} respectively, i.e., its position differs by 60 cm^{-1}.

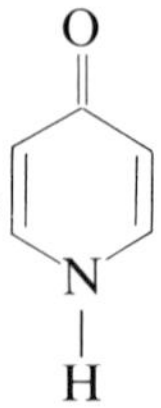

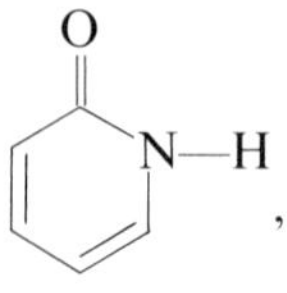

Another example of the sensitivity of the C=O vibration to the structural groups surrounding it is provided by the two general series of compounds (I) and (II). For series (I), where R is CH_3,

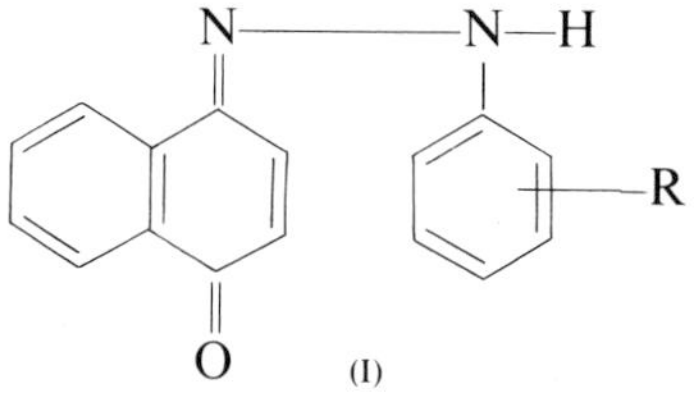

(I) 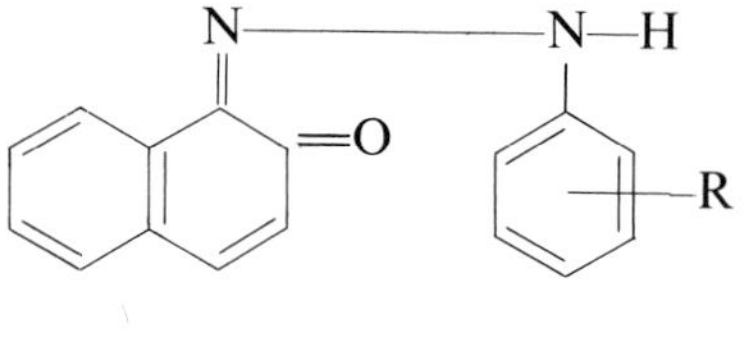 (II)

OCH_3, NO_2, or Cl and the spectrum has been determined in the solvents C_6H_{12}, C_2Cl_4, CCl_4, CH_2Cl_2, $CHCl_3$, or $CHBr_3$, the C=O vibration appears in the range from 1655 to 1620 cm^{-1}, while for series (II) with the same substituents and solvents the vibration appears between 1635 and 1610 cm^{-1}

Thus, it can be seen that the C=O vibration, while somewhat sensitive to environment, still maintains a fairly constant frequency position. It is therefore a very useful group frequency. If the experimental conditions are stated, then its position becomes a very exact parameter for a given compound.

Only a few structural groups have group frequency vibrations in the region where the C=O vibration appears. These include such groups as —C=C—, >NH, aromatics including pyridines, and —OH. In most cases the C=O vibration can be distinguished from vibrations of these groups, since it is usually stronger and more distinct. An example of an exception to this is afforded by some compounds with the structural group —C—NH. For some of these compounds the NH vibration can be more intense than the C=O vibration.

For compounds with two C=O groups both a symmetric and an asymmetric stretch can occur. These compounds usually have bands assignable to each of these vibrations unless symmetry factors make one of the vibrations infrared inactive. Also, if the compound enolizes, only one vibration will be seen.

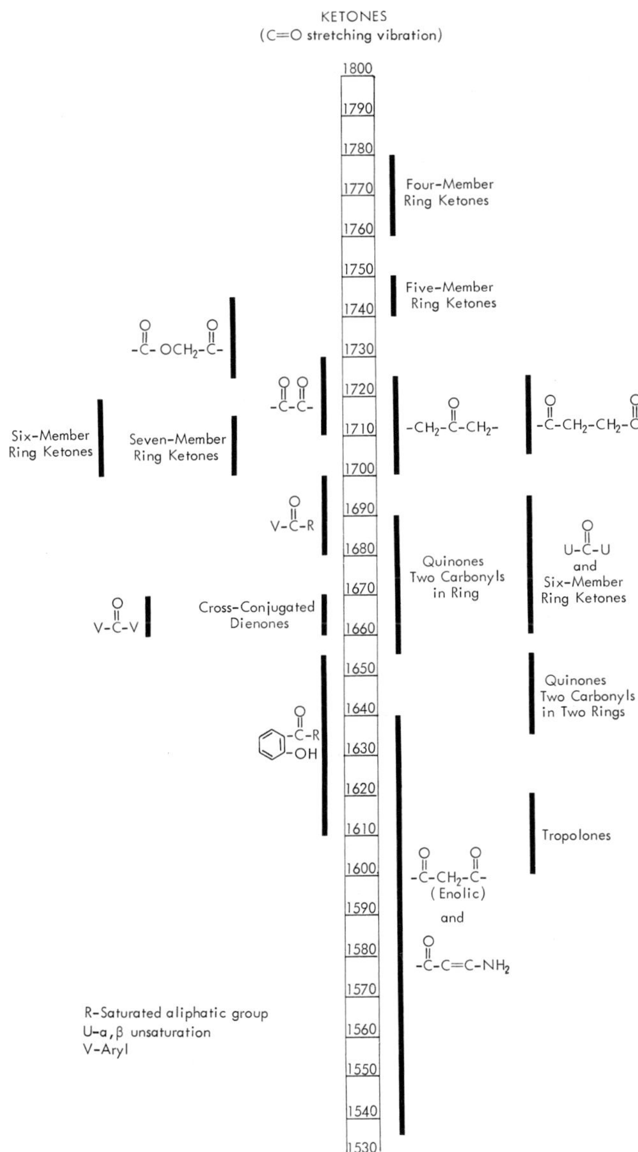

Figure 5-27. Group frequencies for carbonyl compounds.

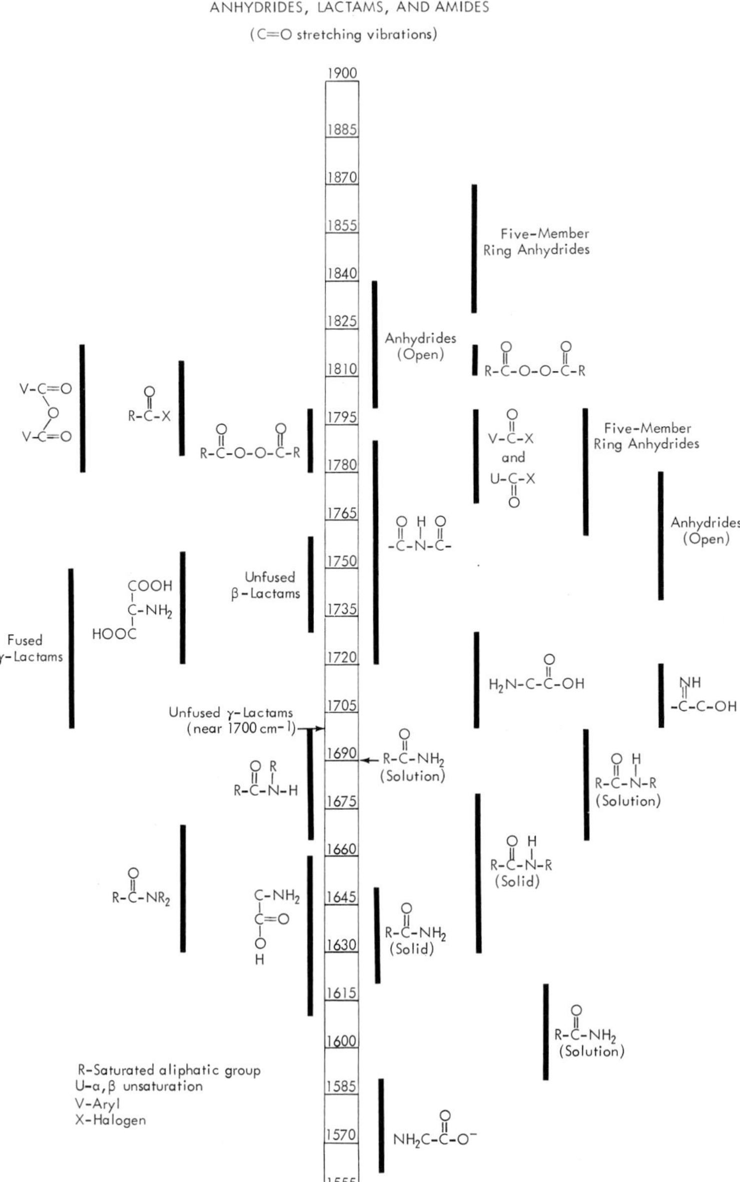

Figure 5-27 (continued).

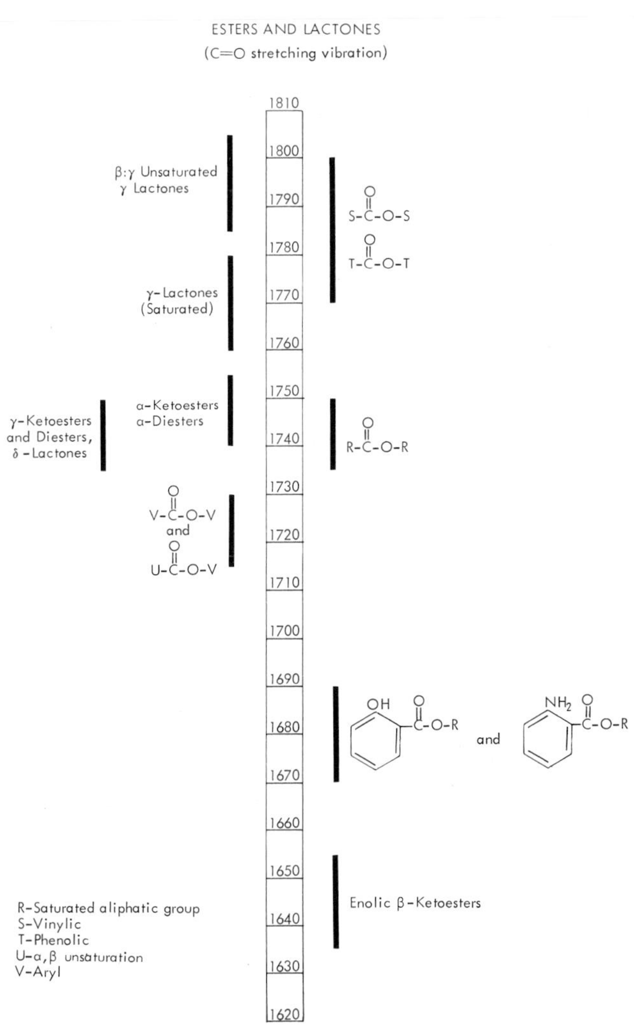

Figure 5-27 (continued).

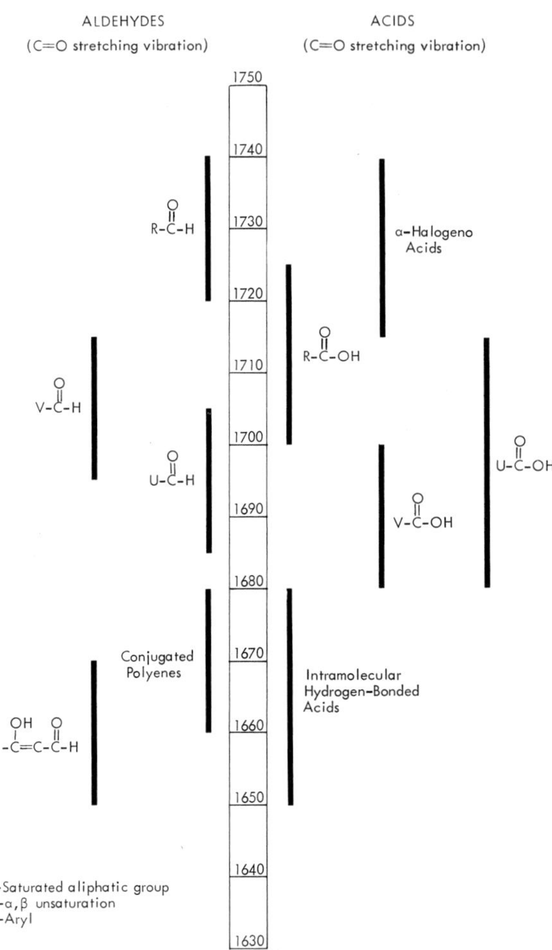

Figure 5-27 (continued).

It is possible that there may be a deformation vibration for the structural unit $R—\overset{\overset{\displaystyle O}{\|}}{C}—R$ which could be a characteristic group frequency indicative of the C=O group. However, data on this are inconclusive and will not be described here.

The general group frequency assignments suggested for C=O compounds are given in Figure 5-27. A selected list of references is included in the Appendix. The effect of solvents on the carbonyl stretch and on other vibrations is thoroughly discussed in References 85 and 86 at the end of this chapter.

The following general observations are offered:

1. The intensity of the C=O stretching frequency can be correlated with the position of the carbonyl group in a molecule. The intensity also can be correlated with the presence of halogen atoms or conjugation of the carbonyl with other groups.

2. Electronegative atoms near the carbonyl group can shift its frequency to higher values than are observed for similar carbonyl compounds with less electronegative atoms.

3. The position of the C=O stretching frequency will be shifted to lower values if an NH group is next to it.

4. A salt such as $CH_3COO^-Na^+$ has two $C—O^-$ stretching frequencies, one at 1580 cm^{-1} and the other at 1425 cm^{-1}.

5. While a deformation frequency involving only a C=O unit cannot occur, the presence of this group will shift the normal deformation frequencies that can occur for the groups next to the C=O unit. For example, the CH_3 deformation frequency in the structural unit $CH_3—\overset{\overset{\displaystyle O}{\|}}{C}—$ appears near 1360 cm^{-1}, while the unit $—CH_2—\overset{\overset{\displaystyle O}{\|}}{C}—$ has a CH_2 deformation frequency near 1420 cm^{-1}; the normal positions of these bands in hydrocarbons are 1380 and 1450 cm^{-1}, respectively.

6. Association of a carbonyl molecule with a solvent molecule or with other carbonyl molecules will shift the position of the carbonyl frequency to lower frequencies.

7. The coplanarity of the carbonyl group with a group such as C—X, where X is a halogen, appears to shift the carbonyl frequency to a higher position than that found for a similar compound where coplanarity is absent.

8. An overtone of the C=O stretching frequency can often be observed in the spectra of carbonyl compounds, and this band is helpful in assigning the carbonyl band.

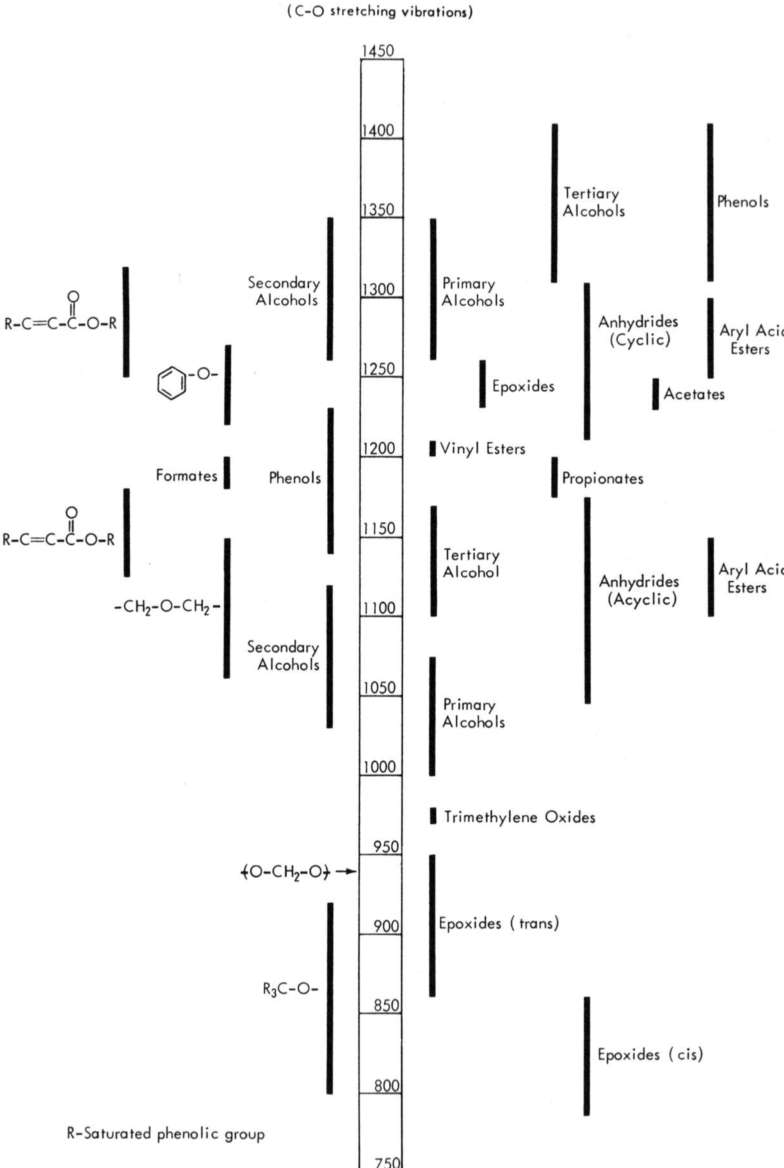

Figure 5-28. Group frequencies for C—O—C and C—O—H groups.

5.10. GROUP FREQUENCIES ASSOCIATED WITH THE C—O—C AND C—O—H STRUCTURAL UNITS

Since the masses of carbon and oxygen are similar and the force constant of the single bond between them is equal to that of many other bonds, the C—O—C frequencies tend to couple with other frequencies and therefore are not good group frequencies.

The region between 1150 and 1060 cm^{-1} is generally assigned to this group, with aryl and conjugated ethers absorbing between 1270 and 1230 cm^{-1}. Cyclic ethers show bands at 1250, 890, or 830 cm^{-1} for small-ring epoxy compounds and at 1140–1070 cm^{-1} for larger ether rings. Alkyl peroxides have a band in the 890–820 cm^{-1} region.

The structural unit C—O—H, which occurs in alcohols, has group frequencies which can be recognized fairly well in the spectra of many alcohols in the 1200–1050 and 1400–1250 cm^{-1} regions. The O—H stretching frequency, useful in recognizing the presence of the C—O—H structural unit, appears in the 3650–2500 cm^{-1} region. A summary of these group frequency assignments is given in Figure 5-28.

5.11. GROUP FREQUENCIES ASSOCIATED WITH NH AND CN STRUCTURAL UNITS

The structural unit NH can occur in such compounds as amides ($R—\overset{\overset{O}{\|}}{C}—NH_2$), amines ($RNH_2$ and R_2NH), and substituted ammonium salts. In each case there are distinct group frequencies associated with the NH group. In addition, in most of these structures, the CN structural unit also occurs. This, however, does not show as distinct a set of group frequencies as the NH unit, although in some cases the CN frequencies can be assigned in the spectra of nitrogen compounds.

A brief summary of the group frequencies for each of the structural units is given in Figure 5-29. In general, the NH stretching frequencies can be found in the region of the OH and CH stretch, the range running from 3500 to 2400 cm^{-1}.

Sandorfy and his co-workers have summarized a series of distinct bands which appear near 2500 cm^{-1} for amine hydrochlorides [71]. These bands, while not strong, are quite distinct for each type of amine hydrochloride.

The aliphatic and aromatic amines have a number of distinct bands which can be used to identify the type of amine. For these amines the structural units NH and CN both have fairly good group frequencies. A brief résumé of these vibrations is presented in

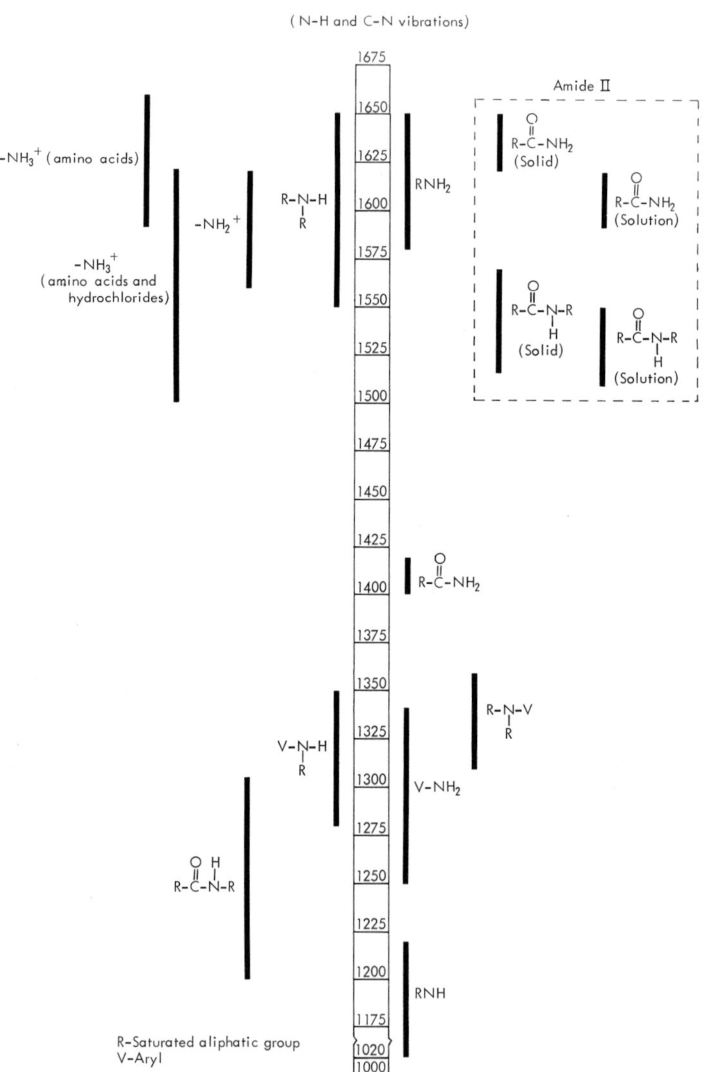

Figure 5-29. Group frequencies for NH, CN, and OH groups.

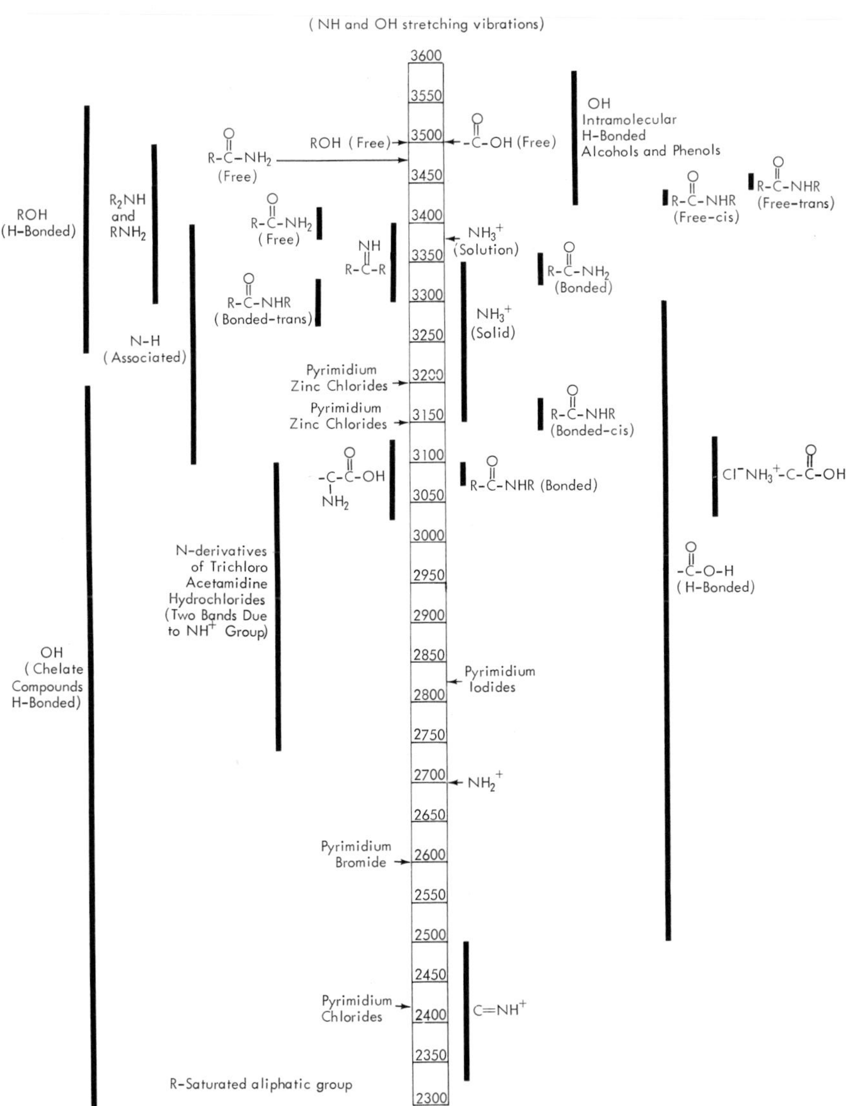

Figure 5-29 (continued).

TABLE 5-XVII. Vibrations for Amines

Vibration	Frequency (cm^{-1})	Remarks
	A. Primary Amines	
N—H sym. stretch N—H asym. stretch	3400 ± 100	Asymmetric band the higher of two bands
NH_2 deformation	1620 ± 30	
NH_2 torsion	290	
Overtone of NH_2 torsion	496 ± 24	
C—N stretch	A. Primary α carbon 1079 ± 11 B. Secondary α carbon 1040 ± 3 C. Tertiary α carbon 1030 ± 8	
	B. Secondary Amines	
N—H stretch	3400 ± 100	One band
C—N—C bend	427 ± 14	
C—N asym. stretch	A. Primary α carbon 1139 ± 7 B. Secondary α carbon 1181 ± 10	
N—H asym. bend	A. Primary α carbon 739 ± 11 B. Secondary α carbon 718 ± 18	
	C. Aromatic Amines	
C—N vibrations	A. Primary 1350–1250	
	B. Secondary (mono-aryl) 1260 ± 4	On deuteration of NH group band is found at 1344 ± 22
	C. Secondary (diamyl) 1241	On deuteration of NH the band is above 1370
	D. Tertiary 1360–1310	
	D. Ethylenediamine Derivatives	
N—C—N vibrations	A. Ethylenediamine 1096, 1052	
	B. Sym. dimethyl ethylenediamine 1147, 1118, 1107, 1093	
	C. Asym. dimethyl ethylenediamine 1023, 1099, 1042	

For the series of methyl, dimethyl, and trimethyl amines, vibrations associated with the CN structure are presented in Table 5-XVIII. The bands for the corresponding hydrochlorides are also listed for comparison.

Table 5-XVII. The vibrations listed are, to a large extent, those suggested by Stewart [72].

The spectra of amine hydrohalides will differ from those of the corresponding amines in a number of ways. First, there will be new vibrations due to formation of the amine ion. For example, a secondary amine hydrochloride will have NH_2^+ vibrations. The NH_2^+ deformation vibrations will result in bands in the 1620–1560 cm^{-1} region which were not present in the secondary amine. The formation of the hydrohalide will also modify the NH stretching vibrations observed for the amine. The CN vibrations of some amines and their hydrochlorides are given in Table 5-XVIII.

Since hydrogen bonding occurs for amine hydrohalides, the spectra for various physical states will differ if varying degrees of hydrogen bonding occur in these states. Other variations in the spectra of hydrohalides can be caused by the interaction between the sampling medium and the hydrohalide, and will be especially great if the sample is dispersed in halide disks, since then ionic equilibrium will be present.

Some references to spectra of amine hydrohalides and compounds containing CN units are included in the Appendix. Also, Table 5-XIII includes some assignments.

5.12. AMIDE I BAND

A band which appears to be primarily a C═O stretching vibration in amides has been called the amide I band. In N-methylacetamide the band is near 1650 cm^{-1} and has been shown to be essentially a C═O stretching vibration [29]. In the solid state the band is lower than in solutions. For primary amides in the solid state the band is near 1650 cm^{-1}, while for secondary amides it is

TABLE 5-XVIII. Vibrations for Amines and Their Hydrochlorides*

Compound	C—N symmetric stretch	C—N asymmetric stretch	C—N bend
Methylamine	1044		
Methylamine hydrochloride	995		
Dimethylamine	930	1024	
Dimethylamine hydrochloride	895		
Trimethylamine	826	1043	425, 365
Trimethylamine hydrochloride	817	985	465, 406

* In some heterocyclic ring compounds having the nitrogen in a ring with a CH_3 group attached to it, a band near 1050 cm^{-1} appears, indicative of the CH_3—N structure.

in the 1680–1630 cm^{-1} range. For tertiary amides in the solid state or in solution it is in the 1670–1630 cm^{-1} range. Figures 5-27 and 5-29 include assignments for a number of amides.

5.13. AMIDE II BAND

The amide II band is a group frequency found for primary amides ($RCONH_2$) and secondary amides (RCONHR). The band occurs in the region from 1600 to 1500 cm^{-1}. Apparently the amide II band is due to the vibration of several structural units such as NH and CN. An extensive study of this band in N-methylacetamide has led to the suggestion that the band is 60% NH bend and 40% CN stretch [29].

The position of the band depends on the physical state and on the degree of association of the molecules, since hydrogen bonding can occur for amides.

In a polyamide such as nylon the band is at 1545 cm^{-1}. An extensive study has shown that the conformation of polypeptides determines the position of the band [73]. In polyacrylamide the amide II band coincides with the carbonyl vibration [74].

In methylthiourea complexes with metals such as platinum, palladium, copper, zinc, and cadmium the amide band appears in a range from 1565 to 1580 cm^{-1}, which is shifted from its normal position at 1550 cm^{-1} in methylthiourea [75]. In thiourea complexes of platinum, palladium, zinc, and nickel the band appears between 1625 and 1615 cm^{-1}, compared to 1610 cm^{-1} in thiourea [76]. In boron halide complexes of acetamides the band appears in the 1555–1525 cm^{-1} range [77]. In trifluoroacetamides the band appears in almost the same positions as in the corresponding acetamides, near 1580 cm^{-1}.

In glucose ureide urea and lactose ureide urea, three strong bands appear in the region from 1695 to 1585 cm^{-1} [78]. Since the three bands may also be due to C=O and NH_2 structural groups, however, it is difficult to assign them correctly. Figure 5-27 includes a number of assignments of amide II bands. A selected list of references to the amide II band is presented in the Appendix. Raman spectra of biological compounds can be measured in water solutions and a number of studies have been reported [87,88].

5.14. AMIDE III, IV, V, AND VI BANDS

The amide III band is a vibration near 1290 cm^{-1} for secondary amides. For N-methylacetamide it has been shown that the band is made up of 40% CN, 30% NH, and 20% $CH_3{-}C\lessgtr$ vibrations [29].

Figure 5-29 includes a number of group frequency assignments of amide II and III bands.

The amide IV band for secondary amides appears between 630 and 600 cm^{-1}. For N-methylacetamide it is near 630 cm^{-1} and has 40% O=C—N and 30% CH_3—C$\lessgtr$ character [29].

The amide V band occurs for hydrogen-bonded secondary amides near 720 cm^{-1}. The band usually is broad and of medium intensity.

The amide VI band occurs for primary amides in the 1420–1400 cm^{-1} region. It usually has medium intensity.

5.15. GROUP FREQUENCIES FOR FLUOROCARBONS

Group frequency correlations for fluorocarbons are quite limited. It appears that interaction between vibrations occurs more extensively for fluorocarbons than for the corresponding hydrocarbons. Because of this it is difficult to find vibrations for fluorocarbons which remain constant for a series of compounds. Even though the extensive interaction of frequencies of fluorocarbons does not lead to good group frequency correlations, this interaction results in quite distinct spectra for each different type of fluorocarbon structure, so that small structural differences can often be detected by studying the infrared spectra of these compounds.

Weiblen, in an extensive survey of the infrared spectra of fluorocarbons, listed the band positions for a number of structural groups to which fluorine was attached [82]. For example, fluorine substitution on a carbon atom bonded to a hydrogen atom will raise the CH stretching frequency. A fluorine atom substituted next to a C≡C structural unit can raise the frequency of the C≡C group, but only slightly. Similarly, a fluorine substituted near a C≡N can raise the frequency of this group. The influence of fluorine atoms substituted near a C=C unit has been discussed in Section 5.5. In general, the C=C stretching frequency is raised by such a substitution. The C=O stretching frequency appears to be raised by 25 to 75 cm^{-1} whenever a fluorine atom is substituted near this unit. This statement would apply to aldehydes, ketones, acid halides, anhydrides, and esters.

5.16. GROUP FREQUENCIES FOR THE C=N, N=O, C≡N, AND OTHER STRUCTURAL UNITS

The C=N and N=O stretching vibration can occur in the region from 1690 to 1480 cm^{-1}, the exact position depending on the structural units attached to the group in question. In allenic-type

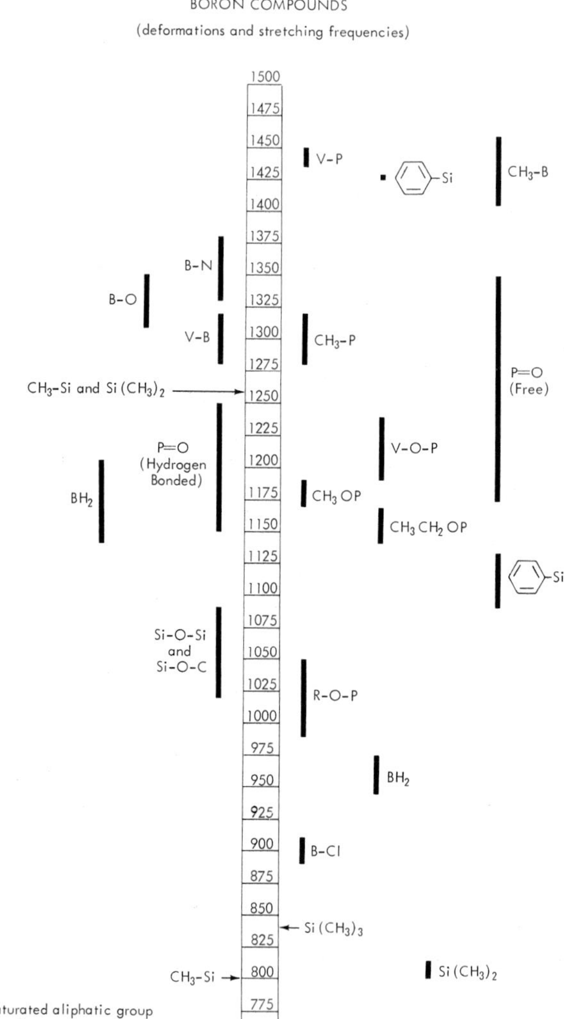

Figure 5-30. Group frequencies for a number of groups.

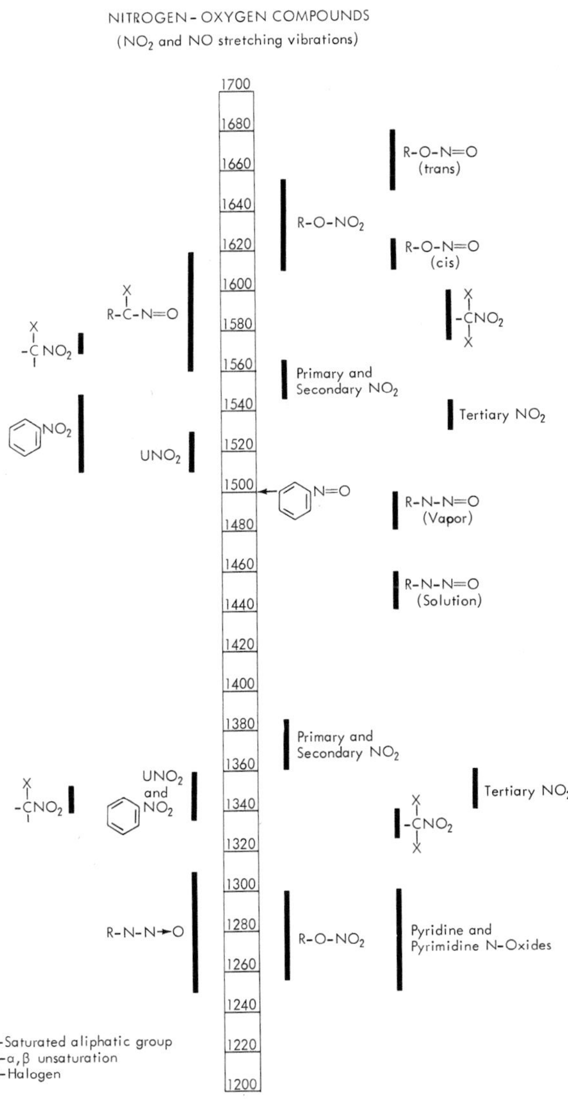

Figure 5-30 (continued).

SOME INORGANIC GROUP FREQUENCIES

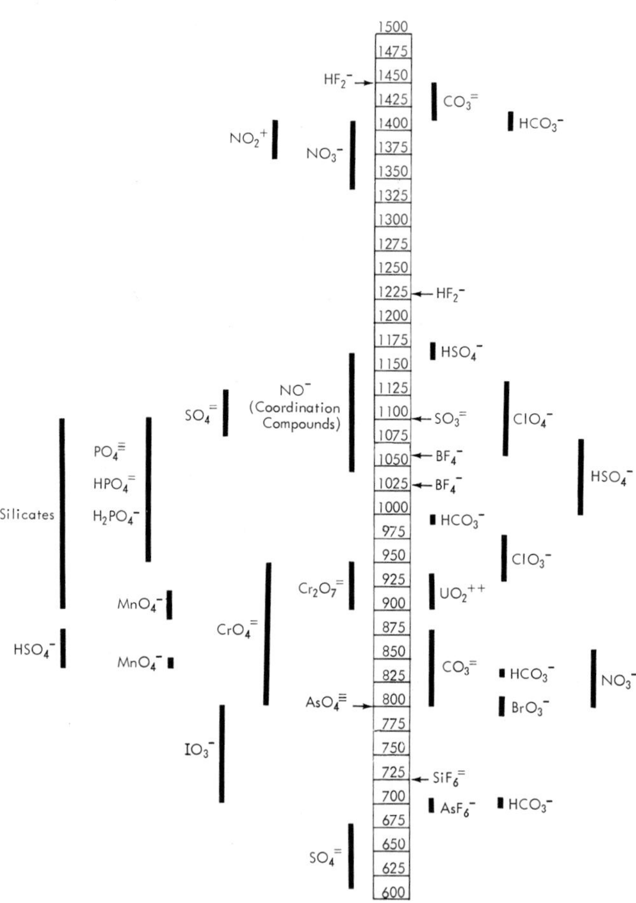

Figure 5-30 (continued).

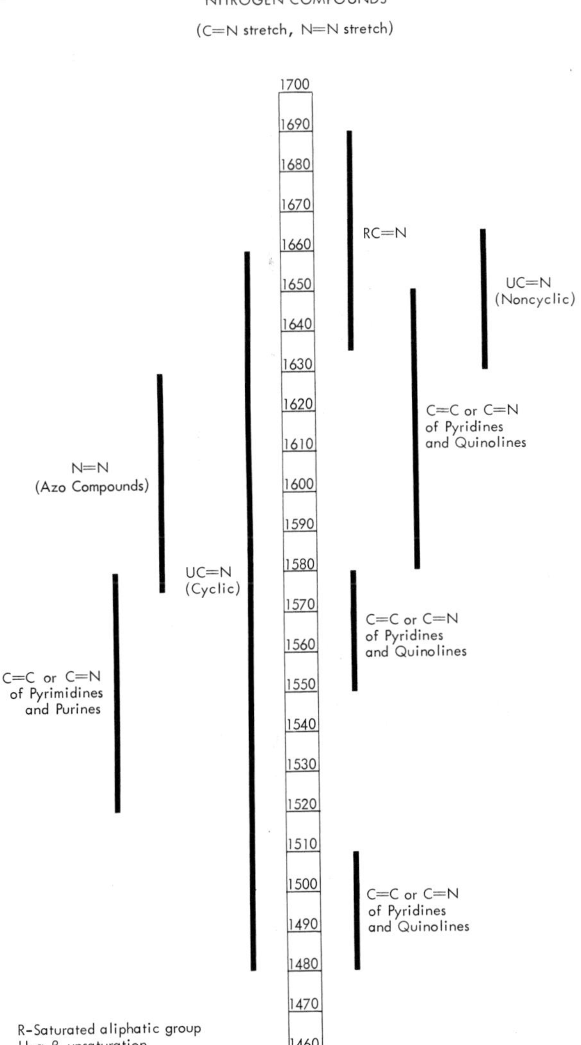

Figure 5-30 (continued).

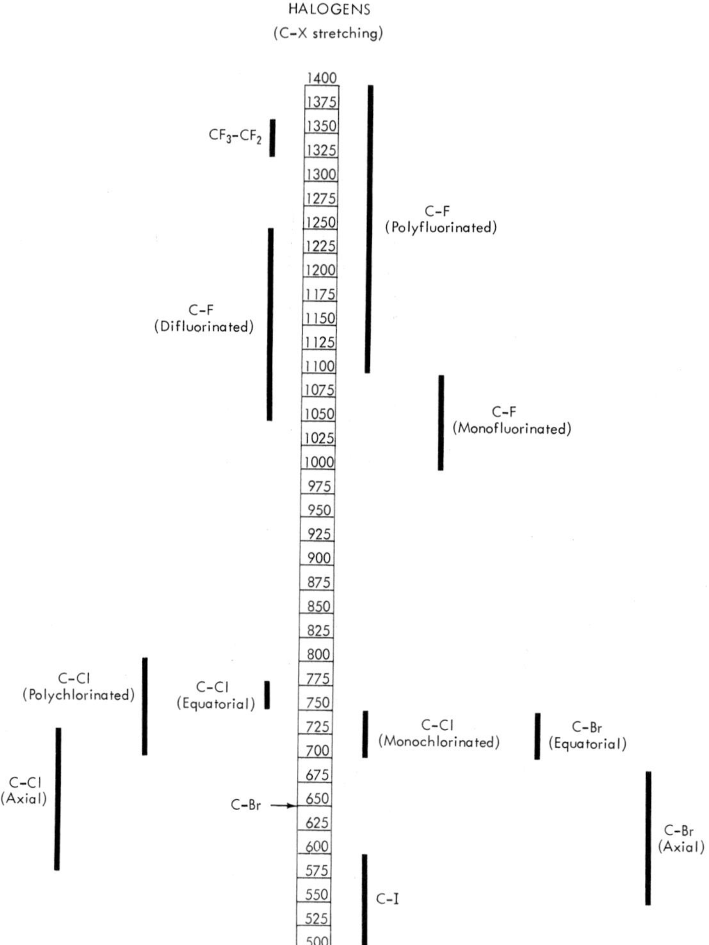

Figure 5-30 (continued).

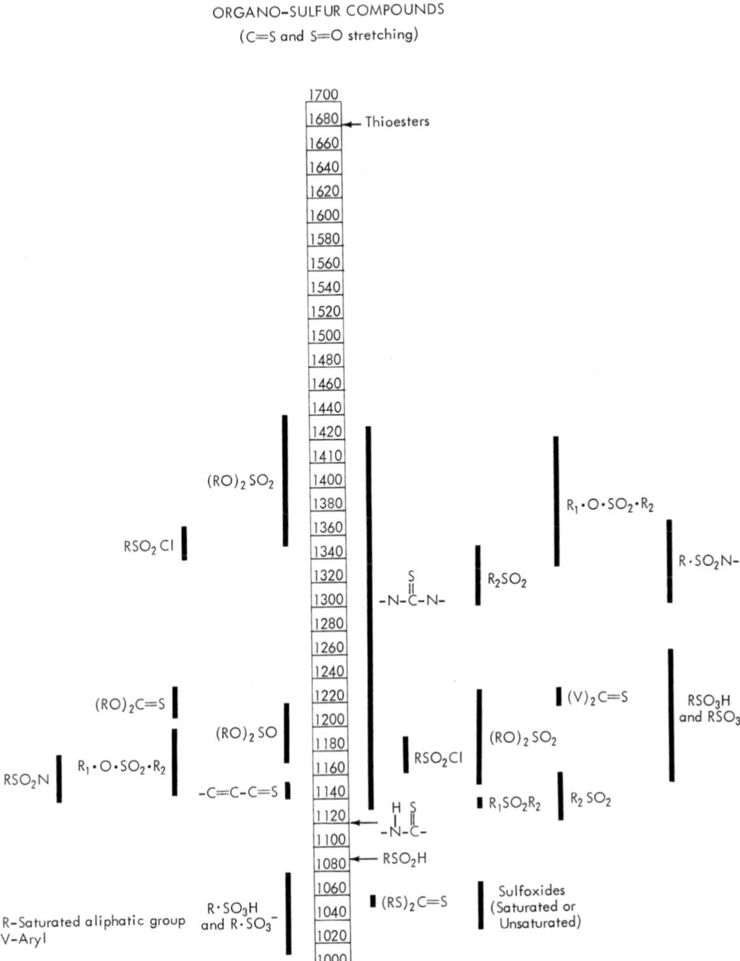

Figure 5-30 (continued).

structures, where the structural unit X=Y=N occurs, strong bands can be found in the 2280 to 2120 cm^{-1} range. Some of these group frequencies are given in Figure 5-30.

The C≡N stretching frequencies appear in the region from 2260 to 2120 cm^{-1} with variable intensity.

Other structural units, such as C—X (where X is a halogen) and N=O, show characteristic group frequencies which may or may not be distinct, depending on the structures attached to these units. These will not be discussed in detail; however, a brief summary is given in Figure 5-30. In addition, group frequencies which have been found for many inorganic compounds are also listed. Good summaries of group frequencies for inorganics have been given by F. Miller and his co-workers [79,80], and by J. Ferraro [81] and others [89–92].

5.17. ORGANIC MOLECULES COORDINATED TO INORGANIC SALTS

Many organic molecules will form complexes with inorganic salts. A compound such as ethylenediamine forms strong complexes with most metallic salts. Even benzene can form weak complexes with such materials as $Cr(CO)_3$ and $SbCl_3$. The interaction of the organic molecule with the inorganic salt can be detected by the changes which occur in the infrared spectra of the organic and inorganic components. For weak interactions we should expect the spectrum of the complex to resemble those of the original organic and inorganic compounds. When strong complexes are formed, as expected, the spectrum of the complex will differ radically from those of the original compounds. An example of the type of spectral changes which can occur when an organic molecule is coordinated to an inorganic salt is illustrated in Table 5-XIX. Here the spectrum of the organic molecule ethylenediamine is compared to the spectrum of ethylenediamine complexed with platinum(II)chloride. While the two spectra were not obtained in the same medium, some general conclusions can be drawn which are of interest. It can be noted that the stretching, wagging, twisting, and rocking frequencies of the NH_2 and CH_2 groups appear shifted to a great extent when ethylenediamine forms a complex, while the NH_2 and CH_2 scissoring frequencies appear to be the least affected. Thus, the scissors would be classified as a "good" group frequency. The NH_2 stretching frequency apparently is perturbed by the anions, while the CH_2 stretching frequencies appear unusually weak. New absorption bands should appear in the complex that are not present in ethylenediamine because a ring has been formed with the platinum and

TABLE 5-XIX. Absorption Bands for Ethylenediamine[1,2] and Ethylenediamine Complex of Platinum(II) Chloride[3]

Compound	Group frequency assignments										
	NH_2 stretch	CH_2 stretch	NH_2 scissors	CH_2 scissors	CH_2 wag	NH_2 wag	CH_2 twist	NH_2 wag	NH_2 twist	CH_2 rock	NH_2 rock
$NH_2CH_2CH_2NH_2$[4]	3335 3246 3171	2917 2891 2858	1597 1608	1469 1456	1360*		1298* 1305*	1249*	920* 951*	775 761	832†
$Pt(NH_2CH_2CH_2NH_2)_2Cl_2$[5]	3141 3307 3064		1610	1454	1395 1373	1326	1311 1275	1154	1138 999	897	831

(1) A. Sabatini and S. Califano, *Spectrochim. Acta* **16**: 677 (1960).
(2) R. J. Mureinik and W. Robb, *Spectrochim. Acta* **24A**: 377 (1968).
(3) D. Powell and N. Sheppard, *Spectrochim. Acta* **17**: 68 (1961).
(4) Based on measurements in the solid, liquid, and gaseous state.
(5) In liquid paraffin and hexachlorobutadiene.
* The CH_2 (and NH_2) wag and twist assignments of Reference 1 have been reversed so as to obtain better agreement with the assignments of Reference 2.
† Band present in liquid state spectrum of $NH_2CH_2CH_2NH_2$ but not considered a fundamental.

the two NH_2 groups. Such bands do occur for the complex at 1050 and 875 cm^{-1}.

It should be remembered that since a comparison is being made here between two compounds of dissimilar symmetry, it cannot be expected that all bands will match.

5.18. PRACTICE INTERPRETATION OF SPECTRA*

One method of teaching interpretation of spectra to students is based on having them memorize a number of generalizations which, when applied to an unknown spectrum, lead the student to postulate several structures. To verify which of the postulated structures is correct, resort is then made to reference spectra or tabular data.

The generalizations are as follows.

1. The strongest bands in the spectrum should be assigned first. Usually their position and relative broadness suggest the class of compound to which the unknown belongs.
2. Carbon–hydrogen absorption bands, in contrast to carbon–oxygen, carbon–nitrogen, carbon–sulfur, and most other carbon–atom bands, do not have a broad appearance. The converse is then true—i.e., broad bands usually are not carbon–hydrogen group frequencies.
3. Two bands of nearly equal intensity and within several hundred wavenumbers of each other may indicate coupled vibrations of a group XY_2. Groups which give strong broad bands of nearly equal intensity are NO_2, SO_2, CO_2^-, NH_2, etc. A bending vibration below the two stretching frequencies is also expected for these groups.
4. The general regions where bands appear is dependent on the bond order and masses of the atoms making up the group. To approximate the positions expected for organic groupings the following reference positions can be used.

F—H:4100	C≡C:2100	S≡O:1400
O—H:3700	C=O:1700	C—O:1050
N—H:3500	C=C:1600	N—O:1100
C—H:3000	N=O:1550	C—S:700
C≡O:2150		

   If these groups are present as XY_2 units, then two bands are observed, one above and one below the position listed.
5. The groups adjacent to the units listed in 4 can change the positions listed. For example, in CH_3OH the C—O stretch is at 1050 cm^{-1}. In CH_3COOH it is at 1250 cm^{-1}. The adjacent

*All the figures in this section will be found in the pocket on the inside back cover.

group shifts can be predicted. In general, adjacent electronegative groups, such as halides or carbonyls, raise the frequency of the unit they are attached to, and electropositive groups lower it.

6. If resonance occurs, so that the bond order of the vibrating unit is reduced, then the frequency is lowered from the position observed when resonance does not occur.
7. In many instances, the greater the number of bands, the more complex the molecule. This rule, however, must be applied with caution, as exceptions occur fairly often.
8. Number and contours of bands for some groups are similar in the spectra of most of the compounds containing them. For example, the SO_3^- group has such a series of bands from 1250 to 950 cm^{-1} when the group is attached to aromatic rings.

We shall now apply these generalities to the interpretation of spectra. All spectra interpreted are in the jacket inside the back cover.

Let us consider first the spectra in Figures 5-31 to 5-34. The class of compound represented is quite obvious, based on four strong, broad bands. The bands are near 3300, 1400, 1050, and 600 cm^{-1}. The class of compound is aliphatic primary alcohol.

For all the compounds in this class the same four bands are seen. The OH stretch is at 3300. The broadness in the 1500–1300 cm^{-1} region is due to the in-plane bend of the OH group. The band near 1050 cm^{-1} is the C—O stretch, and its position identifies that these are primary alcohols. The band at 600 cm^{-1} is the out-of-plane OH bend. The reader can note that these four bands appear at nearly the same position in each spectrum. This is true for all compounds of this class. If a large number of CH_3 and CH_2 groups are present compared to OH, the intensity of these four bands is low, but their broadness helps identify them. There is, however, one important fact that is not discernible from the group frequencies for alcohols. This is that spectra 5-32 and 5-34 are of diols; they are ethylene glycol and 1,4-butanediol respectively. Figures 5-31 and 5-34 are isobutyl and n-butyl alcohol, respectively. Because there are similar vibrating units present in diols and monohydric alcohols, their spectra are similar.

We can now proceed to examine Figures 5-35 and 5-36. Noting bands which are strong or broad, we can again classify the compound represented. Three medium-to-strong and broad bands are seen in both spectra, although the positions differ slightly. For Figure 5-35 the bands are seen near 3400, 1150, and 625 cm^{-1}. In Figure 5-36 they are near 3400, 1100, and 625 cm^{-1}. The experienced interpreter would also note that broadness of the upper parts of bands occurs in the 1600–1500 cm^{-1} region. All these data suggest that the class of compound is alcohol, but not of the primary type. The alcohols represented by the spectra in Figures 5-35 and 5-36 could therefore be secondary or tertiary. It is here that the experience of the interpreter determines if he can choose between these two types of alcohols by viewing the spectra. In correlation tables it is often listed that secondary alcohols have bands near 1150 cm^{-1}, and tertiary near 1200 cm^{-1}. These are C—O stretching frequencies. These positions, however, are found only if the groups attached to the C—O—H unit are alkyl.

Electronegative or unsaturated groups can shift the position of the C—O stretch. The experienced interpreter can, however, decide if such groups are present by noting bands in the spectrum not assignable to the C—O—H group. In Figures 5-35 and 5-36 we find no bands indicative of unsaturation, or of groups such as C—F or C—Cl. A reasonable judgment is then made that only alkyl groups are present, and the compound represented by Figure 5-35 is a secondary, and by 5-36 a tertiary alcohol. The compounds are secondary butyl and tertiary butyl alcohols.

The next two spectra interpreted are Figures 5-37 and 5-38. The absence of broadness in most of the bands of these two spectra; the presence of strong bands in the 900–600 cm^{-1} region; the pattern of weak bands in the 2000–1600 cm^{-1} region; the presence of a sharp band above 3000 cm^{-1}; and the general positions of all the bands, all make the identification of the class of compounds for these two figures very simple. They are both aromatic ring compounds, and probably no groups such as alcohol, ether, or others having O, S, or N in aliphatic structures are present. This latter conclusion is based on the observation that C—O, SO, NO, C—N, etc. groups give rise to broad bands. It would appear that the bands seen in Figures 5-37 and 5-38 are all CH or CC vibrations. The experienced interpreter might add the fact that C=N could be represented here, as well as C=C, in these rings. It is very difficult to distinguish between the C=C and C=N vibrating groups.

The bands in Figures 5-37 and 5-38 appear in similar regions, and for two bands near 700 and 750 cm^{-1} nearly exact coincidence occurs. These two bands are indicative of monosubstituted rings (i.e., there are five adjacent hydrogens on the ring). Thus the rings appear to be monosubstituted six-member rings. Since we have no atoms such as O or S, we can find good group frequencies for any alkyl substituents if they are present. In Figure 5-37, bands near 2900 and 1380 cm^{-1} identify the substituent as alkyl. No such bands appear in Figure 5-38, so the substituent is not alkyl. No bands suggestive of the substituent appear, and this is puzzling until the compound is identified. It is pyridine, and no substituent is present, although the five adjacent hydrogens on the ring are present. The first compound (Figure 5-37) is toluene, and also has five adjacent hydrogens on a ring. Thus all vibrations associated with five adjacent =C—H groups are found in each spectrum.

Figure 5-39 represents a class of compound which usually can be identified from its infrared spectrum. The bands which classify it are near 3500, 1600, and 800 cm^{-1}. At the same time the lack of bands in other regions confirms the identification. The class of compound is primary aliphatic amine, and the compound in Figure 5-39 is n-propylamine. The aliphatic character is recognized by the lack of weak bands in the 2000–1600 cm^{-1} region, and of strong bands in the 900-600 cm^{-1}, which would be due to the out-of-plane bend of the CH ring groups. The aliphatic group has CH_3 and CH_2 vibrations near 3000, 1450, and 1380 cm^{-1}.

The primary amine class is identified by the two NH stretching, and one bending, frequencies near 3500 and 1600 cm^{-1}, respectively. Only the NH_2 group gives these three bands, so the primary character is established. The band near 800 cm^{-1} confirms that this is an amine. It is assigned to a rock of the NH_2 group. A good chemist might note that the odor of this compound might have helped identify it as an amine.

Figure 5-40 is an excellent example of how the strong bands in the spectrum help classify the unknown. The strongest bands appear near 1300 and 1100 cm^{-1}. They are broad and this suggests groups other than carbon and hydrogen. The positions are those expected for SO, and so the assignment of the presence of SO or SO_2 seems logical. That there are two bands of near-equal intensity narrows the choice to SO_2. Finally the lack of strong bands in the 900–600 cm^{-1} region and the lack of weak bands in the 2000–1600 cm^{-1} region, indicate that the sulfone has aliphatic character. The compound is propyl sulfone.

When a medium to strong band appears in the 2500–1900 cm^{-1} region the class of compound can almost always be identified since only a few groups show bands in this region. The groups are X≡Y or X=Y=Z, where X, Y, and Z can be C, O, and N in any combination. In Figure 5-41, a pair of bands appears near 2250 cm^{-1}. The compound therefore contains one of these groups. To ascertain which it is, we can examine other bands in the spectrum. The broad band near 1400 cm^{-1} is certainly important, as is the lack of bands in the 2000–1600 cm^{-1} region. The latter fact suggests that no aromatic rings are present. The band near 1400 cm^{-1} is puzzling, but may be due to the presence of the C≡N group. The compound is identified as acetonitrile.

There are a number of times when the infrared spectrum does not have bands of groups present in the molecule, although these groups are structurally an important part of the molecule. One important case is where the group is present with such symmetry that no infrared band occurs although a Raman band may be detected. For example, in ethylene the C=C stretch is not found in the infrared spectrum but is present in the Raman spectrum. No dipole-moment change occurs in the stretching of this group, due to the symmetrical placement of the group in the molecule. It is not necessary that structural symmetry occur for a group to have an infrared-inactive band. For example, in $BrC{\equiv}CCH_2CH_3$ it is probable the C≡C stretch will be weak or absent, because the change in dipole is small or zero.

Figure 5-42 is an example of a compound where we cannot find bands indicative of one major structural group. Examining the spectrum we can easily identify the presence of a monosubstituted aromatic ring, but we find no indication as to what group the ring is attached to. The assignments are confused of course by the fact that the spectrum was run in nujol so that we could have CH_3 and CH_2 groups present and these bands would be obscured by the nujol bands. This might lead one to assume this is an alkyl benzene compound; however, it is diphenyl acetylene. Thus the C≡C group gives no band which can be used to identify its presence in the molecule.

Figure 5-43 has strong bands in the 1000–400 cm^{-1} region. If these were all sharp we might suspect a polyaromatic compound, but there is a broad band near 780 cm^{-1}. This broadness suggests that groups other than CH are present, and the position suggests these groups are halogen. If the halogen was fluorine attached to carbon, we would expect a series of weak bands in the 2500–2000 cm^{-1} region, and possibly a broad band near 1100 cm^{-1}. The halogen present must therefore be chlorine, bromine, or iodine. The position of 780 cm^{-1} suggests that it is chlorine, but since the position of carbon–halogen frequencies is quite variable we cannot place too high a confidence on this assignment. The compound is 1,1,2,2-tetrachloroethane.

Figure 5-44 should be a simple interpretation problem for the experienced spectroscopist. He would note the lack of CH stretching frequencies in the 3000 cm^{-1} region, and the broad band near 1200 cm^{-1}, which as we pointed out in the previous spectrum discussion can be due to C–F vibrations. The conclusion, then, is that the compound is a fluorinated hydrocarbon. The compound is Fluorolube, a high-molecular-weight fluorocarbon.

Figure 5-45 is presented next, since it is illustrative of how the experienced interpreter uses all bands in the spectrum to identify an unknown. For example, the strong broad band near 1100 cm^{-1} might suggest the presence of a CF group as in the previous spectrum. However, the experienced interpreter also knows that the carbon–oxygen single bond stretch is also found there. Examining the bands at various positions in the spectrum, the interpreter identifies the presence of CH_3 and CH_2, and a lack of unsaturation. The position of

$1100\ cm^{-1}$ is correct for aliphatic ethers, so he would probably suggest this class of compounds. The correct identification is that this is diethyl ether.

All the previous examples could be considered simple examples of spectral interpretation. In most cases, only two or three different structural units were present. We shall now consider examples where more than three structural units can give rise to distinct infrared bands.

Let us consider first Figure 5-46. We notice first the large number of medium-to-strong bands. To begin interpretation, we note strong bands first, and those at positions where their identification is simple. For example, the strong band near $1750\ cm^{-1}$ is almost certainly a carbon–oxygen double-bond stretching frequency. This then is the first group assigned for this molecule. We can also note that aromaticity is present, since bands in the 2000–1600 cm^{-1} region are present, as well as other bands indicative of the ring.

Broadness near 1500 and $1300\ cm^{-1}$ is indicative of an NO_2 group, and this is the third vibrating unit we assign in this spectrum. A broad band near $1200\ cm^{-1}$ together with the position of the carbonyl stretch establishes that the compound is an aromatic aldehyde. If the spectrum had been determined in a medium other than nujol we would have found a band near $2700\ cm^{-1}$ which would have verified the aldehyde structure. Not using nujol would have also allowed one to decide if CH_3 and CH_2 groups are present. They are not in this example.

The final group assigned has a band near $800\ cm^{-1}$. Its presence is difficult to detect in this complex spectrum, and other information on the unknown would have helped in the assignment. The group is C—Cl, and the unknown compound is 5-chloro-2-nitrobenzaldehyde.

Figure 5-47 is another example of a molecule having more than three vibrating groups which give rise to distinct bands. We note as in the previous example that the number of bands is suggestive of complexity. As in the previous example, we first attempt to identify the bands which may be most easily assigned. Certainly a broad band near $3400\ cm^{-1}$ must be an OH or NH group; which it is can often be decided by examining the remainder of the spectrum. The sharp band near $3500\ cm^{-1}$ might be indicative of an OH stretch of a non-associated group, as often phenols have both this type of band and the broad band near $3400\ cm^{-1}$, the broad band representing associated OH groups. If the phenolic group is present, then a broad, strong band is found in the 1300–1250 cm^{-1} region. The region from 1500 to $1100\ cm^{-1}$ has a series of strong bands, and we can tentatively assume that the phenolic structure may be present. Certainly there are aromatic rings present, as all the vibrations of these rings are found. The substitution of the ring, however, cannot be decided, as too many bands are present in the spectrum.

The lack of medium or strong bands between 1600 and 1900 cm^{-1} eliminates the possibility of a carbon–oxygen double bond, and also the possibility that the bands near 3400 cm^{-1} represent NH_2. The NH_2 group has a bending vibration near 1600 cm^{-1}.

The final structural group present does give rise to a band which might be useful in identifying it, but in such a complex spectrum it is difficult to make such an assignment. The group is C–S, and the band due to this group is near 580 cm^{-1}. Its broadness gives a clue that it is not a CH vibration, but this group frequency is often difficult to identify. The compound represented by spectrum 5-47 is bis-3,6-dimethyl-2-hydroxyphenylsulfide.

Figure 5-48 is quickly identified as that of a phenol if we note that the strongest bands occur in the 1500–1100 cm^{-1} region. The aromaticity is seen from bands in the 2000–1600 cm^{-1} region, and the phenolic nature from the bands in the 1500–1100 cm^{-1} region. No other groupings are seen and so the decision that the class of compound is phenolic is made. The compound is catechol.

Our next series of spectra have been chosen to illustrate how the experienced interpreter can quickly classify the compound represented without recourse to tables. Generally this can be done in a very short time by those having viewed a large number of spectra.

Let us consider first Figure 5-49. The interpreter notes three regions of strong bands, and a lack of other bands, suggesting aromaticity. He therefore decides that it is an aliphatic compound, and that the three strong bands in the 3000, 1700, and 1050 cm^{-1} represent CH stretch, C=O stretch, and C—O stretch respectively. Since two C=O stretching frequencies occur, he assigns two of these groups as present. The compound can only be an anhydride and in this case it is valeric.

Figure 5-50 suggests aromaticity, and the large number of bands in the 1000–500 cm^{-1} suggests two sites of unsaturation. No evidence for groups other than CH is present in the spectrum. The pattern of lines in the 2000–1600 cm^{-1} region suggests a distorted series for a monosubstituted aromatic ring, so a second grouping must be giving a band in this region. If we note the very strong bands near 1000 and 950 cm^{-1} we suspect the presence of olefinic structure, since aromatic bands are not so intense in this region. We thus suspect aromatic and olefinic groups. The vinyl group gives bands near 1000 and 950, and since we have bands identifiable for a monosubstituted aromatic ring we put a vinyl group onto a benzene ring and postulate styrene as the compound represented by Figure 5-50. This is correct.

One can become quite proficient in recognizing certain classes of compounds by viewing a large number of spectra of the class. Thus in Figure 5-51, the series of bands from 1250 to 950 cm^{-1}, and their general contours, are indicative of the aromatic SO_3^- group. Both naphthalene and benzene sulfonates have the same general appearance. The compound represented in Figure 5-51 is sodium benzene sulfonate.

The drug chemist, having viewed a number of spectra of compounds having NH_2 and SO_2 groups, usually has no difficulty identifying these groups in unknown spectra. For example, in Figure 5-52, the two sharp bands near 3500 cm^{-1} often appear in the spectra of drugs having the NH_2 group. The compound represented here is sulfacetamide. The large number of bands makes it difficult to assign all bands but the general appearance of the spectrum makes its classification as one of the sulfa drugs fairly straightforward.

There are many other classes of compounds which the interpreter can identify quite easily when he knows that the sample must belong to one of a few classes. For example, the polymer chemist, knowing the sample he has prepared for infrared determination was quite intractable, can suspect the type of polymer before viewing the spectrum. His own background in the field allows him to reduce the number of possibilities to only a few. The infrared spectrum then identifies which of these few is correct. This type of capability comes from viewing a number of spectra of these compounds in his own work, noting certain bands, and that the relative contours of these bands always appear the same. If other compounds are present in minor amounts he can by experience detect these, sometimes by the smallest change in the contours which he noted the polymer generally has. We cannot, of course, cover all the examples the authors have encountered in their studies. We can observe that use of tables of group frequencies without experience is usually a poor method of interpreting spectra. The physical state of the sample, its odor, its hardness, and any other properties, always give clues to the class of compound. The experience of viewing a large number of spectra then allows the interpreter to classify the groups present. Rarely can he identify the exact compound, but having classified it he can use reference spectra to identify it exactly.

We shall illustrate this ability with one more unknown spectrum. Figure 5-53 should indicate an aromatic acid to the experienced interpreter as soon as he views it. The broadness in the 3000 cm^{-1} region, the carbonyl frequency near 1700 cm^{-1}, the band near 1300 indicative of a C—O group adjacent to a carbonyl, and the acid-dimer band near 900 cm^{-1}, make the identification a certainty. The compound is benzoic acid.

BOOKS OF GENERAL INTEREST

D. M. Adams, *Metal–Ligand and Related Vibrations*, St. Martin's Press, New York (1968).

L. Bellamy, *Advances in Infrared Group Frequencies*, Methuen & Co., London (1968).

The Infrared Spectra of Complex Molecules, John Wiley & Sons, New York (1958).

Chemical Applications of Spectroscopy, Vol. 9 of *Techniques of Organic Chemistry*, Inter-Science, New York (1958).

A. D. Cross, *Introduction to Practical Infrared Spectroscopy*, Butterworths, London (1961).

K. Nakamoto and P. J. McCarthy, *Spectroscopy and Structure of Metal Chelate Compounds*, John Wiley & Sons, New York (1968).

H. A. Szymanski, *Interpreted Infrared Spectra*, Vols. 1, 2, and 3, IFI/Plenum Press, New York (1964, 1966, 1967).

A Systematic Approach to the Interpretation of Infrared Spectra, Hertillon Press, Buffalo, New York (1968).

REFERENCES

1. G. Herzberg, *Infrared and Raman Spectra of Polyatomic Molecules*, D. Van Nostrand, Co., Inc., Princeton (1945).
2. L. J. Bellamy and R. L. Williams, *Spectrochim. Acta* **9**: 341 (1957).
3. L. J. Bellamy and R. L. Williams, *Spectrochim. Acta* **9**: 315 (1957).
4. M. Beer, H. B. Kesseler, and G. B. B. M. Sutherland, *J. Chem. Phys.* **29**(5): 1097 (1958).
5. G. E. Hansen and D. M. Dennison, *J. Chem. Phys.* **20**(2): 313 (1952).
6. H. L. McMurry and V. Thornton, *J. Chem. Phys.* **19**(8): 1014 (1951).
7. D. W. Z. Oxford and D. H. Rank, *J. Chem. Phys.* **17**(4): 430 (1949).
8. F. W. Parker and A. H. Nielsen, *J. Mol. Spectroscopy* **1**: 107 (1957).
9. M. El-Sabban, A. G. Meister, and F. F. Cleveland, *J. Chem. Phys.* **19**(7): 855 (1951).
10. L. W. Daasch, C. Y. Liang, and J. R. Nielsen, *J. Chem. Phys.* **22**(8): 1293 (1954).
11. J. L. Lauer, W. H. Jones, Jr., and H. C. Beachell, *J. Chem. Phys.* **30**(6): 1489 (1959).
12. W. J. Lehmann, C. O. Wilson, Jr., and I. Shapiro, *J. Chem. Phys.* **34**(2): 476 (1961).
13. C. W. Young, J. S. Koehler, and D. S. McKinney, *J. Am. Chem. Soc.* **69**: 1410 (1947).
14. L. J. Bellamy and R. L. Williams, *J. Chem. Soc.*, Part 3, 2753 (1956).
15. J. R. Nielsen and H. H. Claassen, *J. Chem. Phys.* **18**(11): 1471 (1950).
16. S. A. Frances, *J. Chem. Phys.* **19**(7): 942 (1951).
17. K. Nakamoto, P. J. McCarthy, and A. E. Martell, *J. Am. Chem. Soc.* **83**(6): 1272 (1961).
18. J. K. Wilmshurst, *J. Chem. Phys.* **25**(6): 1171 (1956).
19. E. R. Shull, T. S. Oakwood, and D. H. Rank, *J. Chem. Phys.* **21**: 2024 (1953).
20. J. K. Wilmshurst, *J. Mol. Spectroscopy* **1**: 201 (1957).
21. W. Lehmann, T. P. Ovak, and I. Shapiro, *J. Chem. Phys.* **30**(5): 1215 (1959).
22. M. Falk and E. Whalley, *J. Chem. Phys.* **34**(5): 1554 (1961).
23. A. E. Parsons, *J. Mol. Spectroscopy* **3**: 73 (1959).
24. N. Jonathan, *J. Mol. Spectroscopy* **5**: 101 (1960).
25. J. G. Pillai and F. F. Cleveland, *J. Mol. Spectroscopy* **5**: 212 (1960).
26. Private communications.
27. G. C. Turrell and J. E. Gordon, *J. Chem. Phys.* **30**(4): 895 (1959).
28. J. R. Barcelo and J. Bellanato, *Spectrochim. Acta* **8**: 27 (1956).
29. T. Miyazawa, T. Shimanouchi, and S. Mizushima, *J. Chem. Phys.* **29**(3): 611 (1958).
30. M. Martinette and S. Mizushima, *Spectrochim. Acta* **14**: 212 (1958).
31. D. A. Dows, *J. Chem. Phys.* **29**(3): 484 (1958).
32. K. Shimizu and H. Murata, *J. Mol. Spectroscopy* **4**: 214 (1960).
33. K. Shimizu and H. Murata, *J. Mol. Spectroscopy* **4**: 201 (1960).
34. J. P. McCullough, W. N. Hubbard, F. R. Frow, I. A. Hossenlopp, and G. Waddington, *J. Am. Chem. Soc.* **79**: 561 (1957).
35. R. J. Philippe, *J. Mol. Spectroscopy* **6**: 492 (1961).
36. W. D. Horrocks, Jr. and F. A. Cotton, *Spectrochim. Acta* **17**(2): 134 (1961).
37. H. R. Linton and E. R. Nixon, *Spectrochim. Acta* **15**: 146 (1959).

38. M. Halmann, *Spectrochim. Acta* **16**: 407 (1960).
39. W. J. Lehmann, C. O. Wilson, Jr., and I. Shapiro, *J. Chem. Phys.* **34**(2): 476 (1961).
40. W. J. Lehmann, C. O. Wilson, Jr., and I. Shapiro, *J. Chem. Phys.* **34**(3): 783 (1961).
41. W. J. Lehmann, C. O. Wilson, Jr., and I. Shapiro, *J. Chem. Phys.* **32**(4): 1088 (1960).
42. W. J. Lehmann, C. O. Wilson, Jr., and I. Shapiro, *J. Mol. Spectroscopy* **7**: 253 (1961).
43. K. Frei and H. H. Gunthard, *J. Mol. Spectroscopy* **5**: 218 (1960).
44. T. Miyazawa, *J. Mol. Spectroscopy* **4**: 155 (1960).
45. A. E. Parsons, *J. Mol. Spectroscopy* **6**: 201 (1961).
46. T. Shimanouchi, M. Tsuboi, T. Takenishi, and N. Iwata, *Spectrochim. Acta* **16**: 1328 (1960).
47. D. M. Adams, J. Chatt, R. G. Guy, and N. Sheppard, *J. Chem. Soc.* 738 (1961).
48. R. C. Lord and B. Nolin, *J. Chem. Phys.* **24**(4): 656 (1956).
49. R. A. Y. Jones and A. R. Katritzky, *J. Chem. Soc.* 4376 (1960).
50. W. E. Fitzgerald and G. J. Janz, *J. Mol. Spectroscopy* **1**: 49 (1957).
51. F. Halverson and R. J. Francel, *J. Chem. Phys.* **17**: 694 (1949).
52. A. R. Katritzky, J. M. Lagowski, and J. A. T. Beard, *Spectrochim. Acta* **16**: 954 (1960).
53. D. W. Aubrey, M. F. Lappert, and H. Pyszora, *J. Chem. Soc.* 5239 (1960).
54. N. Jonathan, *J. Mol. Spectroscopy* **7**: 105 (1961).
55. H. B. Stewart and H. H. Nielsen, *Phys. Rev.* **75**(4): 640 (1949).
56. T. Shimanouchi and I. Suzuki, *J. Mol. Spectroscopy* **8**: 222 (1962).
57. W. West, *Chemical Applications of Spectroscopy*, Vol. 9 of *Techniques of Organic Chemistry*, p. 414, Interscience Pubs., Inc., New York (1958).
58. D. A. Ramsay, *J. Chem. Phys.* **17**(7): 666 (1949).
59. A. Sabatini and S. Califano, *Spectrochim. Acta* **16**: 677 (1960).
60. D. B. Powell and N. Sheppard, *Spectrochim. Acta* **17**: 68 (1961).
61. M. C. Tobin, *J. Mol. Spectroscopy* **5**: 65 (1960).
62. R. H. Biddulph, M. P. Brown, R. C. Cass, R. Long, and H. B. Silver, *J. Chem. Soc.* 1822 (1961).
63. H. D. Kaesz and F. G. A. Stone, *Spectrochim. Acta* **15**: 360 (1959).
64. W. J. Lehmann and I. Shapiro, *Spectrochim. Acta* **17**: 396 (1961).
65. M. Forel, N. Fuson, and M. Josien, *J. Opt. Soc. Am.* **50**(12): 1228 (1960).
66. J. J. Manion and T. S. Wang, *Spectrochim. Acta* **17**: 990 (1961).
67. Y. Kakiuti, *J. Chem. Phys.* **25**(4): 777 (1956).
68. D. Whiffen, *Spectrochim. Acta* **7**: 253 (1955).
69. L. J. Bellamy and R. L. Williams, *Trans. Faraday Soc.* **55**: 14 (1959).
70. C. N. R. Rao, G. K. Goldman, and C. Lurie, *J. Phys. Chem.* **63**: 1311 (1959).
71. C. Brissette and C. Sandorfy, *Can. J. Chem.* **38**: 34 (1960).
72. J. E. Stewart, *J. Chem. Phys.* **30**(5): 1259 (1959).
73. T. Miyazawa and E. R. Blout, *J. Am. Chem. Soc.* **83**: 712 (1961).
74. N. Ogata, *Makromol. Chem.* **40**: 55 (1960).
75. T. J. Lane, A. Yamaguchi, J. V. Quagliano, J. A. Ryan, and S. Mizushima, *J. Am. Chem. Soc.* **81**: 3824 (1959).
76. A. Yamaguchi, R. B. Pentland, S. Mizushima, T. J. Lane, C. Curran, and T. J. Quagliano, *J. Am. Chem. Soc.* **80**: 527 (1958).
77. J. A. Little, *J. Chem. Soc.* **60**: 2144 (1959).
78. L. Segal, R. T. O'Connor, and F. V. Eggerton, *J. Am. Chem. Soc.* **82**: 2807 (1960).
79. F. A. Miller and C. H. Wilkens, *Anal. Chem.* **24**: 1253 (1952).
80. F. A. Miller, G. L. Carlson, F. F. Bentley, and H. H. Wade, *Spectrochim. Acta* **16**(1/2): 135 (1960).
81. J. R. Ferraro, *J. Chem. Educ.* **38**(4): 201 (1961).
82. J. H. Sims (ed.), *Fluorine Chemistry*, p. 449, Academic Press, Inc., New York (1954).
83. J. K. Wilmshurst, *Can. J. Chem.* **35**: 937 (1957).

84. J. R. Durig and D. W. Wertz, *J. Chem. Phys.* **49**: 2118 (1968).
85. H. P. Figeys and J. Nasielski, *Spectrochim. Acta* **23A**: 465 (1967).
86. S. Tanaka, K. Tanabe, and H. Kamada, *Spectrochim. Acta* **23A**: 209 (1967).
87. R. C. Lord and G. J. Thomas, Jr., *Biochim. Biophys. Acta* **142**: 1 (1967).
88. E. L. Elson and J. T. Edsall, *Biochemistry* **1**: 1 (1962).
89. K. Williamson, P. Li, and J. P. Devlin, *J. Chem. Phys.* **48**: 3891 (1968).
90. J. R. Durig and J. S. DiYorri, *J. Chem. Phys.* **48**: 4134 (1968).
91. J. D. S. Goulden and D. J. Manning, *Spectrochim. Acta* **23A**: 2249 (1967).
92. G. E. Kalbus, E. A. Berg, and L. H. Kalbus, *Appl. Spectry.* **22**: 497 (1968).

CHAPTER **6**

Quantitative Analysis

Infrared spectroscopy can be used for quantitative as well as for qualitative analysis, although it does not have the quantitative accuracy of some other analytical techniques—e.g., gas chromatography. In most cases, however, it will do as well as or better than the desired performance on the required quantitative analysis, and frequently it is the only feasible way of doing the job.

There are two approaches to quantitative analysis that can be used: one where there are only a few samples and it is doubtful if any more of this type will be received; the other where there is a series of samples which will arrive over an extended period of time. The method of attacking the problem is very different for each. The following sections of this chapter will outline ways in which these analyses can be accomplished and briefly discuss the various problems involved.

6.1. BEER'S LAW

Beer's law, also referred to as the Beer–Lambert law, or the Beer–Lambert–Bouguer law, expresses the relationship of the absorption of radiation by the sample to the concentration of the desired component and to the path length of the sample. One of the forms in which the law is written follows:

$$A = \log P_0/P = abc \qquad (6\text{-}1)$$

where

A = the absorbance of the sample

P_0 = the radiant power impinging on the sample—the incident radiation

P = the radiant power transmitted by the sample—the transmitted radiation

a = the absorptivity of the component at a particular wavelength (a constant)

b = the path length of the sample

c = the concentration of the component

Theoretically, the constant a is a unique function of the component in question at the frequency for which it is determined, and a plot of absorbance (A) *versus* concentration (c) should give a straight line. Unfortunately, the former statement is never true, and the plot is usually linear only over a short concentration range.

In practice, the constant a is a function of the operating conditions, as well as of the component and the frequency. Whenever it is used, therefore, it must be determined for a specific type of sample, at a specific frequency and under specific operating conditions. It cannot be used for other instruments, and if any changes are made in the instrument for which it was determined (operating conditions or sample composition) the a value must be redetermined.

Failure of the absorbance *versus* concentration plot (called a working curve) to follow a straight line over more than a short concentration range is primarily due to the nonideal behavior of the sample. This nonideality can be traced to molecular interactions, such as association of solute molecules with themselves or association of the solute and solvent molecules. In the case of gases, there is the additional problem of pressure broadening. Causes other than the nonideality of the sample that can contribute to this non-Beer's-law behavior are:

1. intensity data errors due to scattered radiation;
2. the nonlinearity of the optical attenuator wedge; and
3. too broad a slit width.

The scattering problem can be reduced, if not completely eliminated, by the use of short-wavelength filters ahead of the sample. (A look at the blackbody emission curve will show why long-wavelength filters are unnecessary.) Wedge nonlinearity problems may be reduced by operating over the linear portion of the wedge only, or by replacing a nonlinear wedge with one that is linear over at least a large percentage of its range. To eliminate cause 3, it is obvious that the slit width should be decreased.

The best technique to use to get accuracy is to calibrate the system. This operation gives a working curve which will at least partially correct or cancel the errors involved. It is still good practice, however, to minimize all the errors as far as is practical before making the calibration. The calibration should be checked periodically to determine whether it is still valid. The spectroscopist should remember that this is an empirical method and, therefore, must be followed explicitly for good results.

6.2. ANALYSIS OF A SERIES OF SAMPLES

Before starting on a quantitative analysis of any type, it is necessary to have the qualitative results. Stated differently, the analyst must know what components are in the sample prior to determining how much of any component is present. If the qualitative-analysis results are not available, the spectroscopist must determine them.

The next step is to obtain spectrograms of all of the compounds present in the sample, as well as a spectrogram of the sample itself. It is permissible, and probably good practice, to run all components, except the one or ones for which the analysis is to be made, as a mixture rather than as pure materials. At times this can be done by removing the desired component from the mixture and scanning both the extract and the residue.

From these spectrograms the analyst must pick a peak or peaks (one for each component to be determined) which will meet the following three criteria.

1. The most important criterion is that the peak be due to absorption of the component of interest only, with no other component having any absorption at this frequency. If an analysis for several components is desired, a similar peak for each component must be found.

   If the sample was a solid and was put into solution, a change in solvent may be necessary to get a peak free from interference. Extraction of some other component or components may be required to eliminate peak absorption interferences prior to the analysis. If this criterion cannot be met, there is another technique that can be used in such cases. It will be discussed briefly in Section 6.3. However, the simplicity of the solvent-change technique compared to the other makes a thorough investigation to find interference-free peaks well worth the effort.

2. An equally important criterion is the peak intensity. Utilizing the absorbance scale instead of the transmittance scale, the peak should bottom somewhere between 0.05 and 0.40 absorbance units, with the background close to 0.00 absorbance units. In an emergency it is possible to use bands having absorbances as high as 0.60 absorbance units, but no higher. The instrument has its greatest accuracy over the range 0.05 to 0.40 units. This is also the range where the intensity data on a spectrogram is most easily and accurately determined.

If the desired peak does not meet these intensity requirements, there are several correctional possibilities that the operator should investigate. The most obvious of these is changing the cell to a longer-path-length cell for weak bands, or a shorter-path-length cell for strong bands. In going to longer-path cells, care should be taken that the increased intensity in other bands will not cause them to interfere with the chosen band and violate criterion 1. If the sample is a solution, the operator may be able to get the chosen peak into the proper intensity range by varying the concentration. A third possibility, pertaining primarily to liquid samples, is to dilute them with an inert solvent. (The solvent must be chosen with criterion 1 in mind.)

3. The third criterion is the shape of the band. Bands can range anywhere from very narrow, pointed bands to broad, flat bands. If possible, the extremes should be avoided, the best shape being a fairly narrow band with a rounded bottom.

All bands must meet the first two criteria but will not always meet the third. It is also possible that the concentration range to be covered will require that two bands, two cell thicknesses, or, if the sample is put into solution, two different concentrations must be used to cover the desired range.

During the process of picking the peak or peaks, another decision will have been made: the path length of the cell to be used. Variations in nominal cell path length make it unwise to use just any cell of the path length chosen; the cell used in the calibration must also be used to run the samples. If possible, the chosen cell should be set aside and used for this analysis only; if this is not possible, extreme care should be taken with it whenever the cell is used for any other purpose. This is also true when using the cell during the analysis, as anything that changes the path length of the cell will ruin the accuracy of the analysis until the system is recalibrated.

Instrument operating parameters to be used during the calibration and analysis must be investigated and selected. This is a simple operation for those who have low-cost instruments, as instruments of these series have very few variables for the operator to change. The research-type instruments allow the operator to vary a large number of parameters, and values for all of these must be selected at this time. As is true in the case of the cells, once these parameters have been selected they should never be changed unless the system is to be recalibrated. Exceptions to this rule are gain and balance, which should always be checked, and reset if necessary, to maintain proper pen response and eliminate pen drift.

Two of the most important parameters are slit width and scan

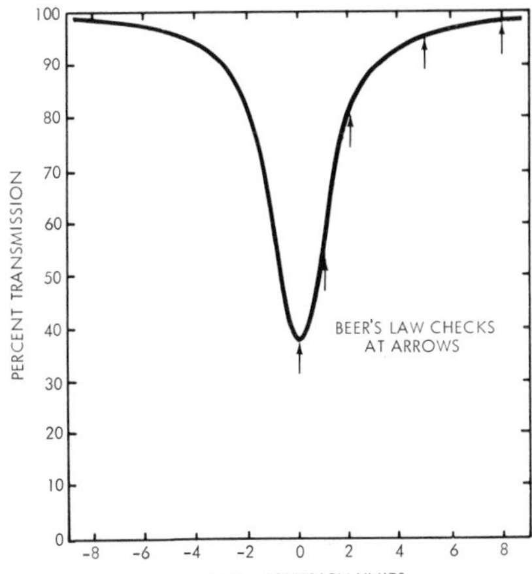

Figure 6-1. The checking of Beer's law at various positions of an absorption band. [Reproduced with permission from *Anal. Chem.* **23**: 273 (1951).]

speed. For low-cost instruments the slit program may be fixed, or the instrument may have a limited number of programs. In the latter case, the slit is usually opened one or two programs above the normal program. On the other hand, research-type instruments have a wide variety of slit programs, and usually require some investigation to find the most useful one. The slits should be opened to give a signal-to-noise ratio of at least 200 : 1, and it may be advantageous to open them even wider. It is possible to open the slits too wide, and as a result the resolution will be so degraded that accurate intensity data can no longer be obtained.

A study of the slit-width–band-width ratio by D. Z. Robinson [1] shows the effects on absorbance intensities of changes in this ratio. Figures 6-1, 6-2, and 6-3 show results from his experiments where different slit-width–band-width ratios were used in determining A as a

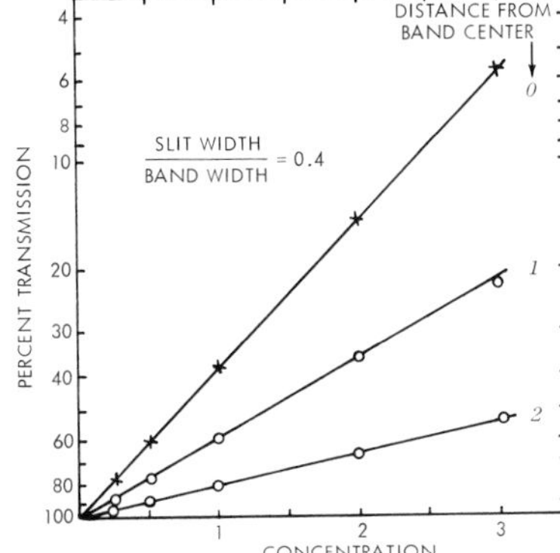

Figure 6-2. The plot of percent transmission *versus* concentration. Data collected at points 0, 1, and 2 of Figure 6-1. Slit-width-to-band-width ratio of 0.4. [Reproduced with permission from *Anal. Chem.* **23**: 273 (1951).]

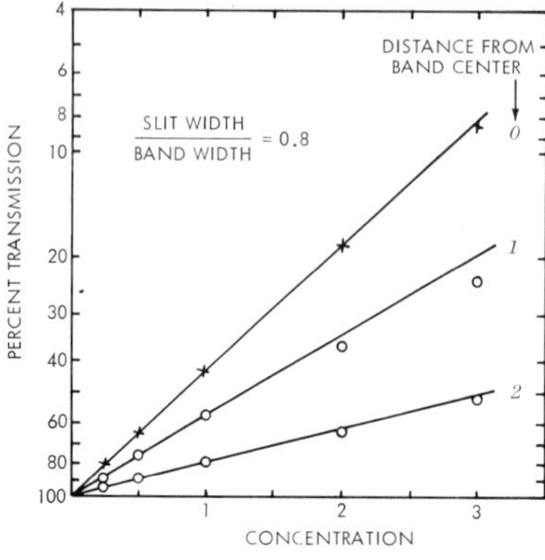

Figure 6-3. The plot of percent transmission *versus* concentration. Data collected at points 0, 1, and 2 of Figure 6-1. Slit-width-to-band-width ratio of 0.8. [Reproduced with permission from *Anal. Chem.* **23**: 273 (1951).]

function of concentration. These measurements were made at several places on the peak (indicated by arrows), including the peak maximum. The results show no deviation from the expected straight line at the smaller ratio, but that deviations occur at low-transmission (high-absorbance) values for the higher ratio. These data reinforce the statement in criterion 2 (above) that intensity-data errors can be reduced by operating the instrument at higher transmission (lower absorbance) values.

In Chapter 2 the requirements for obtaining true intensity data, one of which is scan speed, were discussed (Section 2:2B). On low-cost instruments, the number of scan speeds is very limited, and quantitative analyses are always run at the slowest speed. Research instruments have a wide variety of scan speeds, and as in the case of slit programs it is necessary for the operator to determine which of these is the most suitable. In contrast to the slit-program selection, there is available to the operator a rule of thumb for scan-speed selection. It is the following: to scan over the selected band at ever slower scan speeds until two successive scans show no change in either band shape or band intensity; then to multiply the slowest of these scanspeeds by two or three and scan at that speed. This will provide a good safety factor against variations in the instrument. From a data-accuracy viewpoint it is impossible to scan too slowly, but from an economic viewpoint the time involved in using too-slow scan speeds may be impractical.

A series of calibration standards must be prepared, with the concentration range of the desired component greater than the concentration range expected to be found in the samples. As an example, if the expected range of the desired component it 6–15%, the standard with the lowest concentration should contain less than 6%, while that with the highest concentration should contain more than 15%. The standards should have a composition as close to the samples as possible;

smaller deviations will cause smaller errors in the result due to solute–solvent or solute–solute interactions. The number of standards required is a function of concentration range, desired accuracy, and economics.

In making the calibration, the cell is filled with one of the standards and mounted on the spectrophotometer, and the spectrum is scanned. The scan must cover only a short range of the spectrum, including the peak and its shoulders—scanning the remainder of the spectrum is not necessary or practical. This operation should be repeated until six scans of each standard have been obtained. For calibration purposes it is desirable to average out both short-term and long-term instrument and technique variations. To accomplish this, one scan per day is made on each calibration standard so that the data are collected over a period of a week to ten days. If it is felt that economics preclude taking this much time, the six scans can be run on the same day, but the operator should clean, dry, and refill the cell after each scan.

When all six scans on each of the standards have been completed, the intensity data (in absorbance units) is collected and plotted *versus* concentration to produce a working curve. To work up the data, the interpreter draws a tangent across the shoulders (transmission maxima) of the peak representing his best estimate of where the pen would have drawn the line if the peak were absent. He then erects a line perpendicular to the infinite-absorbance (or $0\%\ T$) line through the peak maximum and intersecting the tangent; the peak maximum shows the total absorbance (A_1), and the intersection of the perpendicular and the tangent shows the background absorbance (A_2). If the scans were made on absorbance paper, he reads A_1 and A_2 on the y-axis scale and calculates the peak intensity by the following formula:

$$A = A_2 - A_1 \tag{6-2}$$

The A values for each standard are averaged and these averages plotted *versus* concentration on ordinary graph paper. The points are connected by the best smooth line, which is rarely straight and seldom passes through the origin.

In running the samples the procedure and operating conditions are the same as used for the standards, except that the six runs are made consecutively rather than over a period of time. A is obtained in the same way as above, and is converted into concentration using the working curve.

6.3. MULTICOMPONENT ANALYSIS

Samples requiring analysis for a number of components can frequently be analyzed by the procedure of Section 6.2. There may be times when it is impossible to find a peak for each component that meets the criteria listed above, so changes have to be made in the technique to correct for this interference.

In this procedure, criterion 1 above is modified to say that each peak selected is primarily caused by one of the components, with the other components contributing only minor or no absorptions at that frequency. If the components do not interact, then the Beer's-law equation becomes:

$$A_1 = a_{11}bc_1 + a_{12}bc_2 + \cdots + a_{1m}bc_m \tag{6-3}$$

where

A_1 = the absorbance at frequency ν_1

a_{11} = the absorptivity constant for component 1 at frequency ν_1

a_{12} = the absorptivity constant for component 2 at frequency ν_1

a_{1m} = the absorptivity constant for component m at frequency ν_1

c_1 = the concentration of component 1

c_2 = the concentration of component 2

c_m = the concentration of component m

b = cell path length

As many peaks must be used as there are components being determined and the absorptivity constant a must be determined for each component at each frequency.

In running the analyses, a series of equations are obtained where A is determined for each peak as discussed above, the a_1, a_2 etc. values will have been determined experimentally and, since the same cell is used for all the work, b will be eliminated from the system.

For a three component analysis, the equations will be:

$$A_1 = a_{11}bc_1 + a_{12}bc_2 + a_{13}bc_3 \tag{6-4}$$

$$A_2 = a_{21}bc_1 + a_{22}bc_2 + a_{23}bc_3 \tag{6-5}$$

$$A_3 = a_{31}bc_1 + a_{32}bc_2 + a_{33}bc_3 \tag{6-6}$$

As can easily be seen, this leads to lengthy, complex calculations, and for large systems a computer may be required. In addition, if Beer's law is not obeyed by the various components, correction factors must be added to the equations.

6.4. SOLID SAMPLES

The best and most accurate method for solid samples is to put them into solution and run them as discussed in Sections 6.2 and 6.3 of this chapter. However, there are times when this cannot be done, and one of the following techniques may be useful.

When the KBr-pellet technique was first introduced it was expected that it would be an excellent alternative to dissolving the sample. However, it rapidly became apparent that there were problems that limited the usefulness of this technique, especially nonhomogenous distribution of the sample in the pellet. Today the Kbr-pellet technique is widely used for solid samples, but normally does not have the accuracy of the solution technique. In using KBr pellets, a larger quantity of the sample–KBr mixture is made up than the 300 mg required for the pellet (both materials are weighed), and from this mixture 300 mg is weighed into the die. In this manner losses due to processing are eliminated, reducing a big potential error. Accuracy can be further improved by the addition of an internal standard to the mixture before pressing the pellet and then plotting the ratio ($A_{\text{sample}}/A_{\text{standard}}$) *versus* concentration, instead of the absorbance of the sample peak *versus* concentration. An internal standard must be chosen that does not have a peak that will interfere with the sample peak, and *vice versa*. In making the pellets, every effort must be made to obtain homogeneous sample distribution and random sample orientation, otherwise the intensity will vary with the pellet position in the holder as well as with sample concentration.

Another technique, used primarily with thin films, is to plot the ratios of the intensities of a peak from each of the two components of the film *versus* concentration. This method is frequently used to determine the ratios of components rather than their absolute concentrations. If the actual concentration of a component is desired, the intensity of the peak used as a standard should be insensitive to concentration but vary with changes in sample thickness, while the peak of the desired component should be sensitive to changes in both sample thickness and concentration and should show large changes in intensity due to small changes in concentration. The intensity of the standard peak is used to correct the desired component peak intensity for small variations in sample thickness, but even with this correction it is desirable to maintain film thickness as close to a particular value as possible. The standard peak may be contributed by another component in the sample or it may be contributed by an added internal standard.

6.5. GAS SAMPLES

In gas analysis the procedure is quite similar to that for liquids, except that the *x*-axis on the working curve will represent partial pressure rather than concentration. These partial pressures are related to concentrations and can always be recalculated to concentrations if necessary.

Quantitative analyses of gases can be very difficult. The factors influencing the results discussed in Chapter 7 (Section 7.1) should be reviewed before running such an analysis. As gas cells have long path lengths compared to liquid cells, and the path lengths are held to small tolerances, cells of the same nominal path length can normally be used interchangeably.

In running quantitative analyses on gases the total pressure in the cell should always be the same for all calibration standards and samples in a series, even if an inert diluent must be added to attain this pressure. The best pressure to use is a parameter that must be determined for each type of sample.

6.6. "ONE-TIME" ANALYSIS

When there are only one or a few samples of the same type, the time and effort involved in setting up a calibration may not be justified, and the following technique can be used. (A qualitative analysis must be made before quantitative analysis can be run.) If the sample is run in a sealed cell of known path length, and the spectrogram compared to the spectrograms of the pure components run in the same cell, very approximate concentrations can be calculated using Beer's law in the form

$$\frac{A_{\text{sample}}}{A_{\text{standard}}} = \frac{C_{\text{sample}}}{C_{\text{standard}}} \tag{6-7}$$

where C_{standard} would be 100%.

Unless the sample is an ideal system where the components follow Beer's law, these results will be only very approximate. The operator can now make a solution based on these approximate concentrations and run it in the same cell. Using the date on this spectrogram, the results can be recalculated and the entire operation repeated until results within the desired accuracy are obtained. A differential run of the unknown sample versus the known mixture can be made as a final check.

There are variations and modifications of this technique, such as using a variable-space cell to determine the path length of an equivalent amount of a component. From the absorptivity constants and Beer's law, an approximate composition can be calculated.

6.7. SOURCES OF ERROR

In several of the previous sections of this chapter some of the sources of error in quantitative analysis were mentioned briefly. A discussion of these and other possible errors, to give a better understanding of their causes and effects on results, follows.

Scattered Radiation. Scattered radiation, also called stray radiation, is radiation of the wrong frequency impinging on the detector. The detector cannot differentiate between frequencies, so it will react to the scattered radiation in the same manner as it will to radiation of the desired frequency. The causes of the scatter are varied and include such things as diffusion of the beam by the cell and sample, optical misalignment, improper radiation baffling (including the motors), and wear on blackened components that makes them more reflecting.

The effects of scattered radiation will show up as false intensity data, usually a decrease in the peak intensity, and the accuracy of the results will obviously suffer. As stated in Chapter 2 (Section 2.2B), the intensity of the scattered radiation can be measured and a correction factor applied to the data. If this is done, the data should be redetermined and the factor recalculated periodically. It is better to eliminate or reduce an error than to correct for it; a short-wavelength filter placed in the sample beam will cut down on the stray radiation.

Zero Setting. The zero of the instrument is a very important factor and must be set correctly to obtain accurate intensity data. It is the reference point against which other points are determined, and an error in the zero will be reflected in these values also. On those instruments where the zero is set electrically, it should be checked, and reset if necessary, for every sample; for those instruments that have a mechanical zero setting, the zero should be checked periodically and reset when necessary. For greatest accuracy in setting the zero, it should be checked by slowly scanning into a totally absorbing band of a compound that shows negligible reradiation at this frequency. If it is necessary to reset the zero, it should be rechecked and the process repeated until no further change is required.

Noncontinuous Sample. This problem usually is caused by improper filling of a cell or poor technique in casting a film. Every place where the sample has a hole, radiation will pass through it without being absorbed. This can easily be illustrated by scanning through a broad totally absorbing band with the cell properly and completely filled and with the cell containing bubbles or being only partially filled. In both cases, the bottom of the band will be flat indicating total absorption but, in the former the flat bottom will be at the infinite-absorbance (or 0% T) line while in the latter it will be some distance above the infinite-absorbance (or 0% T) line. This distance is a measure of the voids in the sample, assuming proper zero setting and no scattered radiation. Obviously, therefore, peak intensities obtained with samples containing voids, whether due to bubbles, holes, or tears, will always be recorded as lower in value than they actually are.

Atmospheric Absorptions. In double-beam instruments, the water vapor and carbon dioxide in the atmosphere should not show on the spectrogram, as they should absorb equally in both beams. However, if there is an appreciable change in path length of one beam, either due to faulty optical alignment or to use of a long-path-length cell in one beam with no compensating cell in the other, these peaks will appear. If the sample peak being used for the analysis is at a frequency where atmospheric absorption occurs, the peak intensity will be a combination of sample absorption and atmospheric absorption.

Even in double-beam instruments operating properly and with both beams equal in length, the atmospheric absorptions can be a problem. Unless the instrument is purged with dry, carbon-dioxide-free air, the water vapor and carbon dioxide will absorb radiation and reduce the energy of the system. This loss of energy increases the possibility of obtaining inaccurate and nonrepeatable intensity data, especially as the concentration of water vapor and carbon dioxide are constantly fluctuating.

In single-beam instruments these absorptions are a bigger problem because the intensities of the sample peak and atmosphere peak will be additive. Again, as the water-vapor and carbon-dioxide concentrations fluctuate, the apparent band intensity will also fluctuate. If it is possible to purge the instrument, the problem will disappear; otherwise these regions should be avoided for quantitative work.

Temperature. Some samples and their peak intensities are temperature sensitive, and the absorbance will vary with change in the temperature of the sample. Quantitative analysis is based on the assumption that the peak intensity will vary with concentration only and not be dependent on other factors.

There are several ways in which this temperature-sensitivity problem can be minimized. One approach is to put the sample cell in the spectrophotometer and wait until it reaches equilibrium before starting the scan. This can be a time-consuming operation and may be economically impractical. Another is to place the sample cell in the spectrophotometer and scan immediately. The sample must be removed and allowed to return to ambient temperature prior to making the next scan. Filters can be placed ahead of the sample to remove all frequencies but those required for the analysis, so that although the sample will still be heated by the spectrophotometer chassis the rise in its temperature will be much less. Another approach is to build a heat sink or heat-dissipation unit around the sample cell to remove the heat before it can raise the sample temperature. Finally, running the sample in a temperature-controlled cell may be a useful solution.

Some materials act as secondary sources when heated above ambient temperatures. This secondary emission has the same effect on peak intensity as scattered radiation. See Chapter 2, Sections 2.3A and 2.5 for a detailed discussion of this problem.

Slit Width. The slit width will determine to some degree the energy available at the detector—the wider the slit the greater the energy.

This increased energy requires less amplification of the detector signal to raise it to a useful level. As stated in Chapter 2, the noise is mostly generated by the detector, and this lower amplification will give a lower overall noise level. The noise-level variations will add to or subtract from the peak intensity and thus affect the accuracy and reproducibility of the peak intensity.

As was stated previously (Section 6.2), the slits should be opened wide enough to give a signal-to-noise ratio of at least 200:1, and at times even higher. The greater this ratio, the smaller the effect the noise will have on the accuracy of the peak intensity. Consideration of another factor, the absorptivity constant, will show that the slit width can be increased to where it is detrimental to obtaining absolute accuracy.

The absorptivity constant is a function of the compound at a particular frequency. Even at very narrow slit widths, the monochromator will pass a band of frequencies resulting in an average absorptivity constant rather than a specific absorptivity constant. The wider the slit width, the wider the band of frequencies, and the larger the number of absorptivity constants present in the average. The absorptivity constant has its highest value at the frequency of the peak maximum and decreases on both sides of the maximum. The wider the frequency bandpass, the lower the value of the absorptivity constant, as the average will include absorptivity constants of smaller and smaller values.

For narrow bands the absorptivity will be further from the true value than for wide bands, as the values will change faster with frequency, a fact that can easily be determined by an examination of a variety of band shapes. As it is desirable to operate with an absorptivity constant as close to the true value as possible, the use of narrow slits is essential.

From these considerations it is obvious that the band shape, the absorptivity constant, and the signal-to-noise ratio all enter into the determination of the most desirable slit width to use. The operator must look for the compromise that will give him the best results, using the above ideas as a guide.

The above discussions deal with the operator's ability to obtain accurate absolute intensity data. For most quantitative analysis it is usually more important to get reproducible intensities rather than accurate absolute intensities. If all other factors are held constant, the reproducibility of the intensities is a function of the reproducibility of the slit width, provided that the random noise is kept to a negligible level. A high noise level will affect both accuracy and reproducibility.

Scan Speed. The selection of proper scan speed has been discussed in Section 6.2 and in Chapter 2. Selection of too fast a scan speed can lead to serious errors in intensity, while too slow a scan speed is economically unprofitable. The operator must select the best compromise between the two.

Gain and Balance. Improper setting of gain or balance can be a very insidious problem, as the resulting errors are not always evident. This is especially true when the errors are very small. Low gain will keep the pen from following the signal and thereby give poor reproducibility. High gain may result in pen overshoot or a higher noise level than is anticipated and can be tolerated. Improper balance means that the amplifier is adding to or subtracting from the signal, and the band intensity is now a composite of the detector and amplifier signals.

In concluding this section, it should be mentioned that many of these sources of error are interacting. A change in a parameter to correct for one error may introduce errors elsewhere and require further changes.

These interactions are discussed in Chapter 2 and must be taken into consideration when the operating parameter settings are investigated and chosen.

6.8. NONINTERCHANGEABILITY OF DATA

Each instrument is an individual and has its own characteristics; it will not necessarily react in the same manner or to the same degree as its neighbors. This has been demonstrated for a series of instruments of the same model made during a short period of time and sitting side by side on the bench. The data collected by Perkin–Elmer Corporation on 50 instruments of the same design serves to illustrate this point. Using two standard samples, the measurements were made as the instruments came off the production line. The results are shown in Figure 6-4.

Instruments of different design and different manufacture as well as instruments located in various laboratories may give intensities varying by even greater amounts. This indicates the difficulty or even impossibility of transferring intensity data from one instrument or (even worse) from one laboratory to another. Various techniques have been proposed to improve this type of data transfer, but none has been very successful.

6.9. MEASURING ABSORBANCE

Section 6.2 gave a brief description of the "baseline" method for determining A, based on the use of chart paper with the ordinate calibrated nonlinearly in absorbance units. This technique can also be

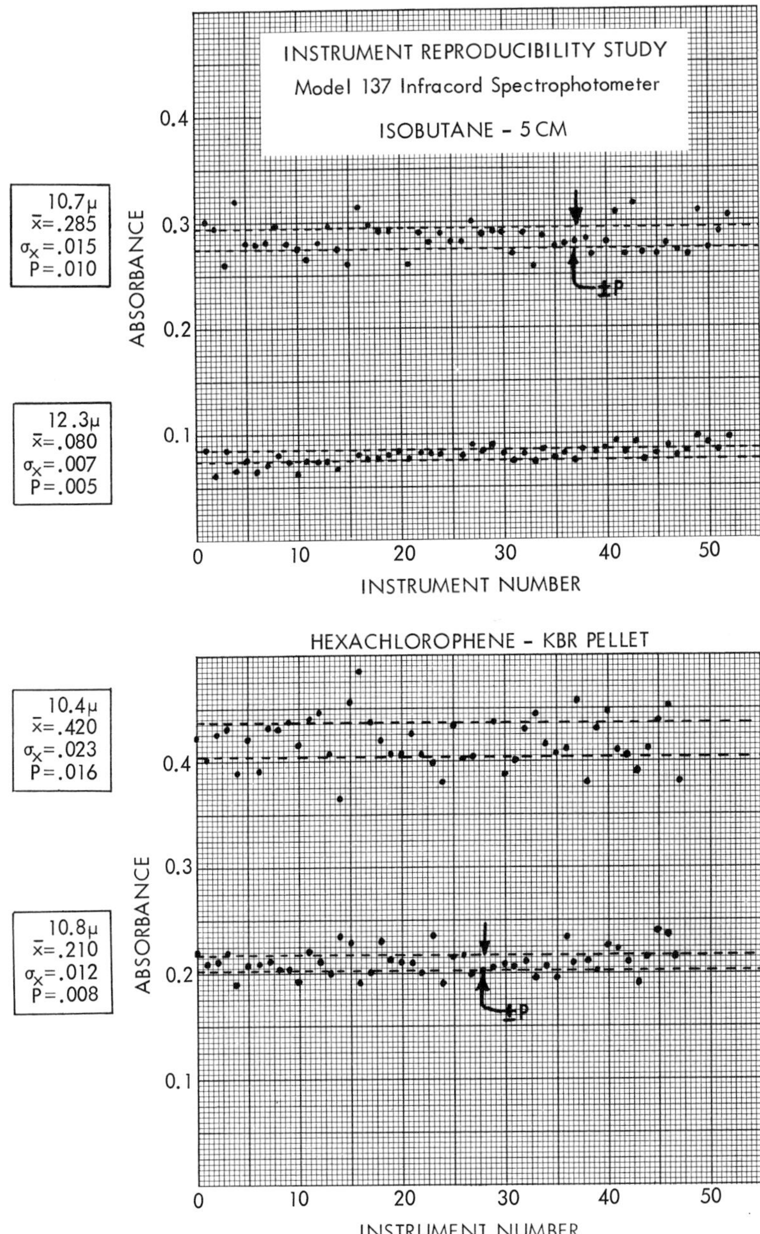

Figure 6-4. Reproducibility of intensity measurements. [Reproduced with permission of the Perkin–Elmer Corporation, Norwalk, Conn.]

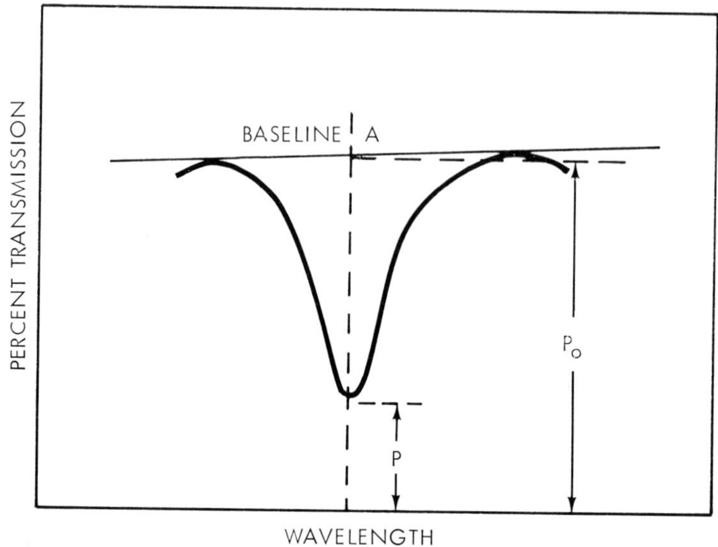

Figure 6-5. Use of a baseline in measuring the absorbance A of an infrared band.

used if the chart paper has the ordinate calibrated linearly in transmittance units. Figure 6-5 illustrates the measurements involved. The way the tangent is drawn for the calibration standards (from shoulder to shoulder or from one shoulder parallel to the infinite-absorbance, or 0% T line) is the way it must be drawn for the samples. As stated in Section 6.2, this line is the operator's estimate of where the background line would be drawn if the peak were not present. For absorbance chart paper, the calculation is a simple subtraction:

$$A = A_{\text{peak}} - A_{\text{background}} \tag{6-2}$$

For transmittance chart paper, the calculation is

$$A = \log (P_0/P) \tag{6-8}$$

where P_0 is the transmittance of the background and P is the transmittance of the peak maximum.

If the chosen peak is not isolated but has one or more peaks adjacent to it (this can occur without violating criterion 1), it may be useful to include these peaks under the baseline. In double-beam instruments, improvements in drawing the baseline can be obtained in some cases by use of a second cell containing solvent in the reference beam. This cell is used to cancel out solvent and other component absorptions which come near to interfering with the chosen peak.

Another method for measuring A is called the point method. The peak used for the analysis is picked in the same manner as in 6-2 and must meet the same criteria. The sample is either scanned over the peak or the monochromator is set to the peak frequency and the transmittance determined at that point. The cell is emptied, cleaned, and refilled with the sample minus the component being determined and, under the same operating conditions, either it is rescanned over the same area, or the transmittance is determined at the same point as above. The true zero of the spectrophotometer must then be determined. Ideally, this is done by replacing the sample with something that is totally opaque at the desired frequency and transparent at all other frequencies. In practice, either a sample with a very high concentration of the desired component is used, or a material that is opaque at the required frequency and transparent at higher frequencies is substituted.

The calculation is as follows:

$$A = \log (P_0/P) \tag{6-9}$$

where

A = the absorbance of the peak

P_0 = the transmittance of the materials minus the component (the background true zero)

P = the transmittance of all materials including the component. (The peak-transmittance true zero.)

In effect this method corrects both P and P_0 for variations in true zero. While (6-9) is the same form as (6-8), the values for P_0 and P should be obtainable with greater accuracy.

6.10. ACCURACY

The accuracy of the results of infrared quantitative analysis will depend on the factors discussed in previous sections of this chapter. For a relatively ideal sample, where the operator does a careful job, the accuracy is expected to be from better than $\pm 1\%$ of the amount present to $\pm 3\%$ of the amount present, depending on the instrument used. Thus, for a research-type instrument in good operating condition, a result of 10% concentration means that the actual value is somewhere between 9.9% and 10.1%.

There are many cases where this type of accuracy is not required and much time and effort can be saved by cutting some of the steps. For example, a valid rough result can be obtained by scanning the peak only once instead of the six times mentioned in Section 6.2. For other samples, the poor quantitative results that will be obtained, even under the best conditions, are useful, and infrared spectrophotometry may be the only way in which any results can be obtained.

As has been mentioned previously, the greatest accuracy is obtained with solutions. Frequently, this accuracy can be improved by the use of internal standards.

6.11. OTHER TECHNIQUES

Some samples may contain components which cannot be separated from a mixture and exist as pure compounds. In this case, it is impossible to either determine absorptivity constants or to make up known calibration standards. As long as it is possible to change the concentrations in the sample and find useful peaks for analysis, it is possible to determine the concentration.

Using a two-component mixture of initial concentrations C_1 and C_2, and changing these concentrations in the second sample to C_3 and C_4,

$$C_1 + C_2 = 1 \tag{6-10}$$

$$C_3 + C_4 = 1 \tag{6-11}$$

Using peaks meeting criterion 1 (Section 6.2) for a useful peak, and Beer's law in the form

$$C = A/ab \tag{6-12}$$

we can substitute for C in (6-10) and (6-11) to obtain

$$(A_1/a_1b) + (A_2/a_2b) = 1 \tag{6-13}$$

$$(A_3/a_1b) + (A_4/a_2b) = 1 \tag{6-14}$$

A_1, A_2, A_3, and A_4 can be determined from the resulting spectrograms, and as the same cell was used for all determinations b will cancel out. Equating the above expressions will give a ratio for a_1/a_2. From these data and the original sample weight, C_1 and C_2 can be calculated.

In some samples the desired component cannot easily be determined in the form in which it is received, and it may be necessary to convert the material to some derivative. The derivative may be required to find a useful analytical peak, or it may provide a better-defined spectrum than the original material.

Samples of low-boiling liquids can frequently be converted to and run as vapors rather than as liquids. Again, this may yield better spectrograms and may also make the sample-handling problems simpler; it has a big advantage for samples containing water, as water vapor has

Infrared Quantitative Analytical Data

Publication is open to any person and is not limited to members of the Coblentz Society or the Society for Applied Spectroscopy. Contributions should be sent directly to the Chairman of the Review Committee, Dr. A. Lee Smith, Spectroscopy Laboratory, Dow Corning Corporation, Midland, Michigan. It is requested that methods be submitted in quintuplicate, using the standard format. Infrared spectra cannot be published as a part of this program.

Determination of C_8 Alkylbenzenes (o-Xylene, m-Xylene, p-Xylene, and Ethylbenzene)

J. A. CASTELLANO, Ultra Chemical Works, Inc., Paterson, New Jersey

CS-158

Component			Range %	Accuracy %	λ or ν B.L. Pts.	Slit (mm) Δλ or Δν	Concn. mg/ml Length (mm)
No.	Name	Formula					
1	1,4 Dimethylbenzene	C_8H_{10}	0-25	±0.5	12.6μ 12.3-12.9	0.384 0.049	50 0.0514
2	1,3 Dimethylbenzene	C_8H_{10}	0-60	±0.5	13.0μ 12.8-13.3	0.426 0.055	50 0.0514
3	1,2 Dimethylbenzene	C_8H_{10}	0-30	±1.0	13.5μ 13.1-13.9	0.506 0.065	50 0.0514
4	Ethylbenzene	C_8H_{10}	0-30	±1.0	14.4μ 14.0-14.8	0.750 0.096	50 0.0514

Instrument: Perkin-Elmer Model 137 Infracord, NaCl prism, normal slit program
Sample Phase: Solution in Cyclohexane
Cell Windows: NaCl
Absorbance Measurements: Base Line.....X..... Point..........
Calculation: Inverted Matrix.....X..... Successive Approx..........
Graphical..........

Relative Absorbances—Analytical Matrix:

Component/λ	12.6μ	13.0μ	13.5μ	14.4μ
1	0.261	0.006	0.000	0.006
2	0.000	0.250	0.000	0.005
3	0.000	0.002	0.432	0.043
4	0.000	0.008	0.000	0.155

Material Purity: Reference compounds 99+% pure.

Comments: Relative absorbances are given as the slope of the Beer's Law concentration curves used and are expressed in terms of absorbances per 100% of constituent.

Determination of γ-Cyanopropyltrichlorosilane in Methyl (γ-cyanopropyl) dichlorosilane

PHILIP J. LAUNER and ALFRED S. CROUSE, Silicone Products Department, General Electric Company, Waterford, New York

CS-159

Component			Range %	Accuracy %	λ or ν B.L. Pts.	Slit (mm) Δλ or Δν	Concn. Vol. % Length (mm)
No.	Name	Formula					
1	γ-Cyanopropyltrichlorosilane	$C_4H_6Cl_3NSi$	0-2	±0.1	16.78μ	0.900	16.7[a] 0.970

Instrument: Perkin-Elmer Model 321, CsBr Prism
Sample Phase: Solution in Carbon Disulfide
Cell Windows: KBr
Absorbance Measurement: Base line.......... Point.....X.....
Calculation: Inverse Matrix.......... Successive Approx..........
Graphical.....X.....

Relative Absorbance—Analytical Matrix:

Component/λ	16.78μ
1	34.0

Material Purity: 99.8%

Comments: These compounds hydrolyze readily in moist air to give hydrogen chloride and high-molecular-weight siloxanes. For special precautions, which should be taken in handling samples, see Anal. Chem. **31**, 1175 (1959) and APPLIED SPECTROSCOPY **14**, 86 (1960). The I_0 measurement is made using a carbon disulfide solution of pure methyl(γ-cyanopropyl)dichlorosilane in each of two matched cells. For the I measurement, the reference cell is left in the reference beam, and the sample cell is filled with a carbon disulfide solution of the unknown. Relative absorbance is given as the slope of the Beer's law concentration curve used and is expressed in terms of absorbance per 100% of constituent.

[a] One ml of sample is added to five ml of CS_2.

Figure 6-6. A typical literature data card on quantitative analysis. [Reproduced with permission of the editor of *Applied Spectroscopy*.]

less intense absorptions than liquid water. As discussed in Section 6.5, there are problems connected with the analysis of gases which may outweigh the advantages to be derived from this technique.

6.12. LITERATURE

The literature contains numerous articles which describe quantitative techniques. The biggest collection of these are in *Applied Spectroscopy* (Figure 6-6) and *Analytical Chemistry*. These methods are quite complete and are the result of an attempt to make available the data that were determined by various company laboratories.

REFERENCE

1. D. Z. Robinson, *Anal. Chem.* **23**: 273 (1951).

CHAPTER 7

Sample-Handling Techniques

To obtain useful spectrograms of the samples, the spectroscopist must do a careful and frequently a lengthy job of sample preparation. The care and effort expended on this phase of the analysis are the biggest contributions to obtaining good results. Even for routine samples, where handling is the same for sample after sample, the prescribed procedures must be carefully followed to obtain the maximum in results.

There are always a number of single-component samples, and occasionally some mixtures, where the sample can be put into a cell or taped to the spectrophotometer and run "as is." Usually, however, the sample must have some chemical or mechanical work done on it prior to its being run.

Separation of mixtures into their components is usually the lengthiest and most difficult part of the sample-handling problem. These separations can be accomplished by one or more of a variety of techniques, such as extractions, distillations, or chromatography; a discussion of these separation techniques does not belong here. However, it should be remembered that during these separations it is possible that changes in the chemical entity of the material may occur, and also that unwanted residues from solvents may be added to the sample. These problems are not necessarily disastrous unless the spectroscopist is unaware of their occurrence.

This chapter will discuss various techniques by which a sample can be run, assuming that the necessary preliminary separations mentioned above have been completed. The spectroscopist should be aware of the possibility that the same sample run by several different techniques may not show identical spectra. In practice, small variations by the operator in processing the sample may produce differences in spectra obtained from the same original sample.

7.1. GASES

Gases usually present few sample-handling problems; the only operation involved is to transfer the gas from the sample container into a previously evacuated gas cell. This is usually done through a manifold system to which both the cell and the sample container are connected. After the connections are made, the entire system, down to the cutoff valve or stopcock on the sample container, is evacuated. The cell is then filled to the desired pressure, closed, and transferred from the manifold to the spectrophotometer. In the case of long-path-length cells, the cell is usually connected to the manifold by rubber tubing, so that it does not have to be removed from the spectrophotometer for evacuating, filling, and cleaning.

The path length of the cell that is needed will depend on whether the compound is a weak or strong absorber, as well as on its concentration in the sample mixture. Pure compounds are usually run in short-path cells (2–10 cm) at low pressures, while mixtures, especially when concentrations in the ppm range are of interest, are run in long-path cells, and frequently at high pressures. Using long-path cells, high pressures, and ordinate scale expansion, it is possible to measure concentrations in the ppb range.

In gas analysis, the pressure of the sample in the cell is an important factor in determining the peak intensity. A change in the partial pressure of a component, or a change in the total pressure of the sample, may enhance or depress the intensity of the absorption peak disproportionately with respect to what is expected from the change in concentration. Increasing the component pressure may cause another problem, called pressure-broadening, which causes poor resolution and faulty band shape, and yields false intensity data. (Under controlled total pressure the data are usually reproducible.) Many times, especially if high resolution is desired, it is better to run the sample in a longer-path-length cell, and at lower pressures, than would normally be used. A good discussion of these and other related problems can be found in Reference 1.

Samples having a high water-vapor content should be run in cells with windows made of a water-insoluble crystal. Frequently, wet samples can be run in cells having sodium chloride or even potassium bromide windows, if the partial pressure of the water vapor is kept low and there is no chance of the water condensing on the windows (keeping the cell at an elevated temperature may be required). Even with water-insoluble crystals, condensation of water on the windows should be avoided, as this will give a spectrum of liquid water rather than the spectrum of the sample. A good rule of thumb is: avoid the condensation of any material onto the cell windows.

Other problems involved in the running of gases, such as cell wall adsorption, are discussed in Chapter 3, Section 3.2.

7.2. LIQUIDS

Running liquid samples also presents few problems, as most liquid samples can be put directly into a sealed or demountable cell. The only determinations the operator has to make are what type of cell, and what cell path length, to use, and whether or not the sample or one of its components will attack the windows.

The type of cell to be used will be determined by the sample to be run and the kind of analysis to be made. The path length of the cell will be determined by the intensity of the bands, the concentration of the components, and the desired results. Low-boiling samples, and samples for quantitative analysis, should always be run in sealed cells, unless the scan will cover only a very short section of the range; viscous or high-boiling liquids are usually run in demountable cells, unless quantitative results are required. A strong absorber run as a pure compound (highly polar compound) will normally be run in a thin cell (0.025 mm), while a weak absorber (nonpolar compound) will require a cell of longer path length (0.100 mm). Thin cells are used to determine the structure at the bottom of strong bands; longer-path-length cells are required for recording weak-intensity bands of pure compounds, or strong-intensity bands of components in very low concentrations in mixtures.

There is a simple technique to determine whether the sample or any of its components will attack the cell windows. Place a drop of the sample on the sanded face of a scrap window (the scrap window must be of the same crystal material as that of the cell windows) and allow it to evaporate. If, after evaporation, the area retains its sanded appearance, the sample will not attack the windows, but, if this area is now clear, the reverse is true.

Occasionally, it is advantageous to run low-boiling or even medium-boiling single-component liquids as vapors. This can easily be done by injecting a small drop of the sample into an evacuated gas cell; normally enough of it will vaporize to provide the partial pressure required to obtain a good spectrum. The cell can be wrapped with heating tape and run at an elevated temperature if it is necessary to increase the vapor pressure.

When small differences between two supposedly similar liquid samples are to be investigated, a differential analysis is useful. In this type of analysis, a cell filled with the standard material is placed in the reference beam of the spectrophotometer and a matched cell filled with the sample is placed in the sample beam. The pen is set at about 50% *T* rather than the usual 100% T prior to running the scan. If the sample has the same composition as the standard, the resulting spectrogram will be a straight line. If the composition of the sample is different from that of the standard, the spectrogram will show peaks going upscale (towards 100% T), indicating a higher concentration of certain components, or the presence of additional components, in the standard; and downscale (toward 0% T), indicating a higher concentration of certain components, or the presence of additional components, in the sample. A qualitative analysis of the materials will allow matching these peaks to specific compounds.

The matched cells can be two sealed cells which have been purchased as a matched pair or they can be a sealed cell and a variable-space cell that has been matched to it. Using a matched pair of sealed cells has the advantage of lower initial cost and does not require the preliminary matching discussed below; however, the match in path length can easily be destroyed (e.g., improper filling of one or both cells may change the path lengths by spreading the window spacing—see Section 3.3). Use of a variable-space cell in conjunction with a sealed cell has the advantage that the path lengths are always matched, and the disadvantages of higher initial costs and the requirement for preliminary matching of the variable-space cell to the sealed cell.

To match a variable-space cell to a sealed cell both cells are filled with the standard, and after the variable-space cell is set for the approximate path length desired (this can be done fairly accurately using the vernier marks on the barrel), a differential spectrogram is run. If the spectrogram shows a straight line, the match is good; if peaks appear on the spectrogram, the cells are not matched. To complete the matching, the monochromator is set to the frequency or wavelength of the peak having the greatest intensity, and the variable-space cell path length is changed until the peak has been eliminated. The spectrum is rerun and again the path length of the variable-space cell is changed if this is required. The spectrum is rerun and the path length changed until the differential yields a straight line, showing that the cells are matched. The sealed cell is then emptied, cleaned, dried, filled with sample, and placed in the sample beam. The differential analysis can now be run.

In running a differential analysis, extra care must be exercised to insure that the gain and balance have been set correctly. If this is not done, false peaks may appear or weak peaks may fail to appear. The first of these errors is usually due to incorrect balance, and the second to incorrect gain. It should also be remembered that in those regions where total absorption occurs no information is obtainable.

If the sample is a solution, the spectroscopist may desire the spectrophotometer to eliminate the solvent peaks from the resulting spectrogram. This can be accomplished by placing an equivalent quantity of solvent in the reference beam prior to making the run. A variable-space cell is most useful for this type of work also, as its path length can easily be changed to whatever path length is required. The operator should remember that he is matching quantities of solvent and not cell path lengths.

As can readily be seen from the chart of solvent windows (Figures 7-1A and 7-1B), there is no solvent that is completely transparent over the entire frequency range. In the areas where the solvent has bands that are totally absorbing, no information can be obtained about the sample (if the solvent absorbs all of the radiation, there is none left to interact with the sample), so that another solution, using a different solvent, must be made and scanned. The most useful pair of solvents is carbon disulfide and carbon tetrachloride (see Figure 7-1).

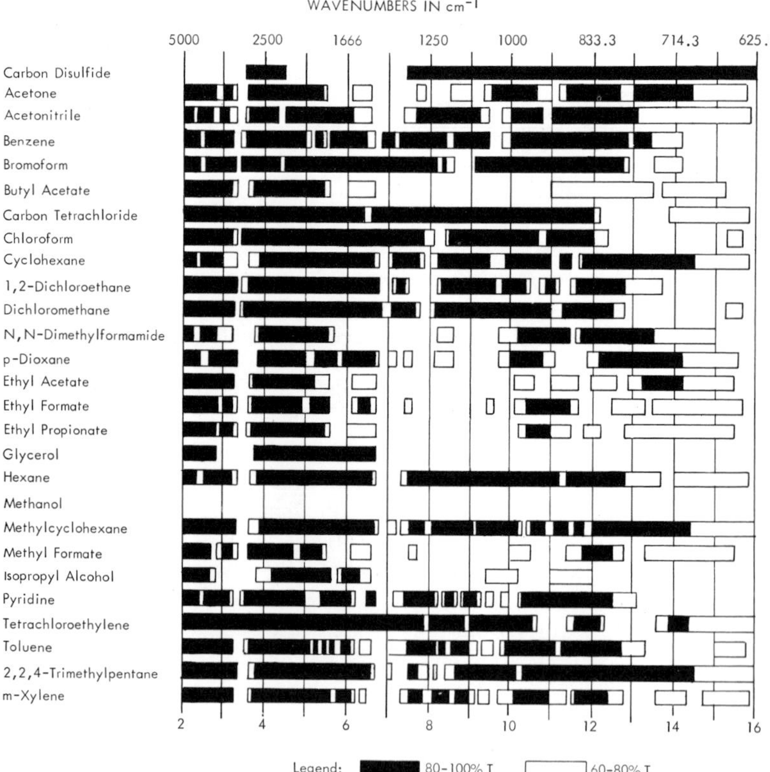

Figure 7-1A. Solvents for the 2–16 μ region. (Reprinted from data of Matheson Coleman and Bell Division of the Matheson Company, Inc.)

Figure 7-1B. Solvents for the 15–35 μ region. The black lines represent useful regions.

This problem and others suggest that the running of samples as solutions is not a good technique to use in infrared. In addition to the solvent absorption problem, there is also the possibility of interactions between the sample and the solvent. These interactions may vary with change of solvent or concentration, and as a result the peaks in the spectra may vary in position, intensity, or shape. The solvent also tends to decrease the energy of the system, and as one of the limiting factors in infrared instrumentation is available energy, this also is an undesirable feature. The energy problem is especially serious with aqueous solutions (they can be run in cells with water-insoluble windows) and limits cell path lengths to less than 0.020–0.025 mm. Except for the following cases, serious consideration should be given to the use of other techniques before a sample is run in solution. Quantitative analyses give the best results using solutions; a study of solvent-solute interactions and concentration effects must be done in solution; and there are other problems that can be dealt with only by the analysis of solutions.

From the above discussion it is obvious that there are times when it is necessary or desirable to calibrate sealed cells. This is usually done by running the clean, dry cell as the sample, and calculating the path length from the resulting interference-fringe patterns (Figure 7-2). Either maxima or minima can be used for this purpose, but it must be remembered that the first maximum or the first minimum is counted as zero and not one. The data are substituted into one of the following equations.

For linear wavelength instruments

$$t_\mu = \frac{n\lambda_1\lambda_2}{2(\lambda_2 - \lambda_1)} \tag{7-1}$$

where

t_μ = the cell path length in microns

n = the number of fringes

λ_1 = the wavelength of the zeroth maximum or minimum

λ_2 = the wavelength of the nth maximum or minimum

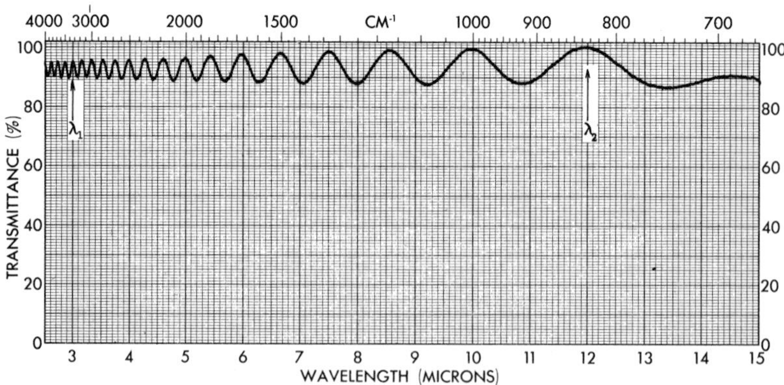

Figure 7-2. Spectrogram indicating the interference-fringe pattern of a clean, dry cell.

For linear wavenumber instruments

$$t_{\text{cm}} = \frac{n}{2(\nu_2 - \nu_1)} \tag{7-2}$$

where

t_{cm} = the path length in cm

n = the number of fringes

ν_2 = the wavenumber of the zeroth maximum or minimum

ν_1 = the wavenumber of nth maximum or minimum

Figure 7-2 can be used to illustrate this calculation. The spectrogram is linear in wavelength, and therefore the equation to use is (7-1), where

$$n = 15$$

$$\lambda_1 = 3.00$$

$$\lambda_2 = 11.95$$

$$t_\mu = \frac{n\lambda_1\lambda_2}{2(\lambda_1 - \lambda_2)}$$

$$= \frac{15 \times 3.00 \times 11.95}{2(11.95 - 3.00)}$$

$$= 30.0\ \mu$$

If for some reason the cell cannot be calibrated by this technique, it can be filled with some standard material of known absorbance, and the path length calculated from the resulting intensity data. Several reasons why the fringe-pattern technique will not always work are: the windows are not flat or polished; the windows are not parallel to each other in the cell; the path length is too long, and the fringes will not appear within the frequency or wavelength range of the instrument.

7.3. SOLIDS

In contrast to gases and liquids, solids present a number of sample-handling problems, as there are few occasions when they can be run without some preliminary work being done on them. In transmission work, the easily run samples are usually limited to thin films; in reflectance work, they are normally limited to small samples for specular-reflection, and flexible materials for ATR or FMIR accessories.

Nujol Mulls. One of the original methods of handling solid materials and one that is still widely used, is the nujol-mull technique. In this technique a small quantity of sample is placed in a mortar and ground into fine, uniform particles—preferably less than 2 μ. A small drop of nujol (mineral oil) is added to the powder, and the material is reground to uniformly disperse the sample in the nujol. The proper ratio of nujol to sample will vary from sample to sample, but a useful rule of thumb to follow is: the completed mull should have a viscosity approximately that of a good grade of whipped cream. When the mulling is completed, the mull should be checked for the presence of large gritty particles, by spreading it around on the head of the pestle with a microspatula. As the spatula moves through the mull, the operator can readily feel whether any such particles are still present.

When the operator is satisfied that the mull has been properly prepared, it is transferred to a salt window with either a microspatula or a rubber policeman (the former is more likely to scratch the window while the latter is more likely to contaminate the mull), and spread in a narrow line down the center of the long axis of the window. A second window is placed on top of the mull, and the two windows are slowly rotated in opposite directions under finger pressure to spread the mull in a thin, even layer between the two windows. The window–mull sandwich is then placed in a demountable cell which is mounted on the spectrophotometer, and the sample is ready to scan.

The spectrogram obtained will show a number of things about the quality of the sample preparation.

1. If the particle size is too large, two effects can be detected. The first is the high background absorption at short wavelengths, due to the particle size being about the same as or larger than the wavelength of the radiation (this scattering decreases as the instrument scans to longer wavelengths or lower frequencies). The second is the Christiansen effect. This effect is noted at lower frequencies (longer wavelengths) where bands that are normally symmetrical become skewed on the lower frequency (long wavelength) side. If, under these conditions, the desired information cannot be obtained from the spectrogram, the mull should be returned to the mortar, reground, and rescanned. There are times when it may be more practical to start over with a fresh sample rather than try to reclaim a poor mull.

2. If the sample-to-nujol ratio is too high in nujol (strong nujol bands and weak sample bands), the mull should be returned to the mortar, more preground sample added, and the whole remulled. If a good spectrum is obtained, there is no such thing as too small a quantity of nujol. However, if more nujol must

be added to get a useful spectrogram, this can frequently be done by removing the windows from the cell, separating them, and adding a microdrop of nujol to the mull on one of the windows. The windows should be rotated back and forth over each other under finger pressure to disperse the additional nujol into the mull before the cell is reassembled.

3. If the spectrogram shows that there is too little mull between the windows, the window–mull sandwich should be removed from the cell, the windows separated, and more mull added. (For those mulls where there is a shortage of sample, the available mull can be scraped into a small rectangle in the center of the window). If there is too much mull between the windows, the window–mull sandwich should be removed from the cell and the windows rotated in opposite directions under firm pressure from the fingers. This will squeeze out some of the mull and leave a thinner layer between the windows. If the sample thickness cannot be sufficiently reduced by this technique, the windows can be separated and one of them cleaned by wiping it very gently with a soft, lint-free tissue. The cell is then reassembled and the sample rerun.

Not all materials (e.g., elastomers, leathers, plastics) can readily be ground; such materials can usually be handled more easily by other techniques. However, if a mull of this type of material is required, it is frequently useful to cool the samples with liquid nitrogen until they become brittle, and then grind them at this temperature.

Normally the sample is ground before adding the mulling fluid, but there are materials where reversing this procedure is advantageous (e.g., a material where the heat generated during the grinding operation is high enough to cause an unwanted change in the sample).

Nujol, like other materials, has absorption bands; carbon–hydrogen stretching, carbon–hydrogen bending, and the long-chain CH_2 rock. Normally these bands do not interfere with the analysis unless CH bands are being studied, in which case some other mulling fluid must be used. The usual alternate material is a completely fluorinated hydrocarbon (referred to as "fluorocarbon oil") having about the same viscosity and refractive index as nujol. Completely chlorinated hydrocarbons, or hydrocarbons where all of the hydrogen has been replaced partially by chlorine and partially by fluorine, are also useful alternates to nujol. These materials have their own absorption bands, but they are transparent at those frequencies (wavelengths) where nujol has its absorption bands. Frequently the literature refers to a "split-mull" spectrogram; here the mull has been run in fluorocarbon oil in the high-frequency (short-wavelength) region (to about 1250 cm^{-1}), and in nujol in the remainder of the range. The ranges may also overlap.

KBr Pellets. In many laboratories nujol mulls have been replaced to a large extent by the KBr-pellet technique. In this technique the sample is again ground to a fine, uniform particle size (preferably under 2 μ), but is mixed with an alkali halide -usually KBr) instead of being dispersed in nujol. This sample–KBr mixture is pressed into a pellet. In this technique both the sample and the KBr are weighed, thus making the technique useful for quantitative analysis for many types of samples. Normally pellets are made to have a total weight of 300 mg; of this total, 0.5–1 % is sample.

The sample is weighed out, transferred to a clean dry mortar, and ground. The mortar must be thoroughly cleaned before it is used, or enough material from previous samples may remain in it so that some of the bands of these samples will appear in the spectrogram. The cleanliness of the mortar and pestle can be checked by making and running a blank pellet. A mortar that has retained a small quantity of water will contribute this water to the next several pellets. Experience has shown that when a mortar has been water-washed this pickup will be excessive, even though the mortar has been thoroughly dried in a vacuum oven.

The sample may be dry-ground, but in many cases it is helpful to add 1 ml of either ethanol or methanol to the mortar and then grind the sample until it is dry. If the operator feels that the particle size is still too large, the above step can be repeated. This wet-grinding technique not only makes the operation easier, but also reduces the amount of work done on the sample. After the sample has been ground to the desired particle size, the KBr is added in small quantities—each addition should be about equal to the amount of total material in the mortar—using the pestle to thoroughly mix the KBr and sample after each addition. The purpose of this grinding operation is to achieve thorough, homogeneous mixing of the sample and the KBr, and not to reduce particle size, even though some grinding may occur. Experience has indicated that the more grinding done on the mixture after the KBr has been added, the greater the quantity of water absorbed from the atmosphere by the KBr.

After all the KBr has been added there are two alternate techniques that can be followed. The traditional one is to transfer the mixture to a small vial, add several steel ball bearings (with certain vibrators, plastic balls are used instead), place the vial on a vibrator, and vibrate it. The purpose of this step is to improve the mixing of the KBr and sample and not to reduce sample particle size. After the mixture has been vibrated for the desired period of time, it is separated from the steel balls and then transferred to the die.

The other technique is to add a small quantity of solvent to the mortar and grind the mixture further under the solvent. With certain types of samples, grinding to near dryness is advantageous, while in others this is unnecessary. In either case, the remaining solvent is evaporated under a heat lamp or on a hot plate, after which the sample is transferred to the die. Any low-boiling solvent, except those absorbed by the KBr or the sample, can be used.

After the sample has been transferred to the die, the operator should spread it evenly over the entire area of the lower ram (an uneven layer of powder frequently results in poor pellets). The die is then assembled, transferred to the press, and evacuated. A pressure of 20,000 psig is applied to press the powder into a pellet. The length of the pressing period can vary in time from 2 to 10 min, depending to some degree on the sample but to a greater degree on the feeling of the operator. After releasing the pressure, the die is disassembled, and the pellet is removed and transferred to a pellet holder. The pellet holder is placed on the spectrophotometer and the spectrum is scanned.

Theoretically, the pellet should be transparent and look like a piece of glass. In practice pellets vary from completely transparent to completely opaque. There are many instances where transparent pellets yield miserable spectrograms while spectrograms from opaque pellets are excellent. If the pellet appears to be completely fused, it should be usable.

As in the case of nujol mulls, the resulting spectrogram will give the operator information about how well the sample preparation was done. The comments on particle size for nujol mulls apply equally to KBr pellets. The KBr at this pellet thickness is completely transparent to beyond $250\ cm^{-1}$ ($40\ \mu$) and will therefore add no bands to the spectrum. It does, however, tend to absorb materials from the atmosphere (primarily H_2O) and these may add unwanted bands to the spectrogram. If the sample concentration is too high, the pellet is broken, part of it is discarded, an equivalent quantity of pure KBr is added to the remainder, and the pellet is remade. If the sample concentration is too low, the pellet can be crushed, the required quantity of preground sample added, and the pellet remade. In general, poor pellets or pellets giving poor spectra should always be remade.

One of the problems in making good pellets is getting a homogeneous dispersion of the sample in the pellet. Frequently this problem can be eliminated by using techniques other than the one discussed above to combine the sample and the halide. The best of these is freeze drying, a technique that gives a fine powder in which the sample is well dispersed. The drawbacks to this technique are the expensive equipment and lengthy preparation time required.

Certain types of samples will give spectra where the intensity, band shape, and frequency of the peaks will vary with the pressing pressure or pressing time. These problems can be held to a minimum by always using exactly the same pressure and time. In certain cases these peak variations can be traced back to the sample-grinding step.

KCl instead of KBr is used with excellent results in a large number of laboratories. They report that KCl is easier to handle and less hygroscopic, and usually gives better results. The one drawback is that it requires a slightly higher pressing pressure (25,000 psig).

The halide used to make the pellet must meet the following requirements.

1. It should have a high transmittance throughout the spectral range to be studied.
2. It should have a low sintering pressure.
3. It should be available in a pure state and be fairly nonhygroscopic.
4. It should have high chemical stability.
5. Its refractive index should be close to that of the sample.

The advantages of using the pellet technique include the following.

1. Spectra are obtained which are free from interfering bands.
2. Less light is scattered.
3. For some materials, spectra with superior resolution can be obtained compared to spectra of the same sample in other media.
4. Pellets can be stored conveniently for long periods of time.

Difficulties encountered with the pellet technique include the following.

1. For ionic samples, exchanges of the halide ions with ions of the sample can occur; the resulting spectra will be representative of the various species present.
2. For some samples the observed spectra do not agree with those determined in other media. This is apparently due to physical isomerization or other changes in the structure of the sample. It also appears that in some instances addition compounds are formed between the halide and the sample [2].
3. Physical or chemical changes can occur in the sample during the grinding or pressing processes.
4. Because of the difficulty of completely removing water from the halide and the sensitivity of pellets to moisture, some water always remains as an impurity.

As stated above the spectrum obtained by the use of a pellet does not always agree with the spectrum of the compound when it was run in some other medium. This anomalous behavior has been investigated extensively [3–6]. One group of workers [5] suggested that in the case of some compounds the anomalous behavior is due to surface-adsorption phenomena, notably the adsorption of a monomeric species of the solute on the halide molecules. For another series of compounds, Baker [7] suggested that the anomalous spectra were due either to polymorphic or amorphous forms of the sample in the halide disk. Tolk [2] has shown that the aging of pellets in a moist atmosphere removes some of the observed anomalies. In some instances heating the pellets achieves the same result.

Baker's studies of this anomalous behavior were quite extensive and he reported that the following factors should be considered in preparing samples in pellet form.

Crystal Energy of the Sample Phase. Compounds with high lattice energy have the same mull and pellet spectra unless polymorphic transitions occur. Normally, compounds which melt above 200°C are stable if the grinding is not too vigorous. Low-melting-point (89–90°C) compounds normally show broadening and shifting of peaks.

Energy of Grinding (Sample and Matrix). If a compound shows broadening and shifting of peaks, vigorous grinding will give a spectrum nearly identical to the liquid spectrum, thus showing that the grinding is merely randomizing the molecular orientation. Hand grinding, because its intensity and uniformity are much harder to control, is inferior to mechanical grinding.

Lattice Energy of the Matrix. The lower the lattice energy of the matrix, the greater the proportion of the grinding done on the sample. To decrease the excess grinding and work done on the sample it may be desirable to have a matrix of high lattice energy. Unfortunately, the higher the lattice energy, the greater the sintering pressure required (KCl < KBr < KI).

Particle Size of the Matrix. As the initial size of the matrix particles is decreased, spectral distortion increases. Baker prefers to start with 20-mesh rather than 250-mesh material, on the basis that the halide absorbs some of the grinding energy and does not permit as much of it to go into the sample. Others [8] have noted that the use of too large a grain size, even though uniform, gives rise to the scattering of an appreciable fraction of the incident light by the pellets.

Ability of the Sample to Recrystallize in the Pellet. (Related to crystal energy.) It is frequently stated that one advantage of the KBr-pellet technique is that the pellets can be stored, repressed, and used as an unchanging reference standard [3]. This is not necessarily correct, because recrystallization in the pellet can occur within a matter of minutes for high-crystal-energy compounds, or over a period of months if the compound has low crystal energy. A possible means of reducing this effect is to heat the pellet below the melting point of the solid, thus allowing stress relaxation and recrystallization.

Relative Stability of Polymorphic Forms. Many compounds can exist in states that are more stable than the original form under the conditions prevailing during the grinding and fusing operations. Accordingly, they can be physically isomerized by vigorous pelleting, and in many organic materials crystalline structures result that are unstable at room temperature. These structures will probably revert to the more stable forms as the pellet is stored at ambient temperature and thus give changing spectra.

Hydrates and samples containing ions such as NO_3 that can undergo low-temperature transformations cannot be studied conveniently by the pellet method [3]. Some oxidation of the KBr may also occur when it is mixed with a strong oxidizing agent.

Baker mentions two other effects that are of lesser importance but may be involved: the surface adsorption of the sample on the matrix powder, and dielectric forces. Only the latter effect will be discussed here.

Dielectric Forces of the Alkali Halide. Several investigators [3,8–11] feel that the most serious difficulty with the pellet technique lies in the possibility of an anion exchange between the compound (especially in the case of inorganic salts) and the alkali halide, causing the appearance of new bands and the shifting of others. The magnitude of these changes depends on the amount of anion exchange involved, with the exact position of the new bands depending on which alkali halide is used [10]. The exchange is promoted and accelerated by water adsorbed on the surface of the sample and the halide. All the halide pellet materials give rise to this phenomenon [3] to a greater or lesser degree, depending on the quantity of water adsorbed and the pressure required for sintering. Because the effect is greatly dependent on the adsorbed water, the degree of exchange is quite variable. There is a possibility that this effect can be eliminated [9] in some cases by evacuating the die for a long period of time and by maintaining a low water-vapor pressure in the work area. However, as yet, no definite solution to the problem has been found.

The above discussion was based on the use of an alkali halide as the matrix. In the case of some materials the operator may improve the results by using some matrix other than an alkali halide—e.g., polyethylene. In the event that such a matrix substitution is made, it is likely that the matrix material may add bands of its own to the spectrogram.

Films. Many solid materials can be laid down as thin films or as thin layers of finely divided particles. Other materials can be microtomed or pressed to give thin films.

Films can be cast from a dilute solution of the sample in some solvent onto a substrate such as plate glass. The solvent is completely evaporated, the film peeled off and run as a free film. The films must have sufficient strength for the operator to peel them from the substrate in pieces large enough to cover the illuminated area of the spectrophotometer. At times this problem can be avoided by casting the film on a liquid substrate, for example a pool of mercury.

The big drawback with free films is that they will probably add an interference fringe pattern on top of the spectrum of the sample. This is true of any thin layer of material with flat, polished, parallel sides. The fringe pattern makes it difficult to distinguish the peaks that belong to the spectrum from the fringes, gives false intensity data by adding to or subtracting from actual band intensities, and at times completely eliminates the band (when a maximum of the fringe pattern has the same intensity and frequency as a band). Running the sample at two thicknesses helps resolve the problem (the fringes are dependent on film thickness only, and will move to different frequencies) but at an added expense in time and effort. Crumpling the film, roughing the surface, or casting the film as a wedge may eliminate or reduce the problem.

A more useful technique for casting films is to cast them directly on the salt windows. (Those samples that are soluble only in water can be cast on silver bromide or other water-insoluble windows.) These films are cast from very dilute solutions (10–12% is a good rule of thumb) and are built up to the desired thickness in a series of thin layers. The solvent is flashed off with a heat lamp or hot plate after each addition of solution. When completely dry the window is mounted in a demountable cell, the cell placed on the spectrophotometer, and the spectrum scanned. If the film is too thin (insufficient band intensity of the spectrum), more layers can be added; if it is too thick, it must be removed from the window and the casting process repeated. The need to recast films can be greatly reduced by casting the film as a wedge rather than with a uniform thickness. This can be accomplished by placing one side of the window on a thin object lying on the table rather than placing it flat on the table. Films cast directly on windows do not give interference fringes.

Some materials form a layer of a fine powder or tiny crystallites instead of the expected film, a certainty if the sample is dispersed instead of being dissolved in the solvent. To yield useful spectrograms these particles must meet the following requirements:

1. the particle size must be very small ($<2\,\mu$); and
2. the particles must adhere to the window.

There are a number of known techniques and commercial types of equipment available for obtaining such dispersions. If a dispersing agent is used, it should be removed prior to or during the deposition of the powder on the plate. If the particle size is too large the radiation will be scattered by the sample. This scatter can be reduced by coating the sample with nujol—in effect, this will be the equivalent of scanning a nujol mull.

Whenever the film-casting technique is used, the operator must be careful of the following problems:

1. All of the solvent must be removed from the film. Usually the more volatile the solvent the more readily this can be accomplished. Also, building up the film in thin layers will facilitate this operation.
2. Some solvents tend to leave solid residues behind when they are evaporated. Unless the spectra of these residues do not interfere with that of the sample, the residues must be removed.
3. Solvent is usually removed with heat, a step which may decompose the sample. If decomposition does occur some other technique must be found to remove the solvent.
4. Windows can be heated without deterioration, but they cannot be placed on a cold surface until they are cool.

Specular Reflectance. Specular reflectance is a useful technique for the analysis of coatings on reflective surfaces. Usually the only sample preparation required is to reduce the physical size of the sample so that it will fit the accessory. The spectrogram obtained by this method will be essentially the same as that obtained by a transmission scan.

The sample area to be analyzed can be varied in size and shape by selection of the proper accessory (different-type reflectance units cover different size sample areas) and the use of masks. Most reflection units will accommodate extra-large-size samples; these can be moved around over the illuminated area, thus allowing the operator to make a profile study of the sample.

The specular-reflectance technique is frequently used to measure layer thickness, especially in the semiconductor industry. The system can be calibrated with known-thickness samples by relating band intensities to these thicknesses and graphing the results. The graph can be used to determine the layer thickness of the samples after their band intensities have been determined. Whenever the composition of the coating or the type of substrate is changed, the system must be recalibrated.

Another, more frequently used approach to layer-thickness determination is possible when the layer whose thickness is to be determined is partially reflecting and partially transmitting. In this case, when the sample is scanned an interference pattern will be recorded, and if the refractive index of the layer and angle of incidence of the radiation are known, the layer thickness can be calculated.

ATR. Attenuated total reflectance and its extension, frustrated multiple internal reflectance (FMIR), are excellent techniques for the analysis of surfaces. In these techniques the sample does not have to be reflecting, as the reflection is done by the crystal–sample interface (Section 3.8).

Sample preparation is normally limited to cutting the material to the required physical size. If the surface is expected to be the same as the bulk material, the sample can be polished flat and smooth, although the operator must be careful during the polishing operation that he does not change the structure or composition of the surface of the sample. (A flat, smooth surface is very helpful in obtaining good results.) Normally, soft, flexible samples can be cut larger than the illuminated area of the crystal, but smaller than the openings in the holder; rigid samples, especially if they have a rough surface, should be cut to the size of the illuminated area of the crystal.

The radiation at the critical angle will penetrate the sample, assuming perfect contact between sample and crystal, to a depth of about one wavelength. This indicates that only a very thin layer of the sample below the surface can be analyzed by this technique, and that the thickness of this layer will increase with increasing wavelength. As the ATR device is set to higher angles, the depth of penetration decreases, allowing investigation of a number of sample thicknesses without changing the sample or resurfacing it. This ability to vary the angle is useful in adjusting path lengths to compensate for types of samples having different absorption intensities. Most samples are run at just above the critical angle but some, especially carbon-black-loaded materials (of no more than approximately 5% carbon-black content) can be run only at the highest available angles.

The path length through a sample is a function of the refractive index of the crystal and the angle of incidence only. Assuming perfect contact between sample and crystal, ATR should furnish an accurate quantitative analysis technique. This is true of liquids and of some soft solids where there is no sample–crystal contact problem.

With most solid materials, sample–crystal contact is a big problem. On the basis that several poor contacts will give the same path length as one good contact, the accessory was changed from a single- to a multiple-reflection unit. There are FMIR crystals in use that have several hundred contact points and which can be used for very thin layers. Unfortunately, these are not commercially available at present.

Another approach to the same problem is to deposit films of the sample from solution onto the crystals. These films do not require control of their thickness.

7.4. MISCELLANY

Other techniques which are used at various times are covered in this section.

Dry Pyrolysis. Dry pyrolysis is used for samples where other techniques are difficult or impossible to use. Examples of this are high-carbon-black-filled materials, abrasive wheels, and insoluble polymers that are difficult to grind.

In this technique the sample is heated to a temperature higher than its breakdown temperature, and the breakdown products, called pyrolyzates, are collected and analyzed. This pyrolysis can be accomplished very simply in a hard glass test tube, or it can be done in a special pyrolysis apparatus under a controlled atmosphere.

The pyrolyzates can be gaseous, liquid, or solid. Most infrared work is done on the liquids, with some done on the gases and very little done on the solids. When the gaseous pyrolyzates are used, they should be dried before they are transferred into the gas cell. If they are transferred from the pyrolysis unit directly into the cell, water vapor tends to condense on the cold cell walls, and even more readily on the windows. This problem is especially serious if the pyrolysis was done in a glass test tube.

The usefulness of this technique is based on the fact that any specific polymer, when heated beyond this breakdown point, will always yield the same products; other polymers will yield their own special breakdown products. This technique is an empirical method, and like all empirical methods should always be done in the same way.

Melts. Many materials melt at temperatures that make the use of this technique very attractive. A small piece of the sample is placed on a window, melted, and allowed to flow in a thin layer over the window. Usually the sample and window are cooled and the sample resolidified; however, there may be times when it is useful to hold the sample in the fused state and run it as a liquid. Seldom is a second window added to make a sandwich, although this addition may be helpful in obtaining the desired sample thickness. Some materials have a melting point that is so low that the heat generated in the spectrophotometer will remelt the sample, and the addition of a second window is required to prevent the now-liquid sample from flowing off the window. A second window is also required if the sample is held and run as a liquid.

Pressed Samples. Some samples are quite soft and can be pressed into a thin film between salt windows. This technique is especially useful for semisolids. Other materials can be pressed into thin films in a press. In this case, the sample is a free film and will show an interference-fringe pattern.

Polarization. This technique is used primarily to determine orientation or crystallinity in a sample, primarily in films and fibers. As stated previously, instruments, especially grating instruments, add varying degrees of polarization to the beam, so that a polarizer will be helpful to get the true spectrogram of the sample. It may be possible to align the sample at 45° to the beam, and thereby get an accurate spectrum without using a polarizer.

Matrix Isolation. Gases can be condensed on cold windows as finely divided deposits, and scanned as such. If the sample is mixed with an inert gas prior to freezing, the sample particles will be separated by the particles of the inert gas in the deposit.

7.5. MICROSAMPLING

In doing microwork the biggest problem is the sample contamination that can occur during the sample-handling operations. This problem is so severe at times that the contamination represents a larger percentage of the material than does the sample. This is especially true of solids that are run as micromulls or as micropellets.

The same general techniques that were discussed above can normally be used with some modifications in microwork. Liquids can be run in micro- or ultramicrocells if a sealed-cell type is required, or in a microdemountable cell for the others. Solids are usually run as micronujol mulls or micro- KBr pellets. These mulls are also run with the microdemountable cell.

Microanalysis finds its greatest use in the biomedical field. Here the final step in sample cleanup is frequently freeze-drying, and time and effort can usually be saved and contamination decreased by adding the proper quantity of KBr to the solution prior to the freeze-drying. The resulting residue requires no further preparation, but is ready to be transferred to the die and pressed into a pellet.

The minimum sample size required to produce a useful spectrogram will depend to some extent on the sample—a strong absorber will require less sample than a weak absorber. For liquids, samples as small as 0.01 μl have been run successfully, while for solids it is possible to run 0.01 μg or less.

7.6. CONCLUSIONS

As infrared spectroscopy is a comparison of unknowns to knowns, the spectroscopist should use the same technique to run the unknowns that was used to run the knowns. The fewer the components contributing bands to the spectrogram, the easier it will be to identify all of them. Saving time on sample preparation frequently leads to spending additional time in correcting errors, and may be highly embarrassing.

Unless the spectrograms are to be published or are to be made part of a report, they may furnish the required information even though they are of poor quality. This is not to be construed as suggesting sloppy work, but is merely intended to point out that obtaining useful information, not beautiful spectrograms, is the desired end.

If the sample contains holes or is too small to fill the beam (as defined by the slit), radiation will pass through the uncovered area without touching the sample and will give false intensity data. This can be demonstrated using a totally absorbing band; the band will not go to zero even though its shape indicates that it should do so if the sample area is too small or contains holes.

Usually the technique to use should be the one that is the least complicated, that requires the least sample handling, and with which the operator normally gets the best and most useful results.

REFERENCES

1. W. Ulrich, Analytical Instrumentation (1963).
2. A. Tolk, *Spectrochim. Acta* **17**: 511 (1961).
3. V. W. Melacke and G. E. Kabus, *J. Inorg. Nucl. Chem.* **6**: 104 (1958).
4. R. D. Elsey and R. N. Hazeldine, *Chem. Ind.* 1177 (1954).
5. V. C. Farmer, *Chem. Ind.* 586 (1955).
6. V. C. Farmer, *Spectrochim. Acta* **8**: 374 (1957).
7. A. W. Baker, *J. Phys. Chem.* **61**: 450 (1957).
8. M. A. Ford and F. Wilkinson, *J. Sci. Instr.* **31**: 338 (1954).
9. F. Vratny, *J. Inorg. Nucl. Chem.* **10**: 328 (1959).
10. W. H. Pleskin and R. P. Euschens, *J. Phys. Chem.* **59**: 1156 (1955).
11. D. J. Millen, C. Polydonopoulos, and D. I. Watson, *Proc. Chem. Soc.* 18 (1957).

CHAPTER 8

Spectra and Reference Library

It is a well-known fact that the ability to perform a thorough search of the literature is of great importance to the research chemist; of similar importance to the spectroscopist is the library of reference spectra that he must acquire and maintain. The ability to locate a spectrum quickly and to identify an unknown by comparison of its spectrum with those on file makes spectroscopy one of the most important analytical and research tools a chemistry laboratory can possess.

The spectrograms for this library may be obtained by purchase of those commercially available. Of greater importance than these commercially available spectra are the reference spectra run in the user's own laboratory. These latter references are more likely to reflect the specific products and research projects of interest to that particular laboratory.

This chapter is designed to acquaint the reader with the various spectral reference systems commercially available, and to present some suggestions that may be valuable in maintaining a useful spectrum library.

8.1. GENERAL REFERENCES TO INFRARED SPECTROSCOPY

A partial listing of books of general interest for spectroscopists follows.

Practical Spectroscopy, G. R. Harrison, R. C. Lord, and J. R. Loofbourow, Prentice-Hall, Inc. (1948). Primarily concerned with optics and electronics.

The Encyclopedia of Spectroscopy, G. L. Clark (editor), Reinhold Publishing Corp. (1960). Articles by various authors in many fields of spectroscopy. The section on infrared spectroscopy gathers together many articles that have appeared in the literature. Several articles were written specially for this text. A useful compendium.

Electronics for Spectroscopists, C. G. Cannon (editor), Interscience Pubs., Inc. (1960). Concerned with the optics and electronics of spectrophotometers.

Infrared Methods: Principles and Applications, G. K. T. Conn and D. G. Avery, Academic Press, Inc. (1960). Concerned with the optics and electronics of spectrophotometers.

Infrared Spectra of Complex Molecules, L. J. Bellamy, John Wiley & Sons, Inc. (1954). Concerned with frequency assignments; one of the most comprehensive monographs on the subject.

Infrared Determination of Organic Structure, H. M. Randall, R. G. Fowler, N. Fuson, and J. R. Dangle, D. Van Nostrand Co., Inc. (1949). Excellent introduction to practical infrared spectroscopy.

Infrared Spectroscopy, Chicago Society of Paint Technology, Chicago, Illinois (1961). Brief text introducing some of the elementary theory and practice of infrared spectroscopy.

Tables of Wavenumbers for the Calibration of Infrared Spectrometers, Butterworths Scientific Publications, London, England (1961). An excellent book for use in wavelength calibration. Calibration points for the 400–600 cm^{-1} region are presented for both high-resolution and prism or small-grating spectrophotometers. Includes brief discussions of effects of pressure, temperature, and also presents material on preparation of samples, reliability of data, and experimental techniques.

Technique of Organic Chemistry, Vol. IX, A. Weissberger (editor), Interscience Pubs., Inc. (1956). An excellent introduction to the theory and applications of infrared spectroscopy. The infrared sections are part of the discussion of spectroscopy in general. Volume XII is also concerned with infrared spectroscopy.

Molecular Vibrations: Theory of Infrared and Raman Vibrational Spectra, E. B. Wilson, Jr., P. C. Cross, and J. C. Decius, McGraw-Hill Book Co. (1955). A theoretical text, intended for those familiar with quantum mechanics.

Infrared and Raman Spectra of Polyatomic Molecules, G. Herzberg, D. Van Nostrand Co., Inc. (1945). The classic text on the history of infrared and Raman spectroscopy.

Introduction to Practical Infrared Spectroscopy, A. D. Cross, Butterworths Scientific Publications (1960). A brief text giving a very short introduction to the theory and practice of infrared spectroscopy; fairly extensive correlation tables.

Absorption Spectroscopy, R. P. Bauman, John Wiley & Sons, Inc. (1962). Covers ultraviolet, visible, infrared, and Raman spectroscopy from a fairly theoretical viewpoint.

Elements of Infrared Technology: Generation, Transmission and Detection, P. W. Kruse, L. D. McGlauchlin, and R. B. McQuistant, John

Wiley & Sons, Inc. (1962). Covers the theory and practice of instrumentation for infrared spectroscopy.

Advances in Spectroscopy, Vols. I & II, H. W. Thomson (editor), (Interscience) John Wiley & Sons, Inc. (1961). Discusses selected topics in all areas of spectroscopy, including emission and absorption, atomic and molecular spectra.

Elementary Introduction to Molecular Spectra, B. Bak, (Interscience) John Wiley & Sons, Inc. (1962). Microwave, infrared, visible, ultraviolet, and magnetic-resonance spectroscopy is discussed.

Fundamentals of Infrared Technology, Holter, Nudelman, Suits, Wolfe, and Zissis, The Macmillan Company (1962). Covers the physics of detectors and instrument optics.

Infrared Absorption Spectroscopy, Nakanishi Koji, Holden–Day, Inc. (1962). Elementary text intended for chemists beginning infrared spectroscopy. 180 pages in four basic sections. Some NMR data are also included.

An Introduction to Infrared Spectroscopy, W. Brugel, John Wiley & Sons, Inc. (1962). A general introductory text. A discussion of American and West German instruments is included.

Introduction to Molecular Spectroscopy, G. M. Barrow, McGraw-Hill Book Co., Inc. (1962). A fairly simple mathematical treatment of molecular spectroscopy. Those desiring a fairly rigorous approach to spectroscopy, but below that presented in Herzberg's classic text, will find this book valuable.

Molecular Spectroscopy, Methods and Applications in Chemistry, G. H. Beaven, E. A. Johnson, H. A. Willis, and R. G. Miller, The Macmillan Company (1962). Ultraviolet, visible, and infrared spectroscopy for beginners in analytical spectroscopy.

Progress in Infrared Spectroscopy, Vol. 1, H. A. Szymanski (editor), Plenum Press (1962). Proceedings of the fifth annual Infrared Spectroscopy Institute held in Canisius College, August 1961. Includes practical chapters on such topics as ultraviolet, far-infrared and Raman spectroscopy, as well as polymer spectra, for the industrial chemist. The question-and-answer section in the Appendix may be helpful in answering many of the beginner's basic questions.

Developments in Applied Spectroscopy, Vol. 1, W. D. Ashby (editor), Plenum Press (1961); Vol. 2, J. R. Ferraro and J. S. Ziomek (editors), Plenum Press (1963). Proceedings of symposia on spectroscopy held in Chicago in 1961 and 1962. Each volume contains a section on infrared and Raman Spectroscopy.

Chemical Applications of Infrared Spectroscopy, C. N. R. Rao, Academic Press, New York (1963). Discusses infrared theory and spectral interpretation. Very extensive bibliography.

Chemical Infrared Spectroscopy, Vol. I (Techniques), W. J. Potts, Jr.,

John Wiley & Sons, Inc., New York (1963). Discusses theory, instrumentation, and sample-handling techniques.

Introduction to Infrared and Raman Spectroscopy, N. B. Colthup, L. L. Daly, and S. E. Wiberley, Academic Press, New York (1964). A discussion of the theory of infrared, instrumentation and the interpretation of spectrograms.

Handbook of Industrial Infrared Analysis, R. G. White, Plenum Press, New York (1964). A discussion of instrumentation and sample handling, plus spectrogram interpretation.

Laboratory Methods in Infrared Spectroscopy, R. G. J. Miller, Heyden and Son, London (1965). Deals with infrared instrumentation and sample-handling techniques.

8.2. GOVERNMENT PUBLICATIONS

A number of government publications on infrared spectroscopy are available, and since it is not practical to list all of them a few will be selected to indicate the general areas and laboratories involved. All those listed may be purchased from the United States Department of Commerce, Office of Technical Services.

Infrared Spectra of Plastics and Resins, PB 111438.

Far-Infrared Spectra of Various Compounds, WADC Technical Note 57-413; WADC Technical Reports 58-198, 57-359, 59-498.

Near-Infrared and Infrared Spectra of Various Compounds, WADC Technical Reports 59-431, 59-344.

General Aspects of Infrared Spectroscopy, SB-466, SB-467.

8.3. HOUSE ORGANS OF VARIOUS COMPANIES AND LABORATORIES

Pertinent house organs are distributed by a number of companies and laboratories. Usually these report new techniques or provide some general information about infrared spectroscopy. They may also list technical-society meetings of interest to the spectroscopist. A partial list of these house organs follows.

Spectroscopia Molecular, F. F. Cleveland (editor). Illinois Institute of Technology, Chicago, Illinois.

Unicam Spectrovision. Unicam Instruments Limited, Cambridge, England.

The Instrument News. Perkin–Elmer Corp., Norwalk, Connecticut.

Hilger Journal. Hilger and Watts Ltd., London, England.

Optica Spectrum Analysis. Optica, Milan, Italy.

In addition, a number of manuals, data sheets, bibliographies and spectrogram compilations are available from some of these firms.

8.4. REFERENCE SPECTROGRAMS AND SPECTRAL RETRIEVAL SYSTEMS

The Sadtler Research Laboratories Spectra

Currently the largest publisher of reference spectrograms is the Sadtler Research Laboratories of Philadelphia. In addition to the spectrograms, Sadtler markets a "spec-finder" which aids in locating spectra in this collection. The spectra are cross-referenced to the ATSM–IBM cards, so that laboratories having the IBM card deck can use this technique to sort the Sadtler spectra. These spectrograms are classified as either standard or commercial.

The standard spectra are run using pure compounds and include the common organic reagents. The commercial series includes dyes, stains, plasticizers, monomers, polymers, polyols, resins, rubbers, soaps, fatty acids and drying oils, solvents, surface-active agents, textile chemicals, waxes, petroleum chemicals, intermediates, and pharmaceuticals. These spectra reflect the materials as they are available commercially, and list their trade as well as chemical names.

The indices available with the spectra are divided into alphabetical, numerical, molecular formula, and chemical classes. In addition, the "spec-finder" can be used to locate spectra by band position. Computer service for locating spectra is also available from the Sadtler laboratories.

Physical data such as melting points are included on the spectrograms. The technique used to obtain the spectrum is stated on the spectrogram.

Coblentz Society Spectra (published by the Sadtler Research Laboratories)

The Coblentz Society has been collecting spectra of high quality from various laboratories, and marketing them through the Sadtler Research Laboratories' marketing facilities. This group of spectrograms, while only a few thousand in number, is of excellent quality.

American Petroleum Institute Research Project 44 and the *Manufacturing Chemists Association Research Project*

This project issues standard reference spectrograms in fiver categories of spectroscopy: infrared, ultraviolet, Raman, mass, and nuclear magnetic resonance. The project deals with petroleum and petroleum products and is limited to this type of compound.

The Documentation of Molecular Spectroscopy (DMS) Spectra (Butterworths Scientific Publications, London, England, and Verlag Chemie, West Germany)

The DMS system consists of punched cards containing the spectrum and physical and structural data on the compound. The cards are sorted with the help of steel needles, or the data can be placed on IBM cards for machine sorting. The coding system used for the cards is quite elaborate. It includes classification by structural characteristics so that groups of compounds having the same structural units can be selected from the pack. As the complete coding is rather extensive, only a brief outline of it will be given. In Figure 8-1, the method of coding is illustrated for 1,6-dimethylpiperid-2-one. The number of carbon atoms in this compound is entered in section A of the card, the basic skeleton is coded in section B, and the substituents are coded in section C. The $N{-}CH_3$ group is entered for this compound. The two sides of the coded card are shown in Figures 8-2A and 8-2B.

Special spectral cards for polymeric materials and natural substances whose structures are not completely known are also issued. The spectral cards are cross-indexed with the literature cards of this system. The literature cards are described in a following section.

An Index to Published Infrared Spectra (H.M. Stationery Office, London, England)

This index to publications of spectra of specific compounds is a continuing series that should prove useful to the spectroscopist. Each volume is indexed according to chemical structure and presents all the pertinent details concerning the spectra reported, including equipment used, spectral range covered, and sampling technique used.

ASTM–IBM Cards (American Society for Testing and Materials, Philadelphia, Pennsylvania)

The American Society for Testing and Materials has available punched cards that can be machine sorted. The codes and systems used on these cards were originated by L. E. Kuentzel and later were somewhat modified by an ASTM subcommittee. The cards used are standard IBM cards, and an IBM sorter must be available for efficient use. The Spectrograms are not printed on the cards, so that a reference spectral library is needed to confirm the identification.

Sources of spectra include the American Petroleum Institute Research Project 44; the Research Laboratories, National Research Council – National Bureau of Standards file; and books, journals, etc.

The system is designed to handle data obtainable with NaCl, KBr, KRS-5, and other prism materials. The cards provide wavelength data to the closest 0.1 μ for the 2–25 μ region, and to the closest 1 μ for the 25–50 μ region. Special cards are available for LiF and CaF_2 prism data, which show better resolution of spectra.

Cards are available punched in either the wavenumber or the wavelength system. The cards are punched to indicate the wavelength or

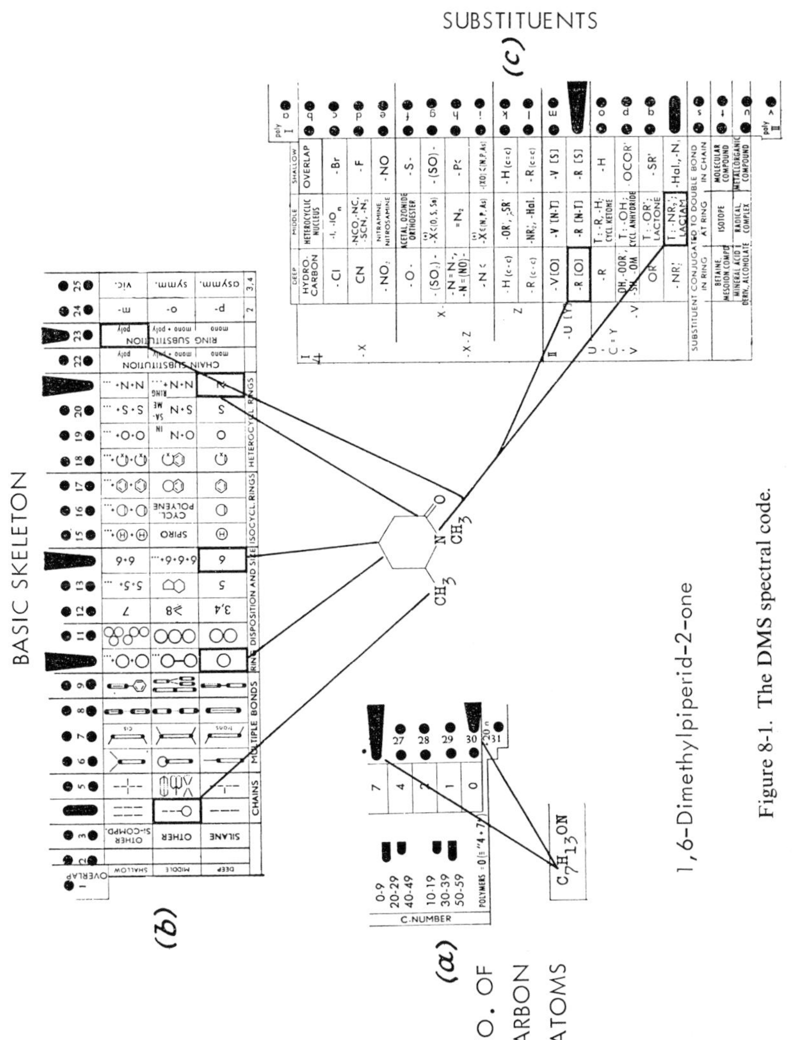

Figure 8-1. The DMS spectral code.

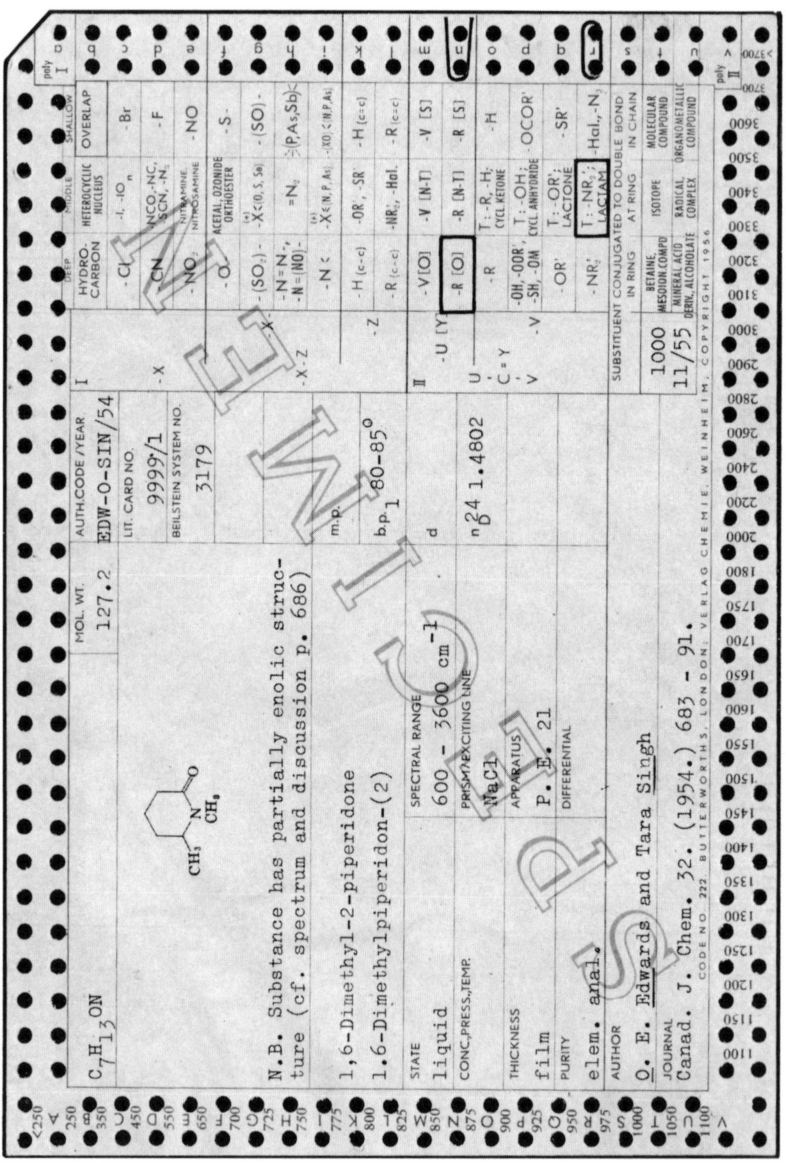
$C_7H_{13}ON$

MOL. WT. 127.2

AUTH.CODE/YEAR EDW-O-SIN/54

LIT. CARD NO. 9999/1

BEILSTEIN SYSTEM NO. 3179

N.B. Substance has partially enolic structure (cf. spectrum and discussion p. 686)

1,6-Dimethyl-2-piperidone

1.6-Dimethylpiperidon-(2)

m.p.

b.p.$_1$ 80–85°

STATE liquid

SPECTRAL RANGE 600 – 3600 cm^{-1}

d

CONC.,PRESS.,TEMP.

PRISM/EXCITING LINE NaCl

n_D^{24} 1.4802

THICKNESS film

APPARATUS P. E. 21

PURITY elem. anal.

DIFFERENTIAL

AUTHOR O. E. Edwards and Tara Singh

JOURNAL Canad. J. Chem. 32. (1954.) 683 – 91.

1000 11/55

CODE NO. 222 BUTTERWORTHS, LONDON; VERLAG CHEMIE, WEINHEIM, COPYRIGHT 1956

Figure 8-2A. A typical DMS spectral card (*front*).

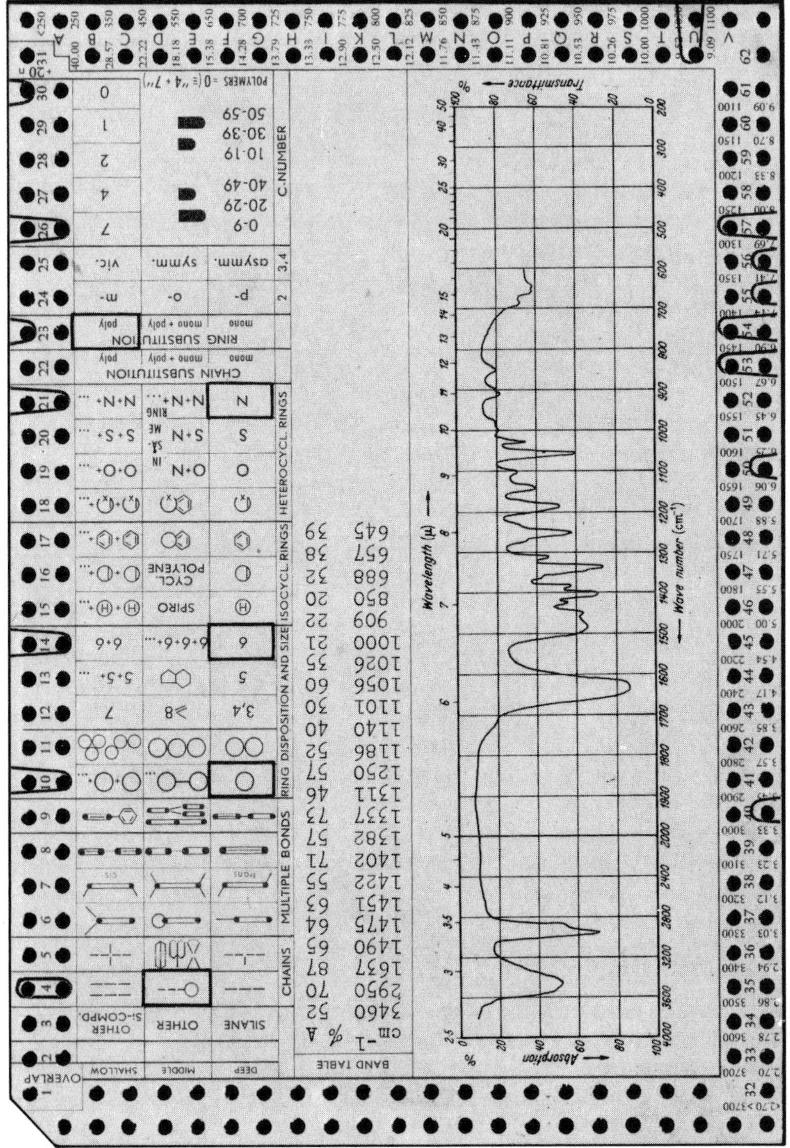

Figure 8-2B. A typical DMS spectral card (*back*).

frequency of bands, some chemical classification, melting or boiling point, number of carbon, nitrogen, oxygen, sulfur, and silicon atoms, and the source of the spectrum.

The 12-position on the card is used to indicate regions of strong absorption, while the 11-position indicates regions of no absorption. A card may then also be sorted for these 11- and 12-punches, and their use can simplify sorting in many cases.

Sorts may be characterized as either positive or negative. The positive sort, as the name implies, involves choosing a band in the unknown spectrum and sorting for all the cards that show this band. (Since the unknown may show shifts in peaks due to temperature and other effects, it is wise to sort for a region about 0.3 μ wide.)

Negative sorts can be used to advantage to eliminate large numbers of cards in the initial sort. They are also useful for impure samples, since a sort could eliminate the desired card if an impurity band is chosen. The negative sort is accomplished by sorting for all cards that show a band where the unknown does not show one. Since the 12-position indicates regions of strong absorption, it may be used to advantage here. Usually, in the case of negative sorts, the sorting is done on a single punch.

Some general rules can be stated concerning the inclusion or omission of absorption bands, but they are applied rather flexibly. Shoulders and weak bands are generally not coded.

In using these cards, it is wise to sort for the least common band first so that a large number of cards may be eliminated in the first sorting. Frequently this can be achieved by selecting the band of longest wavelength.

In selecting band positions, it should be remembered that the standard data are rounded off to the nearest 0.1 μ, so that a broad sorting interval should be used. For instance, a band at 13.46 μ should be checked by sorting both the 13.4 μ and 13.5 μ positions.

Modern sorters may be set to sort for several punches in one column as well as for only a single punch. The use of such sorts may prove quite advantageous in cases where more than one band is present in the wavelength or frequency interval represented by a single column.

Since a small error can eliminate the desired card, it is advisable to keep the various sorts separate, so that it is possible to retrace the sorting steps if an error is made. Obviously, if a sort is made for a weak band that is not coded, the correct card will be automatically eliminated from the deck. The complete deck of punched cards is usually kept in some kind of pre-sort order, so that the entire deck need not be sorted each time.

It is also possible to sort for elements present if these are known. In addition, any of the other physical properties, such as melting point,

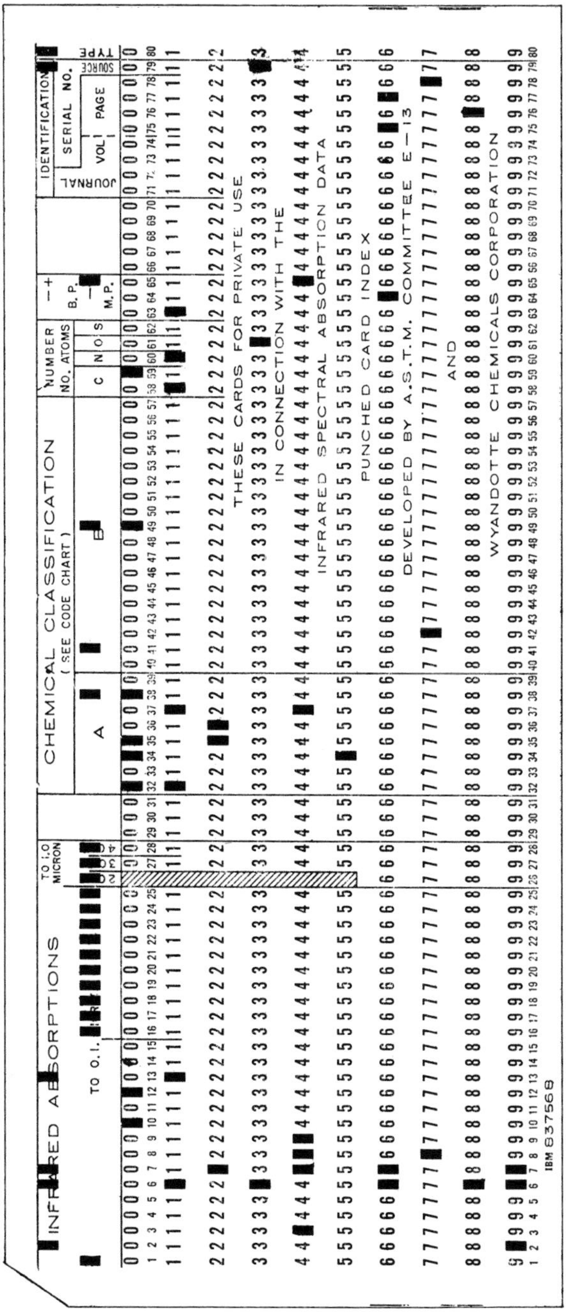

Figure 8-3. A typical IBM spectral card.

can be used as a criterion for sorting. This type of sorting should be done first to decrease the number of cards before frequency or wavelength sorting is started.

A typical card is reproduced in Figure 8-3. The column numbers of the left-hand columns give the wavelength in microns, while the punch position indicates the wavelength to the closest 0.1 μ unit. For example, a punch in column 6 at number 5 would indicate a band at 6.5 μ.

The Documentation of Molecular Spectroscopy (Butterworths Scientific Publications, London, England)

The literature cards of the DMS system can be used to answer the following questions.

1. Which recently published papers deal with some special technical problem?
2. What theoretical work deals with a particular topic?
3. What spectroscopic investigations have been published by a certain author, or in a certain year?
4. What were the main points made by a particular author or in a particular paper?

The cards are coded so that they may be selected from a stack with the aid of steel needles. The coding system includes names of authors, year of publication, general content of paper, spectral region and method, equipment, and sampling. Whenever possible, the cards are linked to the DMS spectral cards. A typical card is shown in Figures 8-4A and 8-4B.

Thermatrex [A "peek-a-boo" system] (Jonker Business Machines, Inc.)

The Thermatrex system is based on the fact that certain characteristics of compounds naturally separate spectra into various categories. The elements present in the compound may determine one classification, the absence of certain peaks can determine another, etc. A particular set of characteristics of the compound is chosen to narrow the search to a certain series. By selecting the characteristics which are to be used for the sort and placing these on a master card (or series of cards) all cards having all these characteristics can be selected. Practically, this is accomplished by representing the data on the index cards by pinholes in a plastic card. There is space for 10,000 pinholes in a $9\frac{5}{8} \times 11\frac{1}{2}$ inch plastic card, arranged in 100 rows of 100 holes. A strong light placed behind the card reveals a series of pinpoints of light wherever a card has been punched. Spectra from the Sadtler, ASTM, API, National Research Council, DMS, Coblentz Society, MCA, and IRDC (Japan) files are included in the Thematrex system.

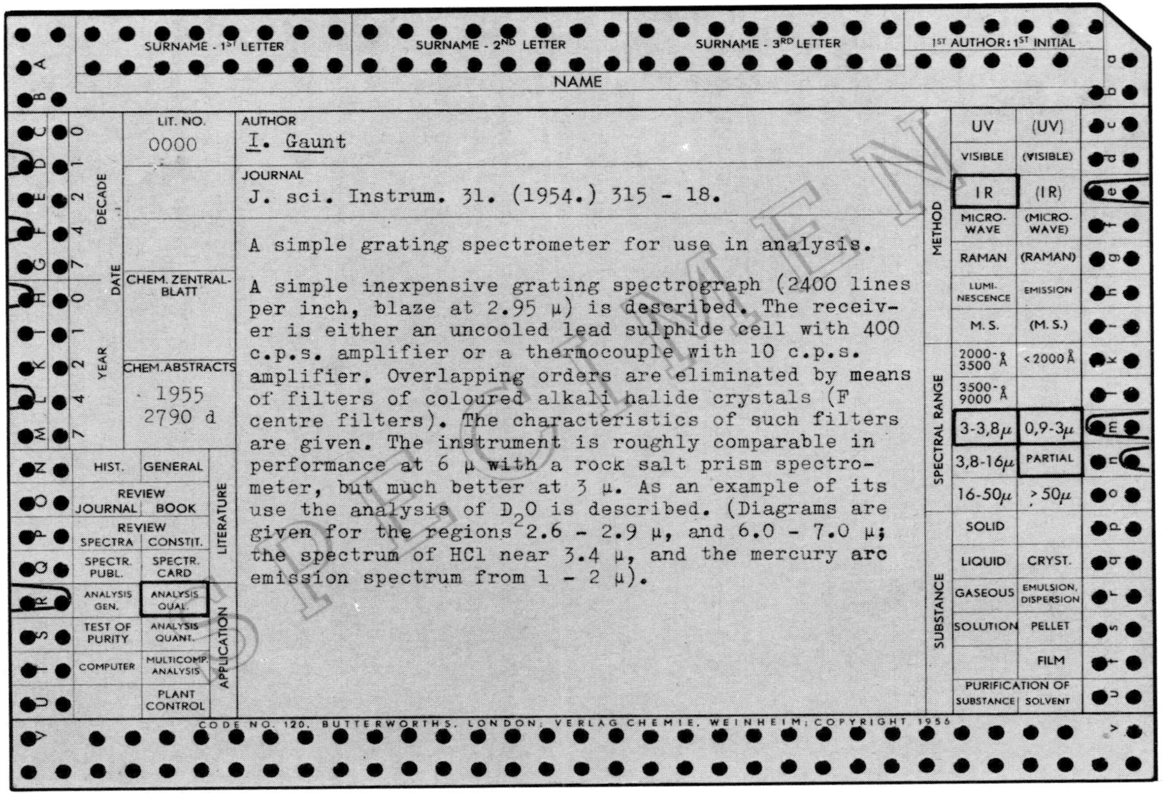

SURNAME - 1ST LETTER | SURNAME - 2ND LETTER | SURNAME - 3RD LETTER | 1ST AUTHOR: 1ST INITIAL

NAME

LIT. NO. 0000

AUTHOR
I. Gaunt

JOURNAL
J. sci. Instrum. 31. (1954.) 315 - 18.

DATE: DECADE 0 1 2 4 7; YEAR 0 1 2 4 7

CHEM. ZENTRAL-BLATT

CHEM. ABSTRACTS 1955 2790 d

A simple grating spectrometer for use in analysis.

A simple inexpensive grating spectrograph (2400 lines per inch, blaze at 2.95 μ) is described. The receiver is either an uncooled lead sulphide cell with 400 c.p.s. amplifier or a thermocouple with 10 c.p.s. amplifier. Overlapping orders are eliminated by means of filters of coloured alkali halide crystals (F centre filters). The characteristics of such filters are given. The instrument is roughly comparable in performance at 6 μ with a rock salt prism spectrometer, but much better at 3 μ. As an example of its use the analysis of D_2O is described. (Diagrams are given for the regions 2.6 - 2.9 μ, and 6.0 - 7.0 μ; the spectrum of HCl near 3.4 μ, and the mercury arc emission spectrum from 1 - 2 μ).

SPECIMEN

LITERATURE: HIST. | GENERAL | REVIEW JOURNAL | BOOK | REVIEW SPECTRA | CONSTIT. | SPECTR. PUBL. | SPECTR. CARD

APPLICATION: ANALYSIS GEN. | ANALYSIS QUAL. | TEST OF PURITY | ANALYSIS QUANT. | COMPUTER | MULTICOMP. ANALYSIS | PLANT CONTROL

METHOD: UV | (UV) | VISIBLE | (VISIBLE) | IR | (IR) | MICRO-WAVE | (MICRO-WAVE) | RAMAN | (RAMAN) | LUMINESCENCE | EMISSION | M.S. | (M.S.)

SPECTRAL RANGE: 2000-3500 Å | <2000 Å | 3500-9000 Å | 3-3,8μ | 0,9-3μ | 3,8-16μ | PARTIAL | 16-50μ | >50μ

SUBSTANCE: SOLID | LIQUID | CRYST. | GASEOUS | EMULSION, DISPERSION | SOLUTION | PELLET | FILM | PURIFICATION OF SUBSTANCE | SOLVENT

CODE NO. 120. BUTTERWORTHS, LONDON; VERLAG CHEMIE, WEINHEIM; COPYRIGHT 1956

Figure 8-4A. A typical DMS literature card (*front*).

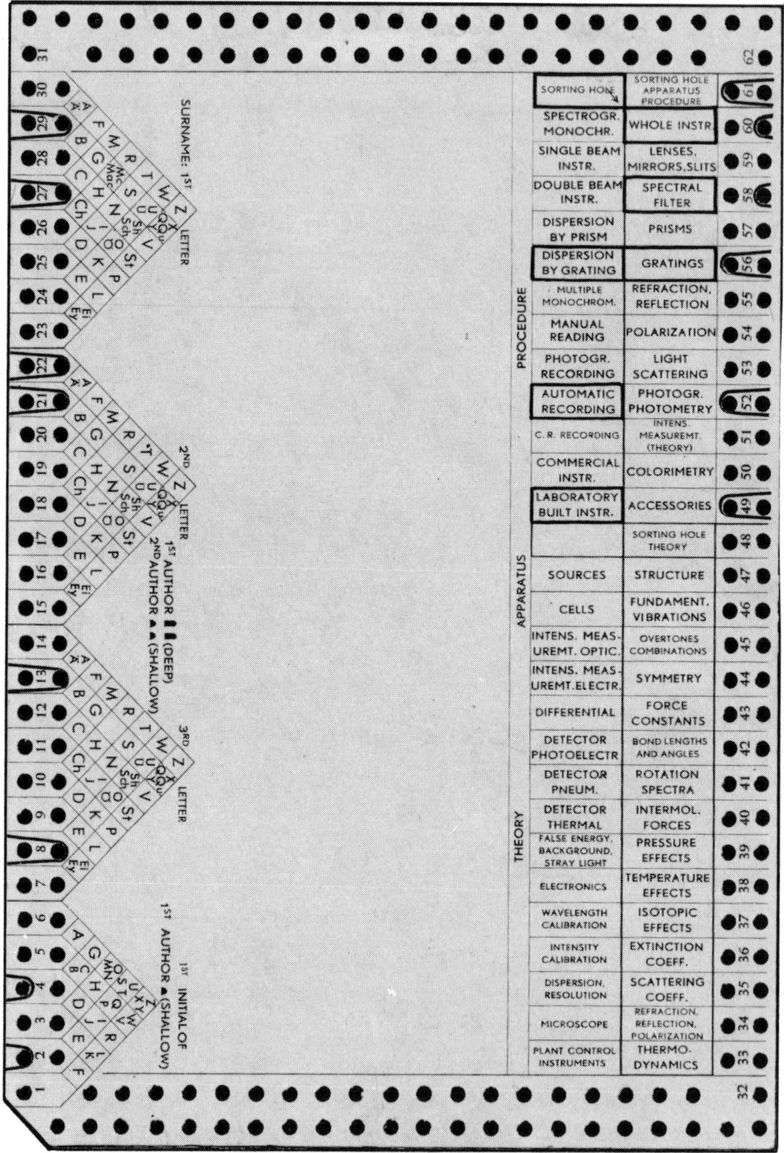

Figure 8-4B. A typical DMS literature card (*back*).

The Documentation of Molecular Spectroscopy System DMS-I Cards [A "peek-a-boo" system] (Butterworths Scientific Publications, London, England)

The DMS-I cards measure 19 × 30 cm, with 5000 possible hole spaces arranged in 50 squares, each subdivided into 100 smaller squares. Each spectral card number has a fixed position on the I card, and each property of the DMS code is represented by one I card. There are 211 cards for the DMS system. The serial number of all spectral cards having the property which the I card represents are punched into the I card. Thus, if a number of I cards representing all the properties to be searched for are selected, the coincident holes will indicate the serial numbers of all the spectral cards for compounds which have all the desired properties. Since the DMS coding system is based on many structural and physical parameters, the "peek-a-boo" system is especially suited for it.

Infrared Band Handbook (H. A. Szymanski, editor, Plenum Press, New York, 1963)

The *Infrared Band Handbook* is based on the assumption that band positions can be reported to 1 cm^{-1}. It consists of a listing by wavelength of all the major bands of a compound. Structural formulas, sample conditions, and group-frequency assignments are given. Correlation tables are also presented in a separate section. The references are quite extensive and should prove useful.

8.5. ABSTRACTING SERVICES AND BIBLIOGRAPHIES

Spectrographic Abstracts, H. M. Stationery Office, London, England. Brief resumés of articles on infrared and Raman spectroscopy are presented.

Infrared—A Library of Congress Bibliography, PB 121998, U.S. Department of Commerce, Office of Technical Services. Titles of articles are presented in this abstracting service.

Infrared Absorption Spectra Index, H. M. Hershenson, Academic Press, Inc. (1969). Index of articles on infrared spectra. Index is arranged according to the compounds whose spectra are reported in the articles.

Titulos Spectroscopic, Theodore H. Zink (editor), 212 Chestnut Hill Drive, Ellicott City, Maryland. This service attempts to abstract all the available material on infrared, Raman, mass, paramagnetic-resonance, nuclear magnetic resonance, emission, flame, x-ray, atomic absorption, and x-ray-fluorescence spectroscopy for both organic and inorganic compounds. The abstracts are presented in both English and Interlingua. The abstracts, on $8\frac{1}{2} \times 1$ inch paper, are classified according to the field of spectroscopy, or if no classification is obvious, the abstract is placed in a general category. The series was discontinued in 1963.

Beckman Bibliography of Infrared Applications. Several issues of a bibliography of infrared applications compiled by W. F. Ulrich of Beckman Instruments, Inc., Fullerton, California, are available from that firm. The practical aspects of infrared spectroscopy are emphasized and the service is by no means a complete abstract of the literature.

APPENDIX 1

Additional References

Compounds with CN Groups

1. B. Holmstedt and L. Larson, *Acta Chem. Scand.* **5**: 1179 (1951).
2. M. Davies and W. J. Jones, *Trans. Faraday Soc.* **54**: 1454 (1958).
3. D. C. Smith, C. Y. Pan, and J. R. Nielsen, *J. Chem. Phys.* **18**: 706 (1950).
4. J. E. Stewart, *J. Chem. Phys.* **30**: 1259 (1959).
5. B. A. Bolto, M. Liveris, and J. Miller, *J. Chem. Soc.* 750 (1956).
6. J. Bellanato, *Spectrochim. Acta* **16**: 1344 (1960).
7. L. Segal and F. V. Eggerton, *Appl. Spectroscopy* **15**: 112 (1961).
8. J. Barceló Matutano and J. Bellanato, *Spectrochim. Acta* **8**: 27 (1956).
9. W. T. Jones, Ph.D. Thesis, University of Wales, 1955. Reference obtained from *Trans. Faraday Soc.* 54: 1458 (1958).
10. T. Miyazawa, T. Shimanouchi, and S. Mizushima, *J. Chem. Phys.* **24**: 408 (1956).
11. N. Jonathan, *J. Mol. Spectroscopy.* **5**: 101 (1961).
12. H. Letaw, Jr. and A. H. Gropp, *J. Chem. Phys.* **21**: 1621 (1953).
13. J. C. Evans, *J. Chem. Phys.* **22**: 1228 (1954).
14. M. Davies and H. E. Hallam, *Trans. Faraday Soc.* **47**: 1170 (1951).
15. W. Pinchas, *J. Chem. Soc.* 3063 (1961).
16. W. Pinchas, *J. Chem. Soc.* 1688 (1961).
17. R. A. McIvor and C. E. Hubley, *Can. J. Chem.* **37**: 869 (1959).
18. A. R. Katritzky and R. A. Jones, *J. Chem. Soc.* 2067 (1959).
19. P. H. Lindenmeyer and P. M. Harris, *J. Chem. Phys.* **21**: 408 (1953).
20. J. Mason and J. Dunderdale, *J. Chem. Soc.* 754 (1956).
21. C. C. Watson, *Spectrochim. Acta* **16**: 1322 (1960).
22. N. S. Ham and J. B. Willis, *Spectrochim. Acta* **16**: 279 (1960).
23. B. L. Crawford, Jr. and W. H. Fletcher, *J. Chem. Phys.* **19**: 406 (1951).
24. D. W. E. Axford, G. J. Janz, and K. E. Russell, *J. Chem. Phys.* **19**: 704 (1951).
25. R. J. Williams, *J. Chem. Phys.* **25**: 656 (1956).
26. J. Mason and J. Dunderdale, *J. Chem. Soc.* 754 (1956).
27. G. B. Barlow and P. J. Corish, *J. Chem. Soc.* 1706 (1959).
28. E. Spinner, *Spectrochim. Acta* **15**: 21 (1959).
29. D. C. McKean, *Spectrochim. Acta* **14**: 161 (1958).
30. D. Hadzi, *J. Chem. Soc.* 847 (1957).
31. R. R. Randle and D. H. Whiffen, *J. Chem. Soc.* 4153 (1952).

Pyridines, Pyrazolones, Isoxazoles, and Quinolines

1. A. R. Katritzky and J. N. Gardener, *J. Chem. Soc.* 2198 (1958).
2. A. R. Katritzky and A. R. Hands, *J. Chem. Soc.* 2202 (1958).
3. A. R. Katritzky and R. A. Jones, *J. Chem. Soc.* 2942 (1960).
4. A. R. Katritzky and R. A. Jones, *J. Chem. Soc.* 2947 (1960).
5. S. Refn, *Spectrochim. Acta* **17**: 40 (1961).
6. D. G. O'Sullivan, *J. Chem. Soc.* 3278 (1960).
7. D. G. O'Sullivan, *J. Chem. Soc.* 3653 (1960).

8. K. J. Morgan, *J. Chem. Soc.* 2343 (1961).
9. A. R. Katritzky and A. J. Boulton, *Spectrochim. Acta* **17**: 238 (1961).
10. A. R. Katritzky and R. A. Jones, *J. Chem. Soc.* 2067 (1959).
11. A. R. Katritzky and R. A. Jones, *J. Chem. Soc.* 3674 (1959).
12. G. L. Cook and F. M. Church, *J. Phys. Chem.* **61**: 458 (1957).
13. H. Shindo and N. Ikekawa, *Pharm. Bull.* (*Tokyo*) **4**: 192 (1956).
14. G. Costa and P. Blasina, *Z. Phys. Chem.* (*Frankfurt*) **4**: 24 (1955).
15. D. A. Lory, F. S. Murphin, J. L. Hales, and W. Kynaston, *Trans. Faraday Soc.* **53**: 1171 (1957).
16. D. G. O'Sullivan and P. W. Sadler, *J. Chem. Soc.* 875 (1959).
17. S. J. Holt, A. E. Kellie, D. G. O'Sullivan, and P. W. Sadler, *J. Chem. Soc.* 1217 (1958).
18. D. G. O'Sullivan and P. W. Sadler, *J. Chem. Soc.* 2202 (1956).
19. D. G. O'Sullivan and P. W. Sadler, *Organic Chem. Bull.* **22**: 283 (1957).
20. A. E. Kellie, D. G. O'Sullivan, and P. W. Sadler, *J. Chem. Soc.* 3809 (1956).
21. A. R. Katritzky and N. A. Coats, *J. Chem. Soc.* 2062 (1959).
22. R. A. Jones and A. R. Katritzky, *J. Chem. Soc.* 1317 (1959).
23. A. R. Katritzky, A. M. Monro, J. A. T. Beard, D. P. Dearnaley, and N. J. Earl, *J. Chem. Soc.* 2182 (1958).
24. A. R. Katritzky and J. M. Lagowski, *J. Chem. Soc.* 657 (1959).
25. A. R. Katritzky and A. J. Boulton, *J. Chem. Soc.* 3500 (1959).
26. A. R. Katritzky, *Quart. Revs.* (*London*) **13**: 535 (1953).
27. S. Califano, F. Piancenti, and G. Speroni, *Spectrochim. Acta* **15**: 86 (1959).
28. A. R. Katritzky, J. M. Lagowski, and J. A. T. Beard, *Spectrochim. Acta* **16**: 954 (1960).
29. R. R. Randle and D. H. Whiffen, Paper No. 12, Report on the Conference of Molecular Spectroscopy, Institute Petroleum (1954).
30. J. G. Hawkins, E. R. Ward, and D. H. Whiffen, *Spectrochim. Acta* **10**: 105 (1957).
31. E. M. Godar and R. P. Mariella, *Appl. Spectroscopy* **15**(2): 29 (1961).
32. L. J. Bellamy and P. E. Rogasch, *Spectrochim. Acta* **16**: 30 (1960).

Amine and Other Nitrogen Hydrohalides

1. E. Baer and H. C. Stancer, *Can. J. Chem.* **34**: 436 (1956).
2. J. Bellanato, *Spectrochim. Acta* **16**: 1344 (1960).
3. C. Brissette and C. Sandorfy, *Can. J. Chem.* **38**: 34 (1960).
4. B. Chenon and C. Sandorfy, *Can. J. Chem.* **36**: 1181 (1958).
5. D. Cook, *Can. J. Chem.* **39**: 2009 (1961).
6. J. C. Grivas, *Can. J. Chem.* **37**: 1260 (1959).
7. R. A. Heacock and L. Marion, *Can. J. Chem.* **34**: 1782 (1956).
8. M. J. Janssen, *Spectrochim. Acta* **17**: 475 (1961).
9. D. M. Kirschenbaum and F. S. Parker, *Spectrochim. Acta* **17**: 785 (1961).
10. R. Mecke and W. Kutzelnigg, *Spectrochim. Acta* **16**: 1225 (1960).
11. R. Mecke and W. Kutzelnigg, *Spectrochim. Acta* **16**: 1216 (1960).
12. D. B. Powell, *Spectrochim. Acta* **16**: 241 (1960).
13. R. Stewart and L. J. Muenster, *Can. J. Chem.* **39**: 401 (1961).
14. C. C. Watson, *Spectrochim. Acta* **16**: 1322 (1960).
15. P. J. Stone, J. C. Craig, and H. W. Thompson, *J. Chem. Soc.* 52 (1958).
16. E. Depas, *Bull. soc. chim. France* 1105 (1953).
17. K. Lehormont, *J. chim. phys.* **49**: 635 (1950).
18. B. Witkop, *J. Am. Chem. Soc.* **76**: 5597 (1954).
19. R. C. Lord and R. E. Marrifield, *J. Chem. Phys.* **21**: 166 (1953).
20. N. N. Greenwood and K. Wade, *J. Chem. Soc.* 1130 (1960).
21. E. Spinner, *J. Chem. Soc.* 1226 (1960).
22. J. D. Waldron, *J. Chem. Phys.* **21**: 734 (1953).

The Carbonyl Vibrations

1. G. Cottis, Ph.D. Thesis, University of Buffalo, 1961.
2. E. W. Abel and G. Wilkinson, *J. Chem. Soc.* 1501 (1959).
3. J. P. Freeman, *J. Am. Chem. Soc.* **80**: 5954 (1958).
4. R. N. Jones, C. L. Angell, T. Ito, and R. J. D. Smith, *Can. J. Chem.* **37**: 2007 (1959).
5. L. J. Bellamy and P. E. Rogasch, *Spectrochim. Acta* **16**: 30 (1960).
6. L. J. Bellamy, *Spectrochim. Acta* **13**: 60 (1958).
7. R. Zbinden and H. K. Hall, Jr., *J. Am. Chem. Soc.* **82**: 1215 (1960).
8. M. L. Josien, J. Lascombe, and C. Vignalou, *Compt. rend.* **250**: 4146 (1960).
9. L. G. Tensmeyer, R. W. Hoffmann, and G. W. Brindley, *J. Phys. Chem.* **64**: 1655 (1960).
10. D. Peltier, A. Pichevin, P. Dizabo, and M. L. Josien, *Compt. rend.* **248**: 1148 (1959).
11. I. Hunsberger, H. S. Gutowsky, W. Powell, L. Morin, and V. Bandurco, *J. Am. Chem. Soc.* **80**: 3294 (1958).
12. L. Gutjahr, *Spectrochim. Acta* **16**: 1209 (1960).
13. T. Miyazawa, *J. Mol. Spectroscopy* **4**: 155 (1960).
14. T. Shimanouchi, M. Tsuboi, T. Takenishi, and N. Iwata, *Spectrochim. Acta* **16**: 1328 (1960).
15. K. Kimuna and T. Kubo, *Bull. Chem. Soc.* (*Japan*) **33**: 1086 (1960).
16. Yu. G. Borodko and Ya. K. Syrkin, *Doklady Akad. Nauk SSSR* **134**: 1127 (1960).
17. D. Cook, *Can. J. Chem.* **39**: 1184 (1961).
18. J. Derkosch and E. Rieger, *Monatsh. Chem.* **90**: 389 (1959).
19. Keishi Kotera, *Yakugaku Zasshi* **80**: 1278 (1960).
20. Keishi Kotera, *Yakugaku Zasshi* **80**: 1281 (1960).
21. L. J. Bellamy and R. L. Williams, *Trans. Faraday Soc.* **55**: 14 (1959).
22. C. N. R. Rao, G. K. Goldman, and C. Lurie, *J. Phys. Chem.* **63**: 1311 (1959).
23. G. Allen, P. S. Ellington, and G. D. Meakins, *J. Chem. Soc.* 1909 (1960).
24. Y. Otsuji and E. Imoto, *Nippon Kagaku Zasshi* **80**: 1199 (1959).
25. C. E. Griffin, *Spectrochim. Acta* **16**: 1464 (1960).
26. P. Mirone and V. Lorenzelli, *Ann. chim.* **49**: 52 (1959).
27. R. Mecke, Sr. and K. Noack, *Spectrochim. Acta* **12**: 391 (1958).
28. H. Hoyer and W. Hensel, *Z. Elektrochem.* **64**: 958 (1960).

The Amide II Band

1. E. R. Blout and M. Idelson, *J. Am. Chem. Soc.* **80**: 4909 (1958).
2. I. Suzuki, M. Tsuboi, and T. Shimanouchi, *Spectrochim. Acta* **16**: 467 (1960).
3. G. R. Bird and E. R. Blout, *J. Am. Chem. Soc.* **81**: 2499 (1959).
4. C. G. Cannon, *Spectrochim. Acta* **16**: 302 (1960).
5. A. Epp, *Anal. Chem.* **29**: 1283 (1957).
6. D. Garfinkel and J. T. Edsall, *J. Am. Chem. Soc.* **80**: 3818 (1958).
7. D. Garfinkel, *J. Am. Chem. Soc.* **80**: 3827 (1958).
8. S. Mizushima, *Sci. Pop. Inst. Phys: Chem. Res.* (*Tokyo*) **29**: 188 (1936).
9. S. Mizushima, M. Tsuboi, T. Shimanouchi, T. Sugita, and T. Yoshimoto, *J. Am. Chem. Soc.* **76**: 2479 (1954).
10. T. Miyazawa and E. R. Blout, *J. Am. Chem. Soc.* **83**: 712 (1961).
11. N. Ogata, *Makromol. Chem.* **40**: 55 (1960).
12. J. H. Robson and J. Reinhart, *J. Am. Chem. Soc.* **77**: 498 (1955).
13. L. Segal, *J. Am. Chem. Soc.* **82**: 2807 (1960).
14. J. E. Stewart, *J. Chem. Phys.* **18**: 248 (1957).
15. A. Yamaguchi, *J. Am. Chem. Soc.* **80**: 527 (1958).

APPENDIX 2

Character Tables of the Most Important Point Groups

C_s	I	$\sigma(xy)$
a'	1	1
a''	1	−1

C_i	I	i
a_g	1	1
a_u	1	−1

C_2	I	$C_2(z)$
a	1	1
b	1	−1

C_{2v}	I	$C_2(z)$	$\sigma_v(xz)$	$\sigma_v(yz)$
a_1	1	1	1	1
a_2	1	1	−1	−1
b_1	1	−1	1	−1
b_2	1	−1	−1	1

C_{3v}	I	$2C_3(z)$	$3\sigma_v$
a_1	1	1	1
a_2	1	1	-1
e	2	-1	0

C_{4v}	I	$2C_4(z)$	$C_4^2 \equiv C_2''$	$2\sigma_v$	$2\sigma_d$
a_1	1	1	1	1	1
a_2	1	1	1	-1	-1
b_1	1	-1	1	1	-1
b_2	1	-1	1	-1	1
e	2	0	-2	0	0

C_{5v}	I	$2C_5$	$2C_5^2$	$5\sigma_v$
a_1	1	1	1	1
a_2	1	1	1	-1
e_1	2	$2\cos 72°$	$2\cos 144°$	0
e_2	2	$2\cos 144°$	$2\cos 72°$	0

$C_{\infty v}$	I	$2C_\infty^\phi$	$2C_\infty^{2\phi}$	$2C_\infty^{3\phi}$	...	$\infty\sigma_v$
σ^+	1	1	1	1	...	1
σ^-	1	1	1	1	...	-1
π	2	$2\cos\phi$	$2\cos 2\phi$	$2\cos 3\phi$	...	0
δ	2	$2\cos 2\phi$	$2\cos 2{\cdot}2\phi$	$2\cos 3{\cdot}2\phi$	...	0
ϕ	2	$2\cos 3\phi$	$2\cos 2{\cdot}3\phi$	$2\cos 3{\cdot}3\phi$	...	0
...	...	...	...	...	...	...

C_{2h}	I	$C_2(z)$	$\sigma_h(xy)$	i
a_g	1	1	1	1
a_u	1	1	-1	-1
b_g	1	-1	-1	1
b_u	1	-1	1	-1

D_2	I	$C_2(x)$	$C_2(y)$	$C_2(z)$
a	1	1	1	1
b_1	1	-1	-1	1
b_2	1	-1	1	-1
b_3	1	1	-1	-1

D_3	I	$2C_3(z)$	$3C_2$
a_1	1	1	1
a_2	1	1	-1
e	2	-1	0

D_{2d}	I	$2S_4(z)$	$S_4^2 \equiv C_2''$	$2C_2$	$2\sigma_d$
a_1	1	1	1	1	1
a_2	1	1	1	-1	-1
b_1	1	-1	1	1	-1
b_2	1	-1	1	-1	1
e	2	0	-2	0	0

D_{3d}	I	$2S_6(z)$	$2S_6^2 \equiv 2C_3$	$S_6^3 \equiv S_2 \equiv i$	$3C_2$	$3\sigma_d$
a_{1g}	1	1	1	1	1	1
a_{1u}	1	-1	1	-1	1	-1
a_{2g}	1	1	1	1	-1	-1
a_{2u}	1	-1	1	-1	-1	1
e_g	2	-1	-1	2	0	0
e_u	2	1	-1	-2	0	0

D_{4d}	I	$2S_8(z)$	$2S_8^2 \equiv 2C_4$	$2S_8^3$	$S_8^4 \equiv C_2''$	$4C_2$	$4\sigma_d$
a_1	1	1	1	1	1	1	1
a_2	1	1	1	1	1	-1	-1
b_1	1	-1	1	-1	1	1	-1
b_2	1	-1	1	-1	1	-1	1
e_1	2	$\sqrt{2}$	0	$-\sqrt{2}$	-2	0	0
e_2	2	0	-2	0	2	0	0
e_3	2	$-\sqrt{2}$	0	$\sqrt{2}$	-2	0	0

D_{2h}	I	$\sigma(xy)$	$\sigma(xz)$	$\sigma(yz)$	i	$C_2(z)$	$C_2(y)$	$C_2(x)$
a_g	1	1	1	1	1	1	1	1
a_u	1	-1	-1	-1	-1	1	1	1
b_{1g}	1	1	-1	-1	1	1	-1	-1
b_{1u}	1	-1	1	1	-1	1	-1	-1
b_{2g}	1	-1	1	-1	1	-1	1	-1
b_{2u}	1	1	-1	1	-1	-1	1	-1
b_{3g}	1	-1	-1	1	1	-1	-1	1
b_{3u}	1	1	1	-1	-1	-1	-1	1

D_{3h}	I	$2C_3(z)$	$3C_2$	σ_h	$2S_3$	$3\sigma_v$
a_1'	1	1	1	1	1	1
a_1''	1	1	1	-1	-1	-1
a_2'	1	1	-1	1	1	-1
a_2''	1	1	-1	-1	-1	1
e'	2	-1	0	2	-1	0
e''	2	-1	0	-2	1	0

D_{4h}	I	$2C_4(z)$	$C_4^2 \equiv C_2''$	$2C_2$	$2C_2'$	σ_h	$2\sigma_v$	$2\sigma_d$	$2S_4$	$S_2 \equiv i$
a_{1g}	1	1	1	1	1	1	1	1	1	1
a_{1u}	1	1	1	1	1	-1	-1	-1	-1	-1
a_{2g}	1	1	1	-1	-1	1	-1	-1	1	1
a_{2u}	1	1	1	-1	-1	-1	1	1	-1	-1
b_{1g}	1	-1	1	1	-1	1	1	-1	-1	1
b_{1u}	1	-1	1	1	-1	-1	-1	1	1	-1
b_{2g}	1	-1	1	-1	1	1	-1	1	-1	1
b_{2u}	1	-1	1	-1	1	-1	1	-1	1	-1
e_g	2	0	-2	0	0	-2	0	0	0	2
e_u	2	0	-2	0	0	2	0	0	0	-2

D_{5h}	I	$2C_5(z)$	$2C_5^2$	σ_h	$5C_2$	$5\sigma_v$	$2S_5$	$2S_5^3$
a_1'	1	1	1	1	1	1	1	1
a_1''	1	1	1	-1	1	-1	-1	-1
a_2'	1	1	1	1	-1	-1	1	1
a_2''	1	1	1	-1	-1	1	-1	-1
e_1'	2	$2\cos 72°$	$2\cos 144°$	2	0	0	$2\cos 72°$	$2\cos 144°$
e_1''	2	$2\cos 72°$	$2\cos 144°$	-2	0	0	$-2\cos 72°$	$-2\cos 144°$
e_2'	2	$2\cos 144°$	$2\cos 72°$	2	0	0	$2\cos 144°$	$2\cos 72°$
e_2''	2	$2\cos 144°$	$2\cos 72°$	-2	0	0	$-2\cos 144°$	$-2\cos 72°$

D_{6h}	I	$2C_6(z)$	$2C_6^2 \equiv 2C_2$	$C_6^3 \equiv C_2''$	$3C_2$	$3C_2'$	σ_h	$3\sigma_v$	$3\sigma_d$	$2S_6$	$2S_3$	$S_6^3 \equiv S_2 \equiv i$
a_{1g}	1	1	1	1	1	1	1	1	1	1	1	1
a_{1u}	1	1	1	1	1	1	−1	−1	−1	−1	−1	−1
a_{2g}	1	1	1	1	−1	−1	1	−1	−1	1	1	1
a_{2u}	1	1	1	1	−1	−1	−1	1	1	−1	−1	−1
b_{1g}	1	−1	1	−1	1	−1	−1	−1	1	1	−1	1
b_{1u}	1	−1	1	−1	1	−1	1	1	−1	−1	1	−1
b_{2g}	1	−1	1	−1	−1	1	−1	1	−1	1	−1	1
b_{2u}	1	−1	1	−1	−1	1	1	−1	1	−1	1	−1
e_{1g}	2	1	−1	−2	0	0	−2	0	0	−1	1	2
e_{1u}	2	1	−1	−2	0	0	2	0	0	1	−1	−2
e_{2g}	2	−1	−1	2	0	0	2	0	0	−1	−1	2
e_{2u}	2	−1	−1	2	0	0	−2	0	0	1	1	−2

$D_{\infty h}$	I	$2C_\infty^\phi$	$2C_\infty^{2\phi}$	$2C_\infty^{3\phi}$	…	σ_h	∞C_2	$\infty\sigma_v$	$2S_\infty^\phi$	$2S_\infty^{2\phi}$	…	$S_2 \equiv i$
σ_g^+	1	1	1	1	…	1	1	1	1	1	…	1
σ_u^+	1	1	1	1	…	-1	-1	1	-1	-1	…	-1
σ_g^-	1	1	1	1	…	1	-1	-1	1	1	…	1
σ_u^-	1	1	1	1	…	-1	1	-1	-1	-1	…	-1
π_g	2	$2\cos\phi$	$2\cos 2\phi$	$2\cos 3\phi$	…	-2	0	0	$-2\cos\phi$	$-2\cos 2\phi$	…	2
π_u	2	$2\cos\phi$	$2\cos 2\phi$	$2\cos 3\phi$	…	2	0	0	$2\cos\phi$	$2\cos 2\phi$	…	-2
δ_g	2	$2\cos 2\phi$	$2\cos 4\phi$	$2\cos 6\phi$	…	2	0	0	$2\cos 2\phi$	$2\cos 4\phi$	…	2
δ_u	2	$2\cos 2\phi$	$2\cos 4\phi$	$2\cos 6\phi$	…	-2	0	0	$-2\cos 2\phi$	$-2\cos 4\phi$	…	-2
ϕ_g	2	$2\cos 3\phi$	$2\cos 6\phi$	$2\cos 9\phi$	…	-2	0	0	$-2\cos 3\phi$	$-2\cos 4\phi$	…	2
ϕ_u	2	$2\cos 3\phi$	$2\cos 6\phi$	$2\cos 9\phi$	…	2	0	0	$2\cos 3\phi$	$2\cos 4\phi$	…	-2
…	…	…	…	…	…	…	…	…	…	…	…	…

T_d	I	$8C_3$	$6\sigma_d$	$6S_4$	$3S_4^2 \equiv 3C_2$
a_1	1	1	1	1	1
a_2	1	1	−1	−1	1
e	2	−1	0	0	2
f_1	3	0	−1	1	−1
f_2	3	0	1	−1	−1

O_h	I	$8C_3$	$6C_2$	$6C_4$	$3C_4^2 \equiv 3C_2''$	$S_2 \equiv i$	$6S_4$	$8S_6$	$3\sigma_h$	$6\sigma_d$
a_{1g}	1	1	1	1	1	1	1	1	1	1
a_{1u}	1	1	1	1	1	−1	−1	−1	−1	−1
a_{2g}	1	1	−1	−1	1	1	−1	1	1	−1
a_{2u}	1	1	−1	−1	1	−1	1	−1	−1	1
e_g	2	−1	0	0	2	2	0	−1	2	0
e_u	2	−1	0	0	2	−2	0	1	−2	0
f_{1g}	3	0	−1	1	−1	3	1	0	−1	−1
f_{1u}	3	0	−1	1	−1	−3	−1	0	1	1
f_{2g}	3	0	1	−1	−1	3	−1	0	−1	1
f_{2u}	3	0	1	−1	−1	−3	1	0	1	−1

Index